T0136962

Lecture Notes in Networks and Systems

Volume 50

Series editor

Janusz Kacprzyk, Polish Academy of Sciences, Warsaw, Poland
e-mail: kacprzyk@ibspan.waw.pl

The series "Lecture Notes in Networks and Systems" publishes the latest developments in Networks and Systems—quickly, informally and with high quality. Original research reported in proceedings and post-proceedings represents the core of LNNS.

Volumes published in LNNS embrace all aspects and subfields of, as well as new challenges in, Networks and Systems.

The series contains proceedings and edited volumes in systems and networks, spanning the areas of Cyber-Physical Systems, Autonomous Systems, Sensor Networks, Control Systems, Energy Systems, Automotive Systems, Biological Systems, Vehicular Networking and Connected Vehicles, Aerospace Systems, Automation, Manufacturing, Smart Grids, Nonlinear Systems, Power Systems, Robotics, Social Systems, Economic Systems and other. Of particular value to both the contributors and the readership are the short publication timeframe and the world-wide distribution and exposure which enable both a wide and rapid dissemination of research output.

The series covers the theory, applications, and perspectives on the state of the art and future developments relevant to systems and networks, decision making, control, complex processes and related areas, as embedded in the fields of interdisciplinary and applied sciences, engineering, computer science, physics, economics, social, and life sciences, as well as the paradigms and methodologies behind them.

More information about this series at http://www.springer.com/series/15179

Oualid Demigha · Badis Djamaa
Abdenour Amamra
Editors

Advances in Computing Systems and Applications

Proceedings of the 3rd Conference on Computing Systems and Applications

Editors
Oualid Demigha
Department of Computer Science
Ecole Militaire Polytechnique
Algiers, Algeria

Abdenour Amamra
Department of Computer Science
Ecole Militaire Polytechnique
Algiers, Algeria

Badis Djamaa
Department of Computer Science
Ecole Militaire Polytechnique
Algiers, Algeria

ISSN 2367-3370 ISSN 2367-3389 (electronic)
Lecture Notes in Networks and Systems
ISBN 978-3-319-98351-6 ISBN 978-3-319-98352-3 (eBook)
https://doi.org/10.1007/978-3-319-98352-3

Library of Congress Control Number: 2018950462

This Springer imprint is published by the registered company Springer Nature Switzerland AG
The registered company address is: Gewerbestrasse 11, 6330 Cham, Switzerland

Preface

Welcome to the proceedings of the 3rd conference on Computing Systems and Applications. This edition of the conference (CSA 2018) was held on April 24–25, 2018, at the Ecole Militaire Polytechnique, Algiers, Algeria. It succeeded the second edition (CSA 2016) that was held at the same place in December 2016. CSA constitutes a leading forum for students, researchers, academics, and industrials to exchange their ideas, research findings, and practical experiences across all computer science disciplines. The conference aims at leveraging modern computer systems technology in fields as varied as: data science, computer networks and security, computer vision, and graphics.

- Besides the peer-reviewed contributions, four keynote talks took place, namely
  1. **Keynote I:** *Formal Modeling of Cyber-Physical Systems: lessons learnt from refinement and proof-based methods*, by Prof. Ait Ameur Yamine, Institut National Polytechnique, Toulouse, France,
  2. **Keynote II:** *Potentials of Intelligence Computational for Big Multi-sensor Data Management*, by Prof. Hadjali Allel, Ecole Nationale Supérieure de Mécanique et d'Aérotechnique, Poitiers, France,
  3. **Keynote III:** *Virtual and Augmented Reality, from Design to Evaluation*, by Prof. Prof. Gherbi Rachid, Université de Paris-Sud XI Orsay, France,
  4. **Keynote IV:** *Data Lakes: New generation of Information Systems*, by Prof. Boussaid Omar, Université de Lyon 2, France.
- As well as three tutorials on hot IT topics:
  1. **Tutorial 1:** *Augmented Reality, how does it work?* by Dr. Bellarbi Abdelkader, CDTA, Algiers, Algeria;
  2. **Tutorial 2:** *Build and Deploy Applications on Secure Linux-IoT Devices*, by M. Harouni Djalal, Open Devices UG, Berlin, Germany;
  3. **Tutorial 3:** *Physics Simulation in Blender*, by M. Cheggour Hamza, Digital Artist and Photographer, Heyday Training & Consulting, Algiers, Algeria.

Central to the conference were the contributions of the participants. Our call for papers resulted in 104 submissions of up to 10 pages. Each paper received at least three reviews. Based on the peer reviews, we selected 30 submissions for presentation amounting to an acceptance rate of 29% out of which 11 were presented as a poster and 19 as plenary talks. One talk is awarded the “CSA 2018 Best Paper Award” based on the scientific value and the technical presentation as assessed by the conference chairs. The present volume contains the reworked papers of the conference. The extended versions of the best papers will be invited to appear in a special issue of EL MIR’AT SCIENCES (ISSN 2170-1555).

We believe the scientific prototypes and the technical content presented at CSA 2018 will contribute to strengthen the great success of computer science technologies in industrial, entertainment, social, and everyday applications.

Finally, we would like to thank all the members of the committees for soliciting and reviewing submissions, and the tutorial proposers, who made it possible to build such a rich supplementary program beside the main scientific plan of the conference. We would like also to thank the team at Ecole Militaire Polytechnique for the professional support and the perfect atmosphere during the seminar. Furthermore, a special thank goes to the participants for their active discussions, their dedication during the seminar as well as for the quality of their timely reviews. The organization of this event would not have been possible without the effort and the enthusiasm of many people. We would like to thank all who contributed to the success of this edition of the conference.

June 2018

Yacine Amara
Conference General Chair
Zakaria Sahraoui
Conference General Chair

Organization

Steering Committee

Oualid Demigha	EMP, Algeria
Billal Merabti	EMP, Algeria
Badis Djamaa	EMP, Algeria
Abdenour Amamra	EMP, Algeria

Website Chair

Salim Khamadja	EMP, Algeria

Exhibitions and Demos Chair

Mohamed-Elarbi Djebbar	EMP, Algeria

Local Organizing Committee

Khoutir Bouchbout	EMP, Algeria
Abdelhamid Boukabou	EMP, Algeria
Kadda Beghdad Bey	EMP, Algeria
Abdelhalim Zaidi	EMP, Algeria
Mohamed Maiza	EMP, Algeria
Tayeb Kenaza	EMP, Algeria
Sofiane Bouznad	EMP, Algeria
Redha Benaissa	EMP, Algeria

Faouzi Sebbak	EMP, Algeria
M'hamed Mataoui	EMP, Algeria
Mustapha Reda Senouci	EMP, Algeria
Ali Yachir	EMP, Algeria
Abdenebi Rouigueb	EMP, Algeria
Malek Nadil	EMP, Algeria
Rafik Mecibah	EMP, Algeria
Lamia Sadeg	EMP, Algeria
Mohamed El Yazid Boudaren	EMP, Algeria
Ahmed Habbouchi	EMP, Algeria
Mohamed Boudali	EMP, Algeria
Sid Ali Hasni	EMP, Algeria
Fairouz Beggas	EMP, Algeria

Technical Program Committee

Abdelhalim Zaidi	EMP, Algeria
Abdelghani Bakhtouchi	EMP, Algeria
Abdelkader Bellarbi	CDTA, Algeria
Abdelkrim Nemra	EMP, Algeria
Abdelouahid Ahmed Derhab	King Saud University, Saudi Arabia
Abdenebi Rouigueb	EMP, Algeria
Abdessamad Réda Ghomari	ESI, Algeria
Adrien Gruson	University of Tokyo, Japan
Aicha Mokhtari Aissani	USTHB, Algeria
Ali Yachir	EMP, Algeria
Allel Hadjali	ISAE-ENSMA, France
Amar Siad	SCCSS, Algeria
Asma Atamna	INRIA, France
Assia Kourgli	USTHB, Algeria
Azzedine Bouaraba	EMP, Algeria
Christophe Lino	IRISA–INRIA, Rennes, France
Djalel Benbouzid	University of Tlemcen, Algeria
Djalel Chefrour	University of Souk Ahras, Algeria
Djamal Benslimane	IUT–Laboratoire LIRIS, France
Djamel Djenouri	CERIST, Algeria
Djamil Aissani	University of Bejaia, Algeria
Emmanuel Grolleau	ISAE-ENSMA, France
Fadila Bentayeb	ERIC, Lyon 2, France
Fairouz Beggas	EMP, Algeria
Faouzi Sebbak	EMP, Algeria
Fouzi Mekhaldi	USTHB, Algeria
Franck Multon	IRISA–INRIA, Rennes, France

Habiba Drias	USTHB, Algeria
Hassina Nacer	USTHB, Algeria
Hayet Belghit	CDTA, Algeria
Hicham Seffendji	British Telecom, UK
Hui-Yin Wu	North Carolina State University, USA
Idir Amine Amarouche	HCA, Algeria
Ilyes Boukhari	CRD-DAT, Algeria
Ismail Boukli Hacene	University of Tlemcen, Algeria
Jihad Itani	AUL University, Lebanon
Kadda Beghdad Bey	EMP, Algeria
Kamel Boukhalfa	USTHB, Algeria
Kasprzyk Rafal	Military University of Technology, Poland
Lakhdar Sais	CRIL–CNRS, France
Lala Khadidja Hamaidi	Technische Universität Darmstadt, Germany
Lamia Sadeg	EMP, Algeria
Louisa Bouallouche-Medjkoune	University of Bejaia, Algeria
Lounis Chermak	Cranfield University, UK
Mahfoud Hamidia	CDTA, Algeria
Mahmoud Belhocine	CDTA, Algeria
Matis Huddon	Trinity College of Dublin, Ireland
M'hmed Mataoui	EMP, Algeria
Mohamed El Yazid Boudaren	EMP, Algeria
Mohamed Salah Azzaz	EMP, Algeria
Mohamed Maiza	EMP, Algeria
Mohammad siraj	King Saud University, Saudi Arabia
Mohammed Messadi	University of Tlemcen, Algeria
Mourad Adnane	ENP, Algeria
Mustapha Reda Senouci	EMP, Algeria
Nadia Zenati	CDTA, Algeria
Nasreddine Lagraa	University of Laghouat, Algeria
Omar Boussaid	ERIC, Lyon 2, France
Omidvar Mathieu Mahmoud	University of littoral Cote d'Opale, France
Oualid Araar	EMP, Algeria
Oualid Djekoune	CDTA, Algeria
Rachid Chalal	ESI, Algeria
Ramdane Maamri	University of Constantine 2, Algeria
Riad Akrour	CLAS–TU Darmstadt, Germany
Ricardo Marques	Universitat Pompeu Fabra (UPF), Spain
Riyadh Baghdadi	MIT, USA
Rochdi Merzouki	Polytech'Lille, France
Sabri Boutemedjet	AppDirect, Montréal, Canada
Saliha Aouat	USTHB, Algeria
Salim Bitam	University of Biskra, Algeria
Samir Benbelkacem	CDTA, Algeria

Samir Chouali	FEMTO-ST, France
Selma Khouri	ESI, Algeria
Sergio Silva	ISEL, Portugal
Shafique Ahmad Chaudhry	Dhofar University, Oman
Sidi Mohammed Benslimane	ESI, Algeria
Tassadit Bouadi	IRISA de Rennes, France
Tayeb Kenaza	EMP, Algeria
Toufik Ahmed	LaBRI, France
Walid Cherifi	EMP, Algeria
Walid-Khaled Hidouci	ESI, Algeria
Wassim Ait Elhara	University Paris 11–Orsay, France
Yacine Challal	ESI, Algeria
Yamine Ait Ameur	INPT-ENSEEIHT–IRIT, France
Youcef Aklouf	USTHB, Algeria
Zaia Alimazighi	USTHB, Algeria
Zohir Irki	EMP, Algeria

Contents

IoT, Computer Network and Security

Formal Modeling of Cyber-Physical Systems: Lessons Learn from Refinement and Proof Based Methods

Yamine Ait Ameur(✉)

Institut National Polytechnique, Toulouse, France
yamine@n7.fr

Abstract. Cyber-Physical Systems refer to the tight integration and coordination between computational and physical resources. Modeling their behavior requires handling continuous and discrete behaviors. The definition of the associated models refers to continuous and discrete systems theories. In this talk, we address the problem of designing correct software to control cyber-physical systems. We recall the necessary basic concepts allowing a designer to model such systems. We also give an overview of the different formal approaches supporting the formal verification of these hybrid systems and we highlight the results obtained using these techniques. In particular, we focus on the use of proof and refinement based methods.

O. Demigha et al. (Eds.): CSA 2018, LNNS 50, p. 3, 2019.
https://doi.org/10.1007/978-3-319-98352-3_1

Scheduling Algorithms for IEEE 802.15.4 TSCH Networks: A Survey

Mohamed Mohamadi(✉) and Mustapha Reda Senouci

Distributed and Complex Systems Lab.,
Ecole Militaire Polytechnique, Algiers, Algeria
med_mohamadi@yahoo.com

Abstract. One of the most promising technologies to enable the future Internet of Things is the Time Slotted Channel Hopping (TSCH) mode, which reveals the robustness of the IEEE 802.15.4 standard by providing high reliability, low latency, and energy efficiency. TSCH has received a lot of attention from the researchers' community. In fact, the TSCH specification had never provided how to build and maintain a schedule; therefore, several researchers devised new scheduling algorithms. This paper focuses on scheduling in TSCH networks. It starts with a gentle introduction to the IEEE 802.15.4 standard and the TSCH mode. It then surveys the State-of-The-Art scheduling algorithms, where algorithms are classified and compared. Each of the algorithms are then presented along with its pros and cons. Finally, weaknesses that need to be addressed are identified.

Keywords: Constrained nodes · Industrial networks
Internet of Things · TSCH · Scheduling · 6TiSCH · IEEE 802.15.4

1 Introduction

We are living a great revolution of the Internet of Things (IoT). It is a widespread and popular technology that is expected to connect billions of devices in few years. IoT breaks through several application domains, including smart cities, environmental monitoring, home automation, health care, and industrial control. The industrial automation refers to the Industrial IoT, where the connected devices are resource-constrained in terms of energy, memory, and processing capabilities. Whereas, industrial networks require high reliability, low latency, energy efficiency, and scalability.

Several international organizations have issued many standards to provide reliable and efficient communications in industrial environments such as WirelessHART [1], ISA-100.11a [2], and IEEE 802.15.4e standard [3]. As an extension of [4], [3] provides a better support for industrial applications by defining a new MAC protocol, the Time Slotted Channel Hopping (TSCH) mode.

A new Working Group (WG) named 6TiSCH[1] was then created to provide IPv6 networking capabilities for TSCH-based low-power wireless networks.

[1] https://datatracker.ietf.org/wg/6tisch/about.

O. Demigha et al. (Eds.): CSA 2018, LNNS 50, pp. 4–13, 2019.
https://doi.org/10.1007/978-3-319-98352-3_2

6TiSCH is a key enabler to fulfill the gap between industrial networks and information technology worlds. In such networks, the link layer would be based on the IEEE 802.15.4-TSCH specifications. In 802.15.4-TSCH medium access control (MAC), nodes synchronize on a frame structure and follow a communication schedule to avoid idle listening, extend the battery lifetime, and improve reliability.

Scheduling is one of the trending topics for the 6TiSCH WG. Indeed, the TSCH specification had never provided means to build and maintain a schedule. Scheduling in 6TiSCH-based wireless networks can be achieved in three different ways. In this paper, we survey existing scheduling algorithms for IEEE 802.15.4e TSCH networks. The main objective is to study each of the existing algorithms, provide its advantages and identify deficiencies that need to be addressed, in order to devise more efficient scheduling algorithms.

The rest of this paper is organized as follows. Section 2 gives an overview of the IEEE 802.15.4e standard. Section 3 provides a background about the TSCH mode. Section 4 presents a State-of-The-Art of the scheduling approaches. In Sect. 5, we establish a comparative study between the different scheduling approaches. Finally, Sect. 6 concludes the paper and draws some future directions.

2 Towards the IEEE802.15.4e Standard

The technical standard IEEE 802.15.4 is the reference standard for Wireless Sensor Networks (WSNs). It defines the operation of low-rate wireless personal area networks (LR-WPANs) and it specifies their physical layer and media access control [4]. Such networks are well known by their ease of installation, extremely low cost, short-range operation, reliable data transfer and a reasonable battery life, while maintaining a simple and flexible protocol stack. Several studies have analyzed the performance of the IEEE 802.15.4 standard in WSNs. Unfortunately, many limitations and deficiencies have been identified that make it unfit for critical applications, which have strict requirements in terms of reliability, latency, energy efficiency, scalability and usually operating in harsh environments. These limitations are: MAC unreliability, unbounded latency, no built-in frequency hopping technique, and poor energy management.

These imperfections make the above-mentioned standard unsuitable for many applications. Especially, when these applications are very demanding in terms of reliability and latency. To overcome these limitations, in 2008, a Working Group dubbed *802.15 Task Group 4e* was created to enhance and add functionalities to the 802.15.4 MAC.

In 2012, the IEEE Standard Association Board approved the *IEEE802.15.4e* [3] as an Enhancement Standard for the IEEE 802.15.4 to better support the industrial markets. The core technology is inspired from the previous industrial networking technologies such as WirelessHart [1] and ISA 100.11.a [2]. The enhanced standard provides new functionalities such as: Low Energy, Information Elements, Enhanced Beacons, MAC Performance Metrics, and Fast Association.

The IEEE 802.15.4e defines five new MAC behavior modes. Each one supports specific application domains. In this work, we focus on the *Time Slotted Channel Hopping (TSCH)* mode that is very effective and useful in the industrial field. Readers are referred to [5] for more details about the other MAC behaviors.

3 TSCH: The New MAC Behavior

The Time Slotted Channel Hopping (TSCH) mode is mainly designed to support industrial automation and process control. The TSCH application domains are, to name a few, defense, robotics, oil and gas industry, health-care, transportation control systems, and green energy production. This MAC protocol combines time slotted access with multi-channel and channel hopping capabilities. Doing so, it provides predictable latency, high network capacity, and high communication reliability, while maintaining very low duty cycles.

Deterministic latency is guaranteed by eliminating collision among competing nodes while using time slotted access to increase chances for the outflow to be achieved. Multi-channel allows many communications at the same time, by using different channel offsets. Consequently, it increases the network capacity. Furthermore, channel hopping mitigates the effects of interference and multipath fading that improves the communication reliability and reduces the energy spent in packet re-transmissions.

In a TSCH network, nodes are synchronized, and time is split into timeslots, each typically 10 ms long. Time slots are grouped into one or more slotframes (Fig. 2) which continuously repeats over time. The slotframe size is not fixed, it could be varied depending on the application needs. The shorter the slotframe, the more often a time slot repeats, resulting in higher reliability, but also in a higher power consumption.

During a timeslot (Fig. 1), one node typically sends a frame to its neighbor, and the neighbor sends back an acknowledgement. If the acknowledgement is not received during the timeout period, re-transmission of the frame waits until the next assigned transmit timeslot.

Basically, there are 16 different channels available for communication where some of these frequencies could be blacklisted because of low quality communication. Each channel is identified by a channel offset. Hence, diversity of channels is got by indicating for each send and receive slot a channel offset. This channel offset is then translated into different frequency on which communication happens at every iteration of the slot frame.

In TSCH all nodes in the network follow a common communication schedule, that instructs each node what to do in each timeslot; send to a particular neighbor, receive from a particular neighbor, or sleep. The network scheduling is one of the greatest challenges identified in [6]; because TSCH specification had never provided how to build, update and maintain a schedule, it was considered out of the scope of the IEEE 802.15.4e standard.

A link is a single element of the schedule characterized by a *timeOffset* and a *channelOffset*, which is reserved to transmit from node A to node B or for node B to receive from node A within a given slotframe.

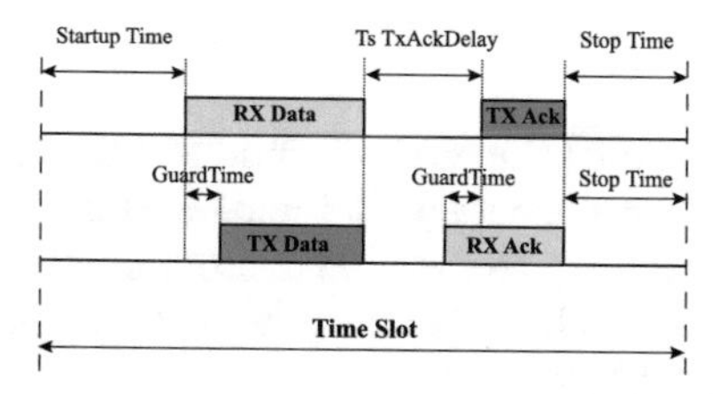

Fig. 1. TSCH timeslot

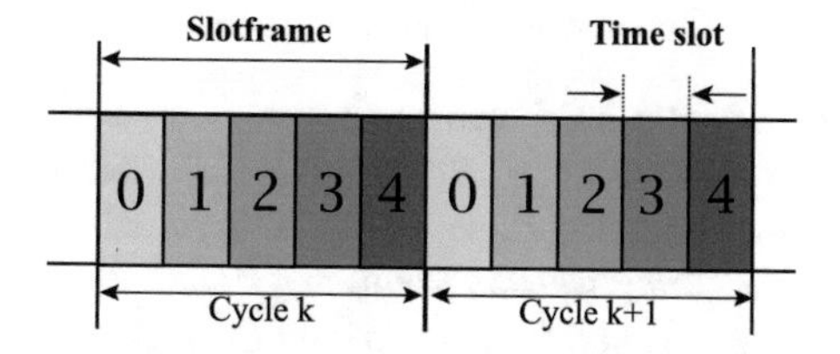

Fig. 2. TSCH slotframe

Figure 3 represents a network with a tree topology of 8 nodes. The communication among these nodes is scheduled as shown in Fig. 4. In this schedule, the slotframe length is equal to 5 slots and only 4 channel offsets are used. TSCH also allows dedicated links, allocated to a single node for transmission (i.e., contention free) and shared links, allocated to more than one node for transmission (i.e., contention-based with TSCH CSMA back-off).

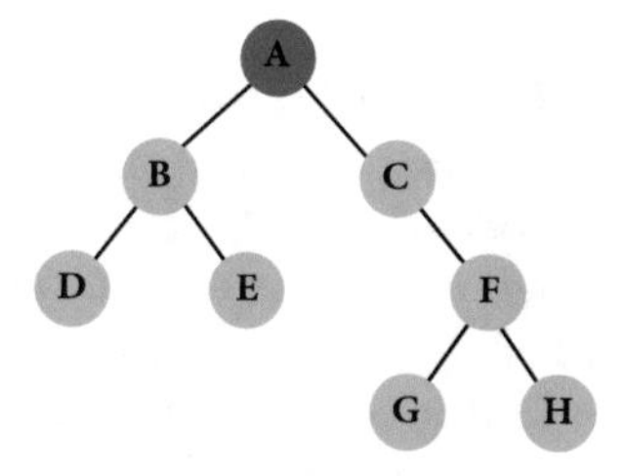

Fig. 3. Tree topology network

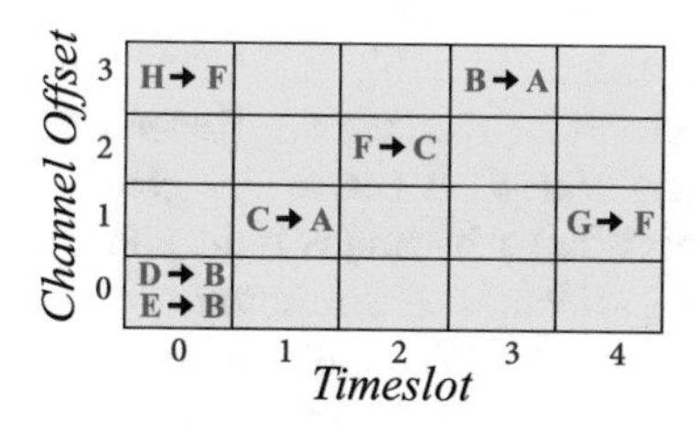

Fig. 4. TSCH schedule

The standard only explains how the MAC layer executes such schedule, but that did not prevent researchers from coming up with solutions. Contrariwise, many solutions were proposed in the literature and will be discussed in the next section.

4 How to Build a Schedule?

The scheduling problem has already attracted research interest in TDMA-based networks as reported in the literature [7]. Even so, most existing multi-channel scheduling schemes are not suitable for TSCH networks. They were not designed for resource-constrained nodes, they do not allow channel hopping per-packet, and they are not efficient in terms of channel utilization. Hence, researchers have designed new scheduling algorithms specially to fit under TSCH networks. Different approaches could be used to build the schedule. They can be broadly classified as centralized, distributed, and autonomous. In the sequel, we will detail the most prevalent scheduling techniques and algorithms for data gathering in multi-channel context.

4.1 Centralized Scheduling

In a centralized approach, only one node schedules the whole communication in the network, the coordinator (i.e., the manager/collector). It builds and maintains the network schedule. The scheduler could be a centralized computer called *Path Computation Element (PCE)* [8]. We discuss here the most important centralized scheduling solutions proposed for TSCH networks.

Palattella et al. [9] proposed the Traffic Aware Scheduling Algorithm (TASA), a centralized scheduling algorithm for IEEE 802.15.4e networks. This solution considers a tree network topology, and focuses on a convergecast scenario, where different amount of data must be delivered to the coordinator. The main aim of TASA is to create the best schedule, minimizing the number of slots needed to deliver all data to the coordinator. Such schedule can be obtained by the *matching and coloring process*. At each iteration, TASA uses the matching algorithm to select a set of eligible links to be scheduled in the same time slot. Then, a vertex-coloring algorithm assigns different channel offsets for each link selected in the previous process. Furthermore, the authors found that using more channels can improve the network throughput, decrease latency, and lower significantly the energy consumption.

Authors of [10], designed the Multi-channel Optimized Delay time Slot Assignment (MODESA). Reversely to TASA, it targets homogeneous traffic conditions where all the nodes generate the same number of packets. In TASA, a conflict-free schedule is built using an iterative procedure. During each iteration, TASA selects a certain number of links and accommodates their transmissions in the same timeslot and using multiple channel offsets if needed. Whereas MODESA selects at each iteration only one node, and chooses a link to accommodate one of its required transmissions. Furthermore, MODESA reduces buffer congestion by scheduling first the nodes that have more packets in their buffers, where TASA does not take into account queue congestion. This approach has been improved and extended to support heterogeneous traffic and multiple coordinators in [11].

Jin et al. [12] proposed a centralized Adaptive Multi-hop Scheduling method (AMUS) based on the latest TSCH MAC. They implemented their solution at the backbone located PCE and use lightweight application layer protocol (i.e., CoAP) to collect required information to compute the schedule. AMUS enables a Multi-hop Scheduling Sequence (MSS) to provide low latency and allocates additional resources to vulnerable links to significantly reduce the delay caused by interference or collisions. This solution outperforms TASA; it enhances communication reliability and achieves ultra-low latency.

A new Centralized Link Scheduling (CLS) for 6TiSCH Wireless Industrial Networks was proposed in [13]. The CLS algorithm constructs efficient multi-hop schedules and reduces the control messages overhead. The allocation and deallocation of slots does not require the entire rescheduling. It leans on the routing protocol for low power and lossy networks (RPL) [14], when a new node joins a network, CLS assigns to it a proper slot and when a nodes preferred parent is changed, CLS deletes the relevant slot and assigns a proper slot. Performance evaluation show that the CLS algorithm requires smaller control messages for scheduling formation than a decentralized approach [15] with almost similar latencies.

In [16], authors formulate the scheduling problem as a throughput maximization problem and delay as a minimization problem. They proposed a graph theoretical approach based on matching theory to solve the throughput maximization problem in a centralized way. Evaluation results show that the proposed solution reaches a very high throughput. The same problem was formulated as an energy efficiency maximization problem in [17], where the authors, proposed a low-complexity Energy Efficient Scheduler (EES) that performs better t han a Round Robin Scheduler (RRS) in terms of energy efficiency, while at the same time yielding a good throughput.

4.2 Distributed Scheduling

When using a distributed approach, each node negotiates with its neighbors and decides locally on which links to schedule with which neighbors. We summarize below the most important distributed solutions.

Tinka et al. [18] proposed the first decentralized scheduling algorithm that deals with mobile nodes. This solution has two variations: Aloha-based scheduling and Reservation-based scheduling. The authors use *advertisement* packets to communicate the available slots and channels on which they can be reached, and *Connection Request* packets to establish a link. While Aloha allocates one channel for broadcasting advertisements for new neighbors, Reservation-based scheduling improves Aloha with a dedicated timeslot for advertisements. Simulation results show that both cope well with dynamic nodes, and that reservation-based algorithm outperforms the Aloha-based algorithm. However, both solutions are not suitable for energy-constrained nodes.

Decentralized Traffic Aware Scheduling (DeTAS) [15], is the distributed version of TASA. It addresses multi-coordinator networks; therefore, it uses combined Micro-scheduling to build the global schedule. To avoid interference each micro-schedule uses dedicated set of channels. DeTAS was compared to TASA, and obtained results show that the former provides a good queue management. In addition DeTAS proves its high performance in terms of duty cycle, end-to-end delay, and link Packet Loss Ratio [19].

In [20], a Label switching over TSCH networks was proposed to build the schedule using a Generalized Multi Protocol Label switching (GMPLS) and a Resource Reservation Protocol-Traffic Engineering (RSVP-TE). Each node sends *PATH* message to establish a route to the sink. This latter sends back a *RESV* message. The Completely Fair Distributed Scheduler (CFDS) uses the information included in the *RESV* message to build the schedule. Simulation results reveal that this method yields a high throughput and a low latency.

Distributed Scheduling for Convergecast in Multi-channel Wireless Sensor Networks (DiSCA) [21]. This solution considers two cases of transmissions: without acknowledgment and with immediate acknowledgment. In this algorithm, in each iteration, a node transmission is scheduled by following a set of rules.

Each iteration provides a micro-schedule. This latter can overlap to reduce the total number of slots. The authors compared DiSCA to [10,22]; obtained results show, that DiSCA was almost close to the optimal schedule with reduced control messages.

Authors in [23] suggested a Distributed Scheduling Algorithm (DIS−TSCH). Their strategy leans on nodes location, where every node knows its local information inherited from RPL (i.e., rank, graph depth...). The main idea of this Distributed Time Slot Scheduling method, is to assign one time slot to the leaf nodes and many slots for internal nodes according to their children. Then, for each assigned time slot, it assigns a channel offset. This schedule may generate some idle slots, but simulation results reveal that this approach attain high throughput, low latency, and low control overhead, which also leads to low power consumption.

Wave [22] is a Distributed Scheduling Algorithm for Convergecast in IEEE 802.15.4e Networks. Every single node within the network knows its conflicting nodes and its parent. A series of waves are flooded in the network, the first one is triggered by a *START* message sent by the coordinator. When a node receives this message and has the highest priority (i.e., depends on the number of packets in its queue) among its conflicting nodes, it assigns to itself a cell in the wave and notifies its conflicting nodes by sending an *ASSIGN* message. This process is yield until all nodes select cells. Once the first wave is propagated, the coordinator transmits a *REPEAT* message to invoke a second wave that improves the first one, this process is repeated until all nodes schedule all their queue packets. Simulation results demonstrate that Wave reduced length of the schedule comparing that with [21] and stands out as an efficient distributed scheduling algorithm.

Demir and Bilgili suggested [24] a distributed divergecast scheduling algorithm (DIVA). Conversely to convergecast, traffic flows in all directions not only from leave nodes to root. Each node starts by broadcasting a Connection Request *CON-REQ*. If it receives a Connection Acknowledgement *CON-ACK* packet as a response, the link will be established between this two nodes. This process is executed by all nodes in the network until achieving the maximum slot frame length. This approach was compared to [18], unfortunately DIVA does not improve Aloha, but supports all slotframe sizes.

A Decentralized Adaptive Multi-Hop Scheduling Protocol was designed for 6TiSCH Wireless Networks (DeAMON) [25]. It is the *de facto* for a variety of industrial control and monitoring applications. DeAMON includes traffic-awareness, sequential scheduling, parallel transmissions, robust over-provisioning, adaptability to topology changes and minimal signaling overhead. This algorithm is similar to [12] but in a decentralized way. Performance evaluation shows that this solution achieves high reliability, reduces latency, maintains high utilization of resources, achieves minimal signaling overhead, and adaptable to traffic topology. Moreover, DeAMON is fully compliant to the 6TiSCH protocol stack.

In the next sub-section we will present another solution which is totally different, it is neither centralized nor distributed and gives very satisfying results.

4.3 Autonomous Scheduling

Orchestra was proposed in [26], as an autonomous scheduling approach where neither a centralized nor a distributed scheduler is required. It is simple, light and flexible. It responds quickly to the sudden network changes leaning on RPL (i.e.: network topology). In Orchestra, no negotiation is needed, nodes calculate and maintain their own schedule locally and autonomously, and update it automatically if the topology evolves without signaling overhead. The schedule is formed of three different slotframes. Each one is exclusively allocated to a specific type of traffic: MAC TSCH beacons, RPL signaling traffic, and application data. The length of each slotframe introduces a trade-off to balance the network capacity, latency, and energy consumption. Simulation results show that it achieves a higher delivery ratio, while keeping a good trade-off between latency and energy consumption.

5 Open Research Issues

Each type of above-mentioned scheduling has strong points and weaknesses. Moreover, solutions for the same class exhibit different performance levels in different scenarios. For a global classification, we put the autonomous approach at the head of the list, followed by the distributed approach then the centralized one. The centralized approach reveals several weaknesses. Usually, it takes a long time for the initialization, what might occur on the expanse of energy consumption. This kind of solutions are not reactive to the sudden changes in the network. Furthermore, the overhead of signaling messages used to create and maintain the schedule reduces the capacity of the network and leads to an additional energy consumption. Nevertheless, centralized built-schedule is certainly conflict-free which increases reliability. Nevertheless, it is still not suitable for dynamic, energy constrained, and large-scale networks. For this aim, the distributed approach is a very appealing solution.

Indeed, decentralized approach responses quickly to the sudden changes of network topology, but it still needs additive communication for the negotiation between neighbors. In addition, some of the decentralized solutions do not consider conflicts, and do not consider energy efficiency as a primary goal. Basically, distributed solutions do not reach the best trade-off between reliability, latency, and energy efficiency. That is what led researchers to design a totally different solution, where there is no need of negotiation between nodes or any central entity to build the schedule.

Autonomous scheduling has proven its efficiency. There is no signaling overhead even locally among neighboring nodes, each node can autonomously decide its own schedule. However, this solution makes nodes unaware of network load, leaving no possibility of optimizing the schedule for different traffic patterns.

It is very crucial to find the best scheduling algorithm that combines all the aforementioned aspects (i.e., *conflict-free schedule, support dynamic topology changes, no signaling messages overhead, reduced slotframe length, support heterogeneous traffic, consider traffic patterns*). Hence, it still is an open research issue, to design a robust solution that ensures high reliability, low latency and energy efficiency in large dynamic networks that fits under the 6TiSCH protocol stack. In addition, the idea of providing a mechanism for these different scheduling approaches to coexist in the same network opens a very wide field of research.

6 Conclusion

In this paper, we have presented the enhanced IEEE 802.15.4 standard that brings new MAC behaviors. Through our study that focuses on the new TSCH mode, we presented the design of this MAC behavior, and discussed thereafter the scheduling as it is one of its most important research issues. We outlined the existing solutions for building a schedule in TSCH networks. We provided a comparative study between these solutions, which were classified into the categories: centralized, distributed, and autonomous. Finally, we identified open issues and laid out areas of possible future research in scheduling in large-scale dynamic networks.

References

1. WirelessHART Specification 75: TDMA data-link layer. HART Communication Foundation Std., Review, vol. 1, no. 1 (2008)
2. ISA, ISA 100: 100.11 a-2009: Wireless systems for industrial automation: process control and related applications. International Society of Automation, Research Triangle Park, NC, USA (2009)
3. IEEE802.15.4e-2012: IEEE Standard for Local and Metropolitan Area Networks. Part 15.4: Low-Rate Wireless Personal Area Networks Amendment 1: MAC Sublayer. Institute of Electrical and Electronics Engineers Std., April 2012
4. IEEE Standard for Information technology – Local and metropolitan area networks – Part 15.4: Wireless MAC and PHY Specifications for LR-WPANs, September 2006
5. De Guglielmo, D., Brienza, S., Anastasi, G.: IEEE 802.15.4e: a survey. Comput. Commun. **88**, 1–24 (2016)
6. Watteyne, T., Palattella, M., Grieco, L.: Using IEEE 802.15. 4e time-slotted channel hopping in the internet of things: problem statement. Technical report (2015)
7. Ergen, S.C., Varaiya, P.: TDMA scheduling algorithms for wireless sensor networks. Wirel. Netw. **16**(4), 985–997 (2010)
8. Farrel, A., Vasseur, J.-P., Ash, J.: A path computation element (PCE)-based architecture. Technical report, August 2006
9. Palattella, M.R., Accettura, N., Dohler, M., Grieco, L.A., Boggia, G.: Traffic aware scheduling algorithm for reliable low-power multi-hop IEEE 802.15. 4e networks. In: 2012 IEEE 23rd International Symposium on Personal Indoor and Mobile Radio Communications (PIMRC), pp. 327–332. IEEE (2012)

10. Soua, R., Minet, P., Livolant, E.: MODESA: an optimized multichannel slot assignment for raw data convergecast in wireless sensor networks. In: Performance Computing and Communications Conference (IPCCC), pp. 91–100. IEEE (2012)
11. Soua, R., Livolant, E., Minet, P.: MUSIKA: a multichannel multi-sink data gathering algorithm in wireless sensor networks. In: 2013 9th International Wireless Communications and Mobile Computing Conference, pp. 1370–1375. IEEE (2013)
12. Jin, Y., Kulkarni, P., Wilcox, J., Sooriyabandara, M.: A centralized scheduling algorithm for IEEE 802.15. 4e TSCH based industrial low power wireless networks. In: 2016 IEEE Wireless Communications and Networking Conference (WCNC), pp. 1–6. IEEE (2016)
13. Choi, K.-H., Chung, S.-H.: A new centralized link scheduling for 6TiSCH wireless industrial networks. In: Internet of Things, Smart Spaces, and Next Generation Networks and Systems, vol. 9870, pp. 360–371, September 2016
14. Winter, T.: RPL: IPv6 routing protocol for low-power and lossy networks, March 2012
15. Accettura, N., Palattella, M.R., Boggia, G., Grieco, L.A., Dohler, M.: Decentralized traffic aware scheduling for multi-hop low power lossy networks in the internet of things. In: 14th International Symposium and Workshops on a World of Wireless, Mobile and Multimedia Networks, pp. 1–6. IEEE (2013)
16. Ojo, M., Giordano, S.: An efficient centralized scheduling algorithm in IEEE 802.15. 4e TSCH networks. In: 2016 IEEE Conference on Standards for Communications and Networking (CSCN), pp. 1–6. IEEE (2016)
17. Ojo, M., Giordano, S., Portaluri, G., Adami, D., Pagano, M.: An energy efficient centralized scheduling scheme in TSCH networks. In: IEEE International Conference on Communications Workshops (ICC Workshops), pp. 570–575. IEEE (2017)
18. Tinka, A., Watteyne, T., Pister, K.: A decentralized scheduling algorithm for time synchronized channel hopping. In: International Conference on Ad Hoc Networks, pp. 201–216, Springer (2010)
19. Accettura, N., Vogli, E., Palattella, M.R., Grieco, L.A., Boggia, G., Dohler, M.: Decentralized traffic aware scheduling in 6TiSCH networks: design and experimental evaluation. IEEE Internet Things J. **2**(6), 455–470 (2015)
20. Antoni, M., Xavier, V., López, V.J., Thomas, W.: Label switching over IEEE802. 15.4 e networks. Trans. Emerg. Telecommun. Technol. **24**(5), 458–475 (2013)
21. Soua, R., Minet, P., Livolant, E.: DiSCA: a distributed scheduling for convergecast in multichannel wireless sensor networks. In: 2015 IFIP/IEEE International Symposium on Integrated Network Management (IM), pp. 156–164. IEEE (2015)
22. Soua, R., Minet, P., Livolant, E.: Wave: a distributed scheduling algorithm for convergecast in ieee 802.15. 4e TSCH networks. Trans. Emerg. Telecommun. Technol. **27**(4), 557–575 (2016)
23. Wang, W.-P., Hwang, R.-H.: A distributed scheduling algorithm for IEEE 802.15. 4e networks. In: 2015 IEEE International Conference on Smart City/SocialCom/SustainCom (SmartCity), pp. 95–100. IEEE (2015)
24. Demir, A.K., Bilgili, S.: DIVA: a distributed divergecast scheduling algorithm for IEEE 802.15. 4e TSCH networks. Wirel. Netw., 1–11 (2017)
25. Aijaz, A., Raza, U.: DeAMON: a decentralized adaptive multi-hop scheduling protocol for 6TiSCH wireless networks. IEEE Sens. J. **17**(20), 6825–6836 (2017)
26. Duquennoy, S., Al Nahas, B., Landsiedel, O., Watteyne, T.: Orchestra: robust mesh networks through autonomously scheduled TSCH. In: Proceedings of the 13th ACM Conference on Embedded Networked Sensor Systems, pp. 337–350. ACM (2015)

Task Scheduling in Cloud Computing Environment: A Comprehensive Analysis

Ali Belgacem[1(✉)], Kadda Beghdad-Bey[1], and Hassina Nacer[2]

[1] EMP, Algiers, Algeria
al.BLgacem@gmail.com, k.beghdadbey@gmail.com
[2] USTHB, Algiers, Algeria
sino_nacer@yahoo.fr

Abstract. Cloud computing plays an important role in the improvement of the Information Technology (IT) industry. However, the cloud Quality of service (QoS) level is considered the biggest challenge facing cloud providers and a major concern for enterprises today. That is why, resource allocation is the optimum solution towards this end, which could be done by the best use of task scheduling techniques. In this paper, we provide a literature analysis of the resource allocation in Cloud computing. As well, we propose a classification of tasks scheduling approaches used in the cloud. Furthermore, our work helps researchers develop existing approaches or suggest a new method in this area.

Keywords: Cloud computing · Resource allocation
Tasks scheduling · Optimization approaches

1 Introduction

Cloud computing is the new face of Information Technology that appears everywhere in modern organizations at storage, networks, servers and applications. It allows resources to be used more appropriately, easily and on demand. At the same time, Cloud reduces the capital expenditure in enterprise (money, energy, time, personnel, etc.). Moreover, Cloud can be integrated with new technologies as Internet of Things or systems based on old technologies because it has several properties as it delivers information in real-time, eliminates security risks and others. Cloud computing consists of three types of service models are: Infrastructure, Platform and Software. Also, Cloud providers (e.g. Google, Amazon and Microsoft) are responsible to manage, allocate and configure the resources in order to satisfy user requirements and maximize profit. Hence, many researchers have been attracted to study resources allocation in Cloud.

The explosive demand in Cloud computing posed a challenge to energy consumption, resource provisioning, waste and ensuring efficient use of resources. To resolve this latter, the organization needs to apply the most suitable approach to manage available computational resources. This notion is generally referred as Scheduling. However, analyzing existing techniques and understanding their

O. Demigha et al. (Eds.): CSA 2018, LNNS 50, pp. 14–26, 2019.
https://doi.org/10.1007/978-3-319-98352-3_3

focus work is necessary for developing some additional applicable technique that can be an improvement of the existing techniques to take advantages from earlier studies [17].

So that, scheduling is a good topic of research to achieve an optimal resources allocation. It aims to enhance the assignment of tasks to resources (resources are an entities where tasks are executed) in order to optimize some objectives that subjected to constraints. Each resource has its own characteristics (Computing power, Memory size, etc.). Also, there are different types of resources: Storage, Power, Networking, and Compute Resources. Since, the scheduling mechanism is an important issue that improves the use of resources and also makes a better performance of Cloud. In this work we show a relative study of the task scheduling in the Cloud computing environment and we propose a classification about the existing methods in literature.

This paper is organized as follows: Sect. 2 gives the state art of resources allocation in Cloud. Section 3 discusses the definition and processes of scheduling problem. Section 4 illustrates a different type of tasks scheduling. Section 5 propose a classification of the tasks scheduling approaches. Section 6 provides a conclusion.

2 Related Works

There are many searches in the scientific literature that surveyed Cloud, more precisely, their main characteristics, features and open issues. Also, they addressed the various approaches were employed to optimize the allocation of resources in this environment.

In [3] authors proposed a classification of Cost optimization approaches for scientific workflow scheduling in Cloud and Grid computing. On the other hand, for Vm placement Challita et al. presented different approaches applied and indicated the motivation behind this type of scheduling as: Energy Management, Resource Usage Optimization and Traffic Engineering [8]. In reference [12] the authors discussed the techniques used for task scheduling founded on Particle Swarm Optimization (PSO), Ant Colony Optimization (ACO), Genetic Algorithm (GA), and two novels League Championship Algorithm (LCA) and BAT algorithm. In the works [17,28] authors gave a detailed study of several aspects of resource allocation and scheduling. Moreover, they mentioned the future challenge and the opportunity for research in this direction.

Whereas, in the work [18], Masdari et al. illustrated an analysis about the workflow scheduling approaches in Cloud computing. In addition to the previous approaches, they pointed to works based Simulated Annealing (SA), and Cat Swarm Optimization (CSO). Besides, they examined both the Heuristic based methods and Hybrid Algorithms. Other work, analyzed job scheduling approaches based primarily on the first-come, first-served (FCFS), Round Robin (RR), Min-Min and Max-Min algorithms [24]. In this paper, we review and analyze the different types of tasks, issues and techniques used in the resource allocation. We also suggest a classification of applied approaches to solve the studied problem.

3 Scheduling Problem Analysis

The main purpose of task scheduling is to find an optimal mapping of tasks to resources that optimize one or more objectives. Besides, the task scheduling issues in Cloud, often considered as an NP-hard problem [18,28].

3.1 Definition

The scheduling explained in terms of these four sets: a set of "m" tasks $T = \{T_1, T_2...T_m\}$, a set of "n" physical resources $R = \{R_1, R_2...R_n\}$, a set of "s" Users $U = \{U_1, U_2...U_s\}$ and a set of "z" functions Objectives $F = \{F_1, F_2...F_z\}$ (Eq. 1) [28].

$$F_z = \sum_{k=1}^{m,n} R_k \times T_k \rightarrow U_k^s \;\; (m\ can\ = n\ or \neq n) \tag{1}$$

It consists of assigning a sets of Tasks T to all available Resources R in Datacenters, in order to complete all tasks under constraints imposed, effectively satisfy user U requests and maximize the revenue of the provider, as shown in Fig. 1.

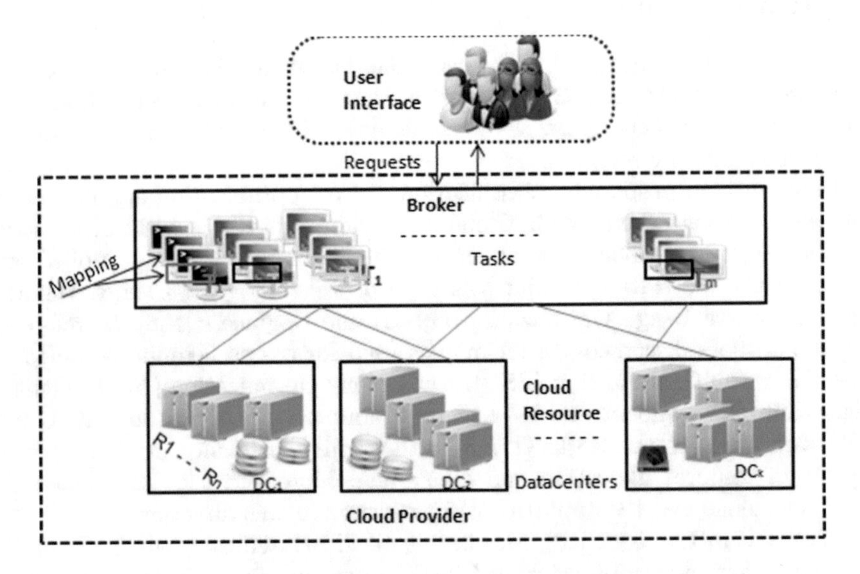

Fig. 1. Resource allocation for task scheduling problem in cloud

3.2 Scheduling Processes

The broker can be a manageable server or a specific datacenter in which the scheduling policy locates. Whereas, the resources are available at the datacenter level. The Cloud provides services to the user using three main steps as illustrated in the Fig. 1:

- The user sends requests;
- The broker receives and analyzes these requests;
- Application of the scheduling approaches to affect each task to the appropriate resource.

4 Task Scheduling Types in Cloud

The diversity of services providing (application, storage, CPU, etc.) and the necessity to make it available to users, as well as the need for a better management of datacenters, led to a multiplicity of task scheduling as we illustrate in Fig. 2.

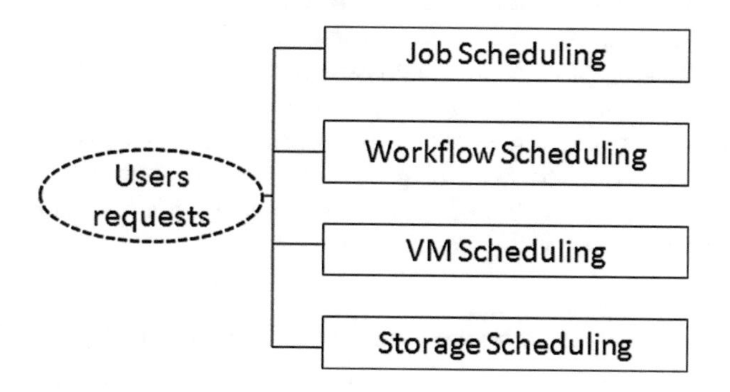

Fig. 2. Type of tasks scheduling in cloud

4.1 Job Scheduling

Usually, job calls on a set of independent tasks [3]. Most approaches begin by evaluating the weight of the tasks in its pool according to (waiting time, deadline, arrival time, execution time, etc.) [2,30] to determine the priority of each task in order to estimate the best mapping.

4.2 Workflow Scheduling

Set of dependent tasks, represented with Directed Acyclic Graph (DAG) where each node indicates the tasks and each edge represents the communication between dependency tasks. Each directed edge e_{ij} states that T_i is the parent task of T_j (Fig. 3). The task can only be executed once all its parent tasks have been completed [31].

4.3 VM Scheduling

Clouds providers create groups of VMs in a resource pool, then each VM should assign to a host. It is one of the main challenges in implementing IaaS because there are many constraints defined in a Service Level Agreement (SLA) document which explain how to associate the VMs with each client. This latter problem can be solved by a VM migration technique.

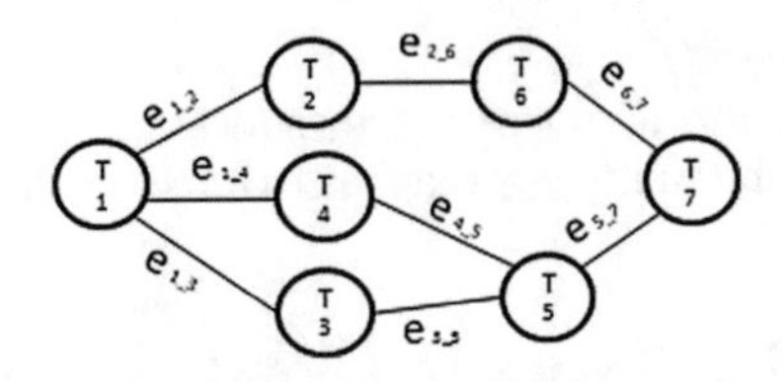

Fig. 3. Example of directed acyclic graph (workflow)

4.4 Storage Scheduling

We mean a set of data block, often with large capacity and heterogeneous type (image, text, sound, etc.) which are stored in clusters that exist in different geographical places. Also, its size increases continually as is the case in the big data. Furthermore, to obtain a good optimization, we require a high processing method for storing and retrieving data, so scheduling is considered as helpful solution to deal with this type of problems. For this purpose, it is generally necessary to create a partition of data, based on the available information about storage sites and resources.

Table 1. Comparison between different tasks scheduling types in cloud

Task type	Main scheduling phases	Main parameters	Example of queries
Job	–Resource allocation phase (allocated pool of tasks to set of resources)	–Weight of tasks or size	–Job requests, –Web applications
Workflow	–Tasks sorting phase, –Resource allocation phase (allocated set of ready tasks to set of resources)	–Data transmission, –Bandwidth, –Subdeadline of tasks	–Workflow applications
VM	–VMs placement (allocated set of VMs to set of physical resources)	–Consolidation of VMs, –Migration of VMs	–VM requests
Storage	–Distribution (Location) of data (allocated blocks of data to Cluster resources)	–Chunk[a]	–Big data application, –Storage requests, –Data mining application

[a]Is a unit for measuring the load of data.

Some works need to improve the task scheduling problem from several aspects, which lead to deal with different types of tasks at the same time. For example, in the work [29], Zhang et al. applied an iterative approach for the scheduling of dynamic workloads in big-data. It allows allocating virtual cluster resources according to the demand. So that, they scheduled both tasks types such as scientific workflow and job. Also, they managed the placement of VMs on top of the physical machines.

The main properties of previous tasks cited in this section are illustrated in Table 1.

5 Tasks Scheduling Approaches Classification

In order to achieve the best solution for resource allocation in Cloud, many categories of scheduling methods are used. Therefore, different classification approaches existed in the literature. Challita et al. classified the dynamic VM placement optimization methods based on four main approaches which are: Constraint Programming, Bin Packing, Stochastic Integer Programming and Genetic Algorithm [8]. However, Masdari et al. classified the workflow scheduling schemes into heuristic, meta-heuristic and hybrid [18]. Instead, Salot et al. classified existing job scheduling algorithms in two modes: Batch mode and online mode [24]. These classifications works are for a particular type of tasks, whereas in our work we take into consideration all types of tasks as listed below (Fig. 4):

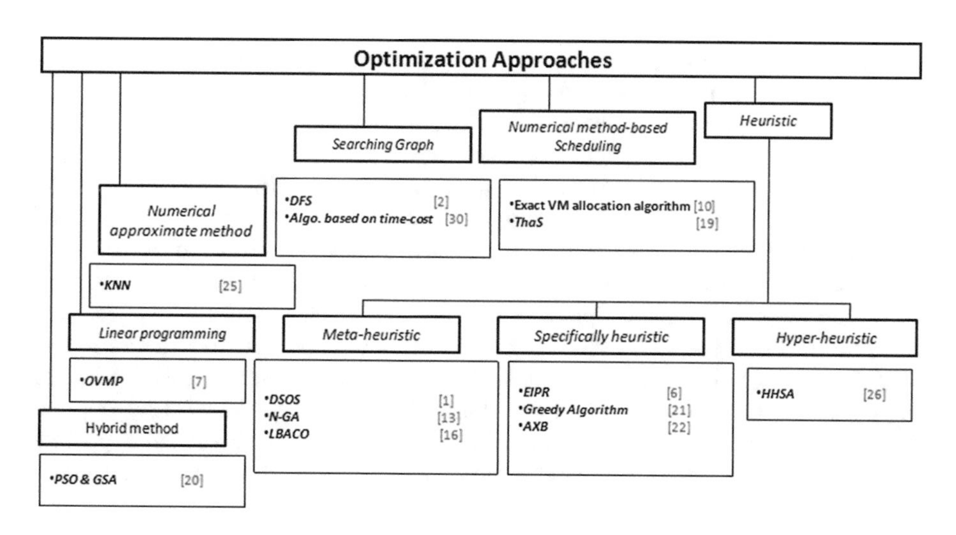

Fig. 4. Classification of tasks scheduling approaches in cloud

5.1 Heuristic

We distinguish different types of heuristics as follows:

Specifically-Heuristic-Based Scheduling: In [6] authors addressed the execution problem of a workflow in the Cloud before deadline-constrained at the smallest possible cost. They used the PCP algorithm to detect the "critical path of Graph" and the IC-PCP based method to examine the available VMs. Also, they applied EIPR algorithm to increase the performance of system. Others, presented a new scheduling strategy based on load balancing (LBE) approach for an independent task in Cloud environment which gave a good results in terms of execution time, makespan and resource utilization [5]. In [21] Nan et al. applied a Greedy Algorithm to achieve the minimal running cost of tasks in resources.

In addition, in the work [22] Panda et al. discussed the optimization problem in a multi-cloud environment and proposed three AXB algorithms based on the traditional Min-Min and Max-Min algorithm. It aims to efficient utilization and collaboration between clouds to get better the makespan. In the work [27] authors defined trust-model bases on the Bayesian theorem, to evaluate the target nodes that have more ability to provide service over the Network. Because, it's difficult to determine the optimal workload scheduling weight for each class of VMs.

Meta-heuristic-Based Scheduling: We distinguish different sorts of scheduling based-Meta-heuristic methods, such as GA, PSO, and ACO [11].

To improve the VMs utilization with reducing the makespan, in the work [1] the authors proposed DSOS approach which scheduling a bag of tasks in a cloud environment. The LBACO algorithm proposed in [16] aimed to optimize the tasks scheduling with load balancing. Further, the integration of the basic ACO principle allows picking out new task scheduling count on the result in the past task scheduling. It aims to minimize the resource wastage.

In the work [15] authors proposed a new approach based on PSO for VM scheduling. They took the characteristic of CPU capacity and memory size of VMs to place on physical servers. Keshanchi et al. improved N-GA based on genetic Algorithm with a heuristic-based HEFT search. The initial population of proposed algorithm is formed by applying three heuristic rank policies (upward rank, downward rank, and a combination of level and upward downward rank) [13].

Hyper-heuristic-Based Scheduling: Other wok aimed to minimize the makespan, by proposing a hyper-heuristic scheduling algorithm (HHSA), to find better scheduling solutions for cloud computing system. It uses a detection operator to automatically define the lowest level heuristic algorithm and a perturbation operator to fine-tune the solutions obtained by each low-level algorithm [26].

5.2 Numerical Method-Based Scheduling

In [10] authors employed Exact VM Allocation algorithm (an extended Bin-Packing approach) to address the power consumption problem. In order to reach a good consolidation of VMs, they used the migration module to handle the state of the unused hosts (sleep, shut down) and therefore determine the VMs placement. To save and reduce energy consumption Mhedheb et al. interpose *ThaS* (Thermal-aware Scheduler) as a mediator between the VMM (hypervisor) and the virtual machines. This, has for purpose to map a VMs request to a physical machines with the consideration of the load and temperature on the hosts [19].

5.3 Numerical Approximate Method-Based Scheduling

Mathematical methods based on probability or operational search, allow estimating a good possible solution of the task scheduling problem. Kong et al. in [14] proposed a scheduling architecture based on Multi-Agents in dynamic Grid and Cloud environments. They suggested a decentralized belief propagation based method (PD-LBP), aimed to enhance the performance and utility obtained from task allocation. This improvement is made in two phases: pruning and decomposition. The first phase focuses on reducing the search space through pruning the resource providers, and the second aims to decompose the network into multiple independent parts. Besides, Sandhu et al. proposed a KNN algorithm to calculate the distance between datacenters location. So that, the distribution of queries on the datacenter is done in an adaptive way, according to the amount of available resources and resources that big data request will consume [25].

5.4 Searching Graph Method-Based Scheduling

The distribution of the resources is represented by a binary tree or acyclic graph. Each resource has different characteristics and is known by its identifier. For mapping the tasks, we search the graph to find the suitable Resource.

Achar et al. presented a novel scheduling algorithm for optimally using the resources, by creating a Tree structure of Virtual Machine (VMT). The mapping is according a prioritizing size for tasks and MIPS for VM. The scheduling algorithm Modified *depth-first-search (DFS)* allows selecting the suitable Virtual Machines to run the tasks [2]. To enhance the reliability of resources, Zhong-wen et al. proposed a scheduling method based on the model of time-cost-trust cloud computing resources, which is based on the subset tree algorithm. They use boundaries function to cut the sub-tree from which they can not get the optimal solution. The optimal solution can be obtained after cutting the branches that do not meet the requirements [30].

5.5 Hybrid Method-Based Scheduling

It is a mixture between different mathematical, heuristic or meta-heuristic methods. Mirzayi et al. [20] introduced new hybrid heuristic algorithm based on PSO and GSA addressed to Workflow scheduling. The algorithm takes place in two parts: the scheduler and the heuristic. So that, according the information taking about VM and workflow the scheduling algorithm determines the ready tasks at each moment. Then, they applied a heuristic algorithm to assign tasks to VMs, in a way that the cost is minimized [7].

5.6 Linear Programming

It is a traditional analytical method based on mathematic formulation. Chaisiri et al. presented an algorithm for optimal placement of virtual machines on host (OVMP). They proposed an architecture system which is composed of user, virtual machine (VM), cloud providers, and cloud broker. The OVMP aiming, to select an optimal solution by solving a stochastic integer programming with two main steps: the first step determines the number of VMs provisioned, while the second step determines the number of VMs assigned for utilization.

Other methods that use the Pareto as a factor to decide the best solution exist. As illustrated in the reference [4], Bessai et al. proposed three complementary bi-criteria approaches (Cost-based approach, Time-based approach, and Cost-time-based approach) for scheduling workflows on distributed Cloud resources. The authors aimed to minimize the completion time and the amount cost incurred by using resources. The cost-time-based approach combined the two others approaches to select only the Pareto solutions.

In this section, we presented our classification of existing approaches in the literature (Fig. 4). Since, Cloud is the result of successive development of technologies from mainframes to the Grid, it is not strange that the type of the tasks and the based approaches can be similar. Because some works propose an improvement and adaptation in the Cloud of existing approaches in other environments [3]; which explains the convergence of our classification with other authors. But, cloud characteristics (virtualization) allows the possibility to achieve new Objectives (efficient resource utilization, SLA requirement, etc.) [17]. This led to predominance of new heuristic and numerical based scheduling approaches such as LCA [12], bin-packing and K nearest neighbor to formulate the Task scheduling problems.

The Table 2 shows a comparison between existing works for task scheduling in Cloud. These approaches aim essentially to minimize: (total energy conception, thermal dissipation costs, resource wastage) and also to enhance (QoS, data distribution over a cluster, etc.) and other criteria.

Table 2. Comparison between the tasks scheduling approaches based on different aspects

Task	Works	Approach classification	Issue
Job	Discrete Symbiotic Organism Search (DSOS) [1]	Meta-heuristic	QoS
	Modified depth-first search (DFS) [2]	Search graph	Utilisation of resource
	Algorithm based on time-cost-trust model [30]	Search graph	Efficiency resource scheduling
Workflow	Cost-based approach, Time-based approach, Pareto selection approach [4]	Pareto	Qos and cost
	Enhanced IC-PCP with Replication (EIPR) algorithm [6]	Heuristic	QoS
	A hybrid heuristic algorithm for workflow scheduling (HSGA) [9]	Meta-heuristics	Load balancing and resource utilisation
	Genetic algorithm N-GA [13]	Meta-heuristics	Qos
VM	Multi Objective GA (MOGA) [23]	Meta-heuristics	Energy Consumption and QoS
	Extended Bin-Packing approach [10]	Numerical method based scheduling	Energy Consumption
	ThaS (Thermal-aware Scheduler) [19]	Numerical method based scheduling	Energy Consumption
Storage	Adaptive K nearest neighbor [27]	Numerical approximate method	QoS

6 Conclusion

Through analyzing the task scheduling problem in Cloud, this paper has illustrated the nature of tasks and presented the taxonomy of scheduling approaches for resources allocation. In the literature, four types of tasks predominant: workflow, job, VM and storage. Besides, we deduce that each work has its specific notation, actors and scenario. Our proposed classification approaches is based on various types of tasks scheduling in Cloud. In each category, we have analyzed existed works, extracted and classified the used methods. The heuristic approaches are the most used technics and given effective results.

Scheduling is an ideal way to achieve a good allocation of resources in Cloud. Hence, it allows improving the services performances provided to consumers. Although the existing of several approaches to overcome the various task scheduling problems, much research remains to be done in this area. Especially with the rapid development of the technologies which led to the emergence of new problems. Our work helps to outline the current research on task scheduling approaches in Cloud paradigm in order to improve actual solutions or solve future problems. So, looking forward to more approaches based on multi-objective scheduling, scheduling for big data analytic and multi-tasks scheduling in Cloud.

References

1. Abdullahi, M., Ngadi, M.A., et al.: Symbiotic organism search optimization based task scheduling in cloud computing environment. Future Gener. Comput. Syst. **56**, 640–650 (2016)
2. Achar, R., Thilagam, P.S., Shwetha, D., Pooja, H., et al.: Optimal scheduling of computational task in cloud using virtual machine tree. In: 2012 Third International Conference on Emerging Applications of Information Technology (EAIT), 30 November–01 December 2012, Kolkata, India, pp. 143–146. IEEE (2012)
3. Alkhanak, E.N., Lee, S.P., Rezaei, R., Parizi, R.M.: Cost optimization approaches for scientific workflow scheduling in cloud and grid computing: a review, classifications, and open issues. J. Syst. Softw. **113**, 1–26 (2016)
4. Bessai, K., Youcef, S., Oulamara, A., Godart, C., Nurcan, S.: Bi-criteria workflow tasks allocation and scheduling in cloud computing environments. In: 2012 IEEE 5th International Conference on Cloud Computing (CLOUD), November 2012, Chicago, IL, USA, pp. 638–645. IEEE (2012)
5. Bey, K.B., Benhammadi, F., Boudaren, M.E.Y., Khamadja, S.: Load balancing heuristic for tasks scheduling in cloud environment. In: Proceedings of the 19th International Conference on Enterprise Information Systems - Volume 1: ICEIS, 26–29 April 2017, Porto, Portugal, pp. 489–495. INSTICC, SciTePress (2017)
6. Calheiros, R.N., Buyya, R.: Meeting deadlines of scientific workflows in public clouds with tasks replication. IEEE Trans. Parallel Distrib. Syst. **25**(7), 1787–1796 (2014)
7. Chaisiri, S., Lee, B.S., Niyato, D.: Optimal virtual machine placement across multiple cloud providers. In: 2009 IEEE Asia-Pacific Services Computing Conference, APSCC 2009, December 2009, Biopolis, Singapore, pp. 103–110. IEEE (2009)
8. Challita, S., Paraiso, F., Merle, P.: A study of virtual machine placement optimization in data centers. In: 7th International Conference on Cloud Computing and Services Science, CLOSER 2017, April 2017, Porto, Portugal (2017). https://hal.inria.fr/hal-01481631
9. Delavar, A.G., Aryan, Y.: HSGA: a hybrid heuristic algorithm for workflow scheduling in cloud systems. Cluster Comput. **17**(1), 129–137 (2014)
10. Ghribi, C., Hadji, M., Zeghlache, D.: Energy efficient VM scheduling for cloud data centers: exact allocation and migration algorithms. In: 2013 13th IEEE/ACM International Symposium on Cluster, Cloud and Grid Computing (CCGrid), pp. 671–678. IEEE (2013)

11. Gupta, A., Garg, R.: Load balancing based task scheduling with ACO in cloud computing. In: 2017 International Conference on Computer and Applications (ICCA), 6–7 September 2017, Doha, United Arab Emirates, pp. 174–179. IEEE (2017)
12. Kalra, M., Singh, S.: A review of metaheuristic scheduling techniques in cloud computing. Egypt. Inf. J. **16**(3), 275–295 (2015)
13. Keshanchi, B., Souri, A., Navimipour, N.J.: An improved genetic algorithm for task scheduling in the cloud environments using the priority queues: formal verification, simulation, and statistical testing. J. Syst. Softw. **124**, 1–21 (2017)
14. Kong, Y., Zhang, M., Ye, D.: A belief propagation-based method for task allocation in open and dynamic cloud environments. Knowl. Based Syst. **115**, 123–132 (2017)
15. Kumar, D., Raza, Z.: A PSO based VM resource scheduling model for cloud computing. In: 2015 IEEE International Conference on Computational Intelligence and Communication Technology (CICT), October 2015, Liverpool, UK, pp. 213–219. IEEE (2015)
16. Li, K., Xu, G., Zhao, G., Dong, Y., Wang, D.: Cloud task scheduling based on load balancing ant colony optimization. In: 2011 Sixth Annual Chinagrid Conference (ChinaGrid), August 2011, Dalian, Liaoning, China, pp. 3–9. IEEE (2011)
17. Madni, S.H.H., Latiff, M.S.A., Coulibaly, Y., et al.: Resource scheduling for infrastructure as a service (IaaS) in cloud computing: challenges and opportunities. J. Netw. Comput. Appl. **68**, 173–200 (2016)
18. Masdari, M., ValiKardan, S., Shahi, Z., Azar, S.I.: Towards workflow scheduling in cloud computing: a comprehensive analysis. J. Netw. Comput. Appl. **66**, 64–82 (2016)
19. Mhedheb, Y., Jrad, F., Tao, J., Zhao, J., Kołodziej, J., Streit, A.: Load and thermal-aware VM scheduling on the cloud. In: International Conference on Algorithms and Architectures for Parallel Processing, October 2013, Liverpool, UK, pp. 101–114. Springer (2013)
20. Mirzayi, S., Rafe, V.: A hybrid heuristic workflow scheduling algorithm for cloud computing environments. J. Exp. Theor. Artif. Intell. **27**(6), 721–735 (2015)
21. Nan, X., He, Y., Guan, L.: Optimization of workload scheduling for multimedia cloud computing. In: 2013 IEEE International Symposium on Circuits and Systems (ISCAS), pp. 2872–2875. IEEE (2013)
22. Panda, S.K., Gupta, I., Jana, P.K.: Task scheduling algorithms for multi-cloud systems: allocation-aware approach. Inf. Syst. Front., 1–19 (2017)
23. Portaluri, G., Giordano, S.: Multi objective virtual machine allocation in cloud data centers. 2016 5th IEEE International Conference on Cloud Networking (Cloudnet), October 2016, Pisa, Italy, pp. 107–112. IEEE (2016)
24. Salot, P.: A survey of various scheduling algorithm in cloud computing environment. Int. J. Res. Eng. Technol. **2**(2), 131–135 (2013)
25. Sandhu, R., Sood, S.K.: Scheduling of big data applications on distributed cloud based on QoS parameters. Cluster Comput. **18**(2), 817–828 (2015)
26. Tsai, C.W., Huang, W.C., Chiang, M.H., Chiang, M.C., Yang, C.S.: A hyper-heuristic scheduling algorithm for cloud. IEEE Trans. Cloud Comput. **2**(2), 236–250 (2014)
27. Wang, W., Zeng, G., Tang, D., Yao, J.: Cloud-DLS: dynamic trusted scheduling for cloud computing. Expert Syst. Appl. **39**(3), 2321–2329 (2012)
28. Zhan, Z.H., Liu, X.F., Gong, Y.J., Zhang, J., Chung, H.S.H., Li, Y.: Cloud computing resource scheduling and a survey of its evolutionary approaches. ACM Comput. Surv. (CSUR) **47**(4), 63 (2015)

29. Zhang, F., Cao, J., Tan, W., Khan, S.U., Li, K., Zomaya, A.Y.: Evolutionary scheduling of dynamic multitasking workloads for big-data analytics in elastic cloud. IEEE Trans. Emerg. Top. Comput. **2**(3), 338–351 (2014)
30. Zhong-wen, G., Kai, Z.: The research on cloud computing resource scheduling method based on time-cost-trust model. In: 2012 2nd International Conference on Computer Science and Network Technology (ICCSNT), December 2012, Changchun, China, pp. 939–942. IEEE (2012)
31. Zhu, Z., Zhang, G., Li, M., Liu, X.: Evolutionary multi-objective workflow scheduling in cloud. IEEE Trans. Parallel Distrib. Syst. **27**(5), 1344–1357 (2016)

Big Data Processing Security Issues in Cloud Environment

Imene Bouleghlimat(✉) and Salima Hacini

Department of Software Technologies and Information Systems,
Abd El Hamid Mehri University, Constantine, Algeria
{imene.bouleghlimat,
salima.hacini}@univ-constantine2.dz

Abstract. Nowadays, large amounts of heterogeneous data are generated continuously from social media, connected devices, and digital process, etc. To extract useful information and insight from these huge data, new management, and processing systems that support the high Volume, high Variety, and high Velocity of data are required. Cloud Computing provides suitable infrastructures to maintain big data processing and storage. However, outsourcing storage and computation of data to Cloud servers make security a big concern. This paper studies the security issues of big data processing in Cloud environment and schemes that are proposed to ensure its privacy, integrity, and availability in order to invasion future research directions.

Keywords: Big data · Cloud computing · Big data processing security issues

1 Introduction

According to Gartner, connected devices will reach 20.4 billion by 2020. Through the increasing use of these devices, continuous data generation is having an explosive increase of digital data that are either structured or unstructured. This data is called "big data" [1]. Big data refers to high Volume, high Velocity, and high Variety of data that require new processing systems in order to enable enhanced decision-making, insight discovery and/or process optimization [2]. Where big information of society is included in big data's, their analysis can deliver useful insights to commercial companies in order to grow their business [2]. The Cloud presents a suitable platform for big data as it provides scalable and flexible computation and large storage capabilities as services. However, the outsourcing resource sharing nature makes security the main concern that hinder users from using Cloud services for big data storage and processing [3]. Where a Cloud provider leads big data processing, the data owner cannot ensure that the processing is done completely and correctly due to the huge amounts of data and computing. Therefore, Cloud provider can cheat to improve their results and make a profit [4]. The big data lifecycle comprises various phases such as collection, ingestion, pre-processing, processing, post-processing, storage, sharing, recording, provenance and, preservation [5]. Data processing phase may include various activities such as computational, data mining, analytics, and visualization [5]. Examples of activities over each phase are listed in Fig. 1. Each phase of big data life cycle presents a research area

O. Demigha et al. (Eds.): CSA 2018, LNNS 50, pp. 27–36, 2019.
https://doi.org/10.1007/978-3-319-98352-3_4

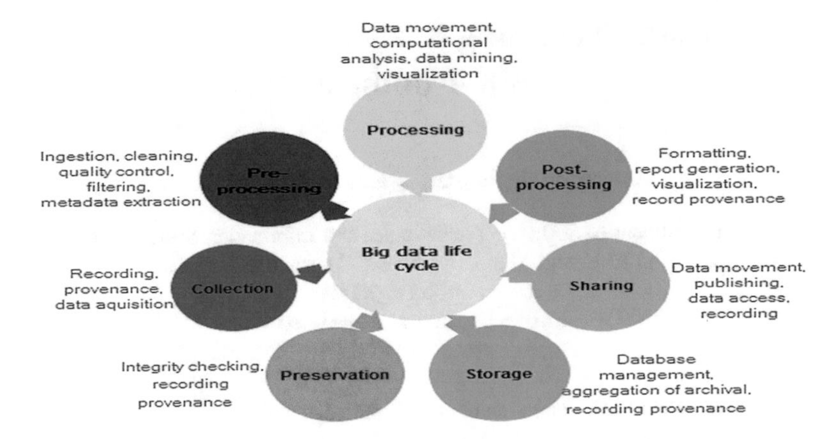

Fig. 1. Big data life cycle [5]

due to the issues and challenges that emerge when the volume of data grows considerably. In the data collection phase, data comes from various sources with different formats: structured, semi-structured, and unstructured [5]. As the collected data may contain sensitive information, big data storage systems need to ensure that the stored data are secure. Big data is collected and stored in order to generate useful knowledge in the processing phase. The processing of data performs data mining algorithms such as classification and clustering to extract new insights for decision-making [6]. The extracted insights may also contain sensitive data. So, data and the outsourced result must be protected from unauthorized access. In this paper, we focus on big data processing phase security.

Our purpose is to provide an overview of research for the security of big data processing in Cloud environment in order to identify the limitations of the existing schemes and motivate future research directions.

The rest of this paper is organized as follows. Section 2 describes big data processing and its phases. The main Cloud services proposed for big data processing are presented in Sect. 3. Section 4 exposes security issues related to big data processing in Cloud environment. Section 5 and 6 describe and discuss some existing schemes for secure big data processing. Finally, Sect. 7 concludes the paper.

2 Big Data Processing

Big data is a concept that can be described as the characteristics of three ‘V’s, Volume, Variety, and Velocity [7]. The term big in big data refers typically to the huge volume of data sets that are generated per day. The generation of data sets from various sources with diverse formats, types, and structures makes big data’s Variety. The Velocity of big data refers to the high speed in which data can be accessed and analyzed to provide the required result. Other Vs are also considered when looking at big data: Veracity and Value. The Veracity of big data describes the quality or trustworthiness of the original data and the acquired information where the Value is related to the potential gain for companies when extracting useful information and insights using this big data [7]. Processing large amounts of data requires numerous computation and storage resources. Thus, the main challenges of big data processing include capture, storage, and analysis [8].

2.1 Big Data Capture

Structured and unstructured data has to be captured from heterogeneous sources like Social Media, emails, mobile devices, video, etc. Since data sets that are generated daily are becoming increasingly diverse and voluminous, the organization must capture information in time [5]. Filtering the data to only keep the significant information is one of the capturing tasks. From a security perspective, ensuring that data are collected from trusted sources is a challenge [6]. Spoofing, spamming, and phishing attacks may hack data provider in order to get an access to the data in this phase [6].

2.2 Big Data Storage

Relational database management systems (RDBMSs) are facing challenges in handling big data [9]. In order to store structured and unstructured data, various storage systems are emerging. Based on Cloud environments, it is possible to store big data in a distributed manner. However, as the collected data may contain sensitive information, it is important to take sufficient security precaution [6].

2.3 Big Data Analysis

Analytics is the process of deriving insight from data [10]. Data mining algorithm used for big data analysis requires high computing and performance processors. To address these two main requirements of big data storage, processing, and analytics, the Cloud is the appropriate platform. Since big data is stored in a distributed way on the Cloud, it should be processed in parallel either inter processing or intra processing so that new knowledge can be discovered. Compared to intra big data processing, inter big data processing is more challenging, as big data sharing should first be executed before processing, and during the time of data sharing, many new security and privacy issues arise [11].

3 Cloud Computing Services for Big Data Processing

To store and analyze large amounts of data, many applications and systems are delivered in Cloud platforms as services. These services may be a file system for the storage of big data, a framework to organize the data and access to them in an efficient way, an execution tool to distribute the computational load among the servers in the Cloud and a query system for knowledge extraction. A common platform used for big data processing is Apache Hadoop. Hadoop [12] is the most used data processing engine in Cloud environments considering its scalability, ease-of-use, and fail-over properties [4]. To perform massive amounts of data, parallel and distributed techniques and models such as MapReduce are required. MapReduce [13] is the programming model of Hadoop that can process large volume of data in parallel. It splits data into chunks. JobTrackers using Map tasks process these chunks in parallel and store intermediate results in Hadoop Distributed File system (HDFS). The Reduce tasks take results from Map tasks as input in order to perform a summary collection. Thus, Hadoop is able to run MapReduce algorithms on an unlimited number of processing nodes [14]. HDFS is the open source counterpart of Google File System [15, 16]. Therefore, it provides a large distributed storage for data.

4 Big Data Processing Security Issues in Cloud Environments

While the volume of big data increases, data management and processing require a distributed storage and processing such as Cloud environment. However, Cloud Computing is still suffering from security issues [17, 18]. Thus, big data has lead to serious security challenges. Big data processing security issues have four aspects: Infrastructure security, Data provenance, Data management and, data/computation integrity and privacy.

4.1 Infrastructure Security

One of the main challenges for securing big data processing is to secure the infrastructures that ensure storage and processing activities. Many of technologies have been developed over Hadoop, MapReduce and Hive; However, they have not got adequate security protections yet [19]. MapReduce is the de facto standard for parallel/distributed processing systems [20]. However, various kinds of attacks such as impersonation, denial of service (DoS), replay and eave-drooping [21] torment MapReduce infrastructure and its distributed processing and storage. By an impersonation attack, a malicious user can execute a MapReduce job to perform wrong computations, or cause data leakage [21]. However, the main purpose of DoS is to make a mapper or a reducer non-functional and interrupt MapReduce computations. A trusted MapReduce must provide an access control scheme, authentication protocols, privacy preserving of storage and computations schemes, data and process integrity, data computations auditing, and data availability for authorized and authenticated users only.

Availability of Data and Nodes. In MapReduce, each job is divided into many tasks that are assigned to different nodes. The input of a job is stored in a distributed file system where multiple replicas are kept to ensure high availability. Techniques that can predict replay, impersonation and denial of service attacks are required [21].

Confidentiality of Data and Computation. Preserving confidentiality and privacy of data or computations refers to their protection from unauthorized users and the Cloud provider itself. Indeed, Cloud providers may not be curious to learn from users' sensitive information. Therefore, techniques that preserve the privacy of outsourced data and computation are needed.

The Integrity of Data and Computations. The protection of big data with a high level of integrity assurance requires ensuring that data storing and processing have not been altered or deleted by unauthorized entities [22]. Accountability and auditing are security requirements for both MapReduce and big data [23]. In MapReduce, accountability is provided when all working machines are held responsible for the tasks they have completed [23, 24]. Furthermore, to make effective decisions, data that are managed by MapReduce must be trustworthy. Assuring data or computations trustworthiness is related to ensuring the integrity of data and computations. Trustworthiness of data requires protection from malicious nodes that aim to derive the user's data or to compromise insights that they will be predicted from these data. Therefore, schemes and techniques that can verify completeness, correctness, and freshness of data and computations are recommended [21].

4.2 Data Provenance

Big data provenance refers to the information that describes the source, or origin of data [25]. As big data is collected from various sources, the volume of the data grows and complexity of provenance security will be more difficult. Hence, the malicious provenance of data leads to integrity problems that create a mix of bad and good data. The mixture of bad and good data in an environment that do not validate or control data input, leads up to wrong insights. In addition, the provenance of data might be compromised or deleted by intruders which leads to privacy and confidentiality breaches, data loss and unauthorized access to data. Therefore, ensuring data provenance using powerful protection techniques is an important step before processing of sensitive data.

4.3 Data Management

Data are varied by format, size and, technique of collection [26]. With the evolution of computing technologies, large amounts of data can be managed in parallel, as many tools are available in Cloud environments such as BigTable, Hadoop and MapReduce [14]. Data management using Hadoop and BigTable may provide the availability of data due to the replicated storage of distributed blocks. However, data privacy in these systems is a major concern whenever large amounts of data are processed. Processing such as data mining and predictive analytics can deduce information linkage that can cause the exposition of data [23]. Generally, security of data management refers to secure data storage, transaction logs, auditing, and data provenance [27].

4.4 The Integrity and Privacy of Data and Computation

Computation of big data is outsourced to the Cloud that is characterized by a processing power to perform tasks. However, data and computation integrity and privacy are the main security concerns for Cloud users. While the Cloud distributes the computation across its nodes, a compromised one can corrupt the integrity of many tasks performed by the Cloud. Therefore, data integrity should be checked constantly in order to prove that data and computation are intact [28]. Related to the outsourced data, security is required to gain the user's trust. It involves preserving its integrity and privacy. The Privacy of big data ensures that the data are only known by authorized and authenticated users. To provide data privacy, various techniques are proposed such as encryption, anonymization, and de-identification [29]. Big data integrity guarantees that the outsourced data are not modified or deleted. The users need to verify the integrity of their data and computation without downloading them [30]. Therefore, schemes that verify the integrity while preserving the privacy of data and computation are also required. In addition, while the generated data are dynamic in nature, the verification of integrity research needs to continuously evolve along big data and Cloud evolution in order to deal with the new emerging requirements and address new security issues [30].

5 Related Work

In order to secure big data processing in Cloud environments, many techniques and schemes are proposed. As we are interested in data and computation security schemes, we review the schemes that are suggested to secure outsourced data and computation in Cloud environment and discuss their advantages and shortcomings.

To validate data integrity and error correcting of large files, the authors in [31] present a new scheme called Proof of Retrievability (PoR). In this scheme, special blocks called sentinels are firmed to an encrypted file F. Encryption is used to make sentinels indistinguishable from other file blocks. To check the integrity of the data file F, the verifier challenges the prover by specifying the positions of some sentinels and asking them to return the associated sentinels values. Therefore, any modification or deletion of a substantial portion of F induces with a high probability the suppression of the sentinels [31]. This scheme preserves the privacy of the stored data file. However, it supports the verifiability only for the special blocks and it may generate computational storage overhead due to the encryption process, error correcting code inserted and the sentinels added when it is used for big data files. The authors of [32] proposed a PDP that can ensure the possession of the file on an untrusted storage. Using RSA-based homomorphic verification tags, it provides a public auditing of the outsourced data. This model helps the clients to check that their data, which is stored in the outsourced server have not been modified. However, it cannot deal with dynamic data verification [30]. Authors in [33] proposed a dynamic provable data possession using a rank-based authenticated skip lists to prove and update the remotely stored data. It supports dynamic data but it involves computation overhead. Moreover, it is used to check the integrity of file blocks and does not support public verifiability [30]. To support both public verification and dynamic data, authors in [34] proposed a new model that uses a batch-auditing protocol to audit multiple clients simultaneously. In this scheme, a trusted third-party auditor (TPA) delegates the public verification of dynamic data storage in Cloud. This model supports dynamic data operations such as data deletion [34] but it is not a fully trusted scheme because the verifier is public and assumed to be unbiased, which is not verified at all times. In [35], high availability and integrity layer for Cloud storage (HAIL) is suggested to manage the integrity and availability across independent server storage services. It makes use of PoR as building blocks by which storage resources can be tested and reallocated when failures are detected. A PoR uses file redundancy within a server for the verification and enables the prover to certify a client whose file is available without any loss. HAIL provides high availability of data. However, it is used for static data only [35]. In [4], the authors presented an integrity protection mechanism that is operated through a reputation-based system. The proposed protocol utilizes duplicated computation to locate potential malicious worker nodes and introduces a dynamic reputation system to increase the success probability of finding out malicious nodes [4]. The results of each worker nodes are checked. This Framework can reduce the risk of integrity breach [4]. However, it may generate a high computational cost when the volume of data to be processed is big due to the duplicated computation. In addition, the mechanism does not preserve the integrity of data storage. Thus, the data that is altered before computation can lead to incorrect conclusions. To protect the privacy of big data computation, authors in [36] proposed an encryption scheme to perform tasks over the encrypted data named Fully Homomorphic Encryption (FHE). However, computation over encrypted big data is complex and highly time-consuming [37]. To secure big data storage, authors in [38] proposed an approach that divides big data into sequenced parts based on certain properties and stores each part at a Cloud storage service provider. This approach protects the mapping of data elements instead of protecting the big data itself. In [39], a privacy-preserving big data deduplication approach based on the binary search tree (BST) and referred to as EPCDD is proposed. For three-tier cross-domain architecture, EPCDD achieves privacy-preserving, data accountability and availability, and efficiency. A labeled Homomorphic Encryption approach for privacy-preserving processing of outsourced data referred to as labHE is proposed in [40]. It can accelerate the homomorphic computation where the function submitted for computation on encrypted data is known to the decryptor. It achieves privacy against honest-but-curious Cloud servers. However, it does not address the integrity of computation results.

6 Discussion

Big data owner may outsource their data for storage in a Cloud environment and perform computation over these outsourced data. Therefore, schemes that can ensure both the privacy and integrity of the stored, processed, and shared data are required. In order to identify future research directions related to the integrity verification and privacy preserving of big data processing in a Cloud environment, we focus our discussion on some papers that addressed the privacy and integrity of data and computation as presented in Table 1. The comparison between these works is based on the main security requirements and public verifiability, the overhead, the efficiency, and large data support criteria where ✓ indicates the existence of the criterion, X indicates its absence. From the comparison table, we can infer that the approach that can cover most criteria is labHE [40]. As a future direction, labHE and HAIL can be used to provide privacy-preserving and public verification for big data processing in Cloud services.

Table 1. Comparison of data and computation security schemes

Ref	Addressed security issues					Overhead	Efficiency	Large data support
	Data Privacy	Integrity			Availability			
		Data	Computation	Public verifiability				
[32]	✓	✓	X	✓	✓	I/O, Computational	✓	✓
[33]	✓	✓	X	X	✓	Computation, Communication	X	✓
[34]	✓	✓	X	✓	✓	Computation, Communication	✓	✓
[31]	✓	✓	X	✓	✓	Computational	✓	✓
[35]	X	✓	X	✓	✓	Low	✓	✓
[4]	X	✓	✓	✓	✓	Computational	✓	X
[36]	✓	✓	✓	X	✓	Computational	X	✓
[38]	✓	X	X	X	X	Related to data size	✓	✓
[39]	✓	X	X	X	✓	Low	✓	✓
[40]	✓	X	X	X	✓	Low	✓	✓

7 Conclusion and Perspectives

The security of big data processing in Cloud environments is a topic that requires attention. This paper presented a review of big data processing security in Cloud environments. The main purpose of the present review is to identify the security issues of big data processing in Cloud environments in order to decide our future research direction. We respectively describe big data processing phases, present some Cloud technologies that perform big data processing and emphasize the main issues of big data processing. By considering the public verifiability, overhead, efficiency, and large data support of big data processing privacy, the integrity, and the availability various existing approaches have been discussed. From the comparison table, we find that the most appropriate scheme for big data processing privacy is the labeled Homomorphic Encryption. Our future work is to propose a labeled Homomorphic encryption-based approach that can preserve both the privacy and the integrity of big data processing over Cloud environments while overcoming open issues.

References

1. Chen, M., Mao, S., Liu, Y.: Big data: a survey. Mob. Netw. Appl. **19**(2), 171–209 (2014)
2. Shukla, S., Kukade, V., Mujawar, S.: Big data: concept, handling and challenges: an overview. Int. J. Comput. Appl. **114**(11), 6–9 (2015)
3. Ding, Y., Wang, H., Shi, P., Fu, H., Guo, C., Zhang, M.: Trusted sampling-based result verification on mass data processing. In: 2013 IEEE Seventh International Symposium on Service-Oriented System Engineering, pp. 391–396 (2013)

4. Gao, Z., Desalvo, N., Khoa, P., Kim, S., Xu, L., Ro, W., Verma, R., Shi, W.: Integrity protection for big data processing with dynamic redundancy computation. In: IEEE International Conference on Autonomic Computing, pp. 159–160 (2015)
5. Gómez-Iglesias, A., Arora, R.: Using high performance computing for conquering big data. In: Arora, R. (ed.) Conquering Big Data with High Performance Computing, 1st edn., pp. 13–30 (2016)
6. Alshboul, Y., Wang, Y., Nepali, R.: Big data lifecycle: threats and security model. In: Twenty-First Americas Conference on Information Systems, Puerto Rico, pp. 1–7 (2015)
7. What Is Big Data? - Gartner IT Glossary - Big Data. Gartner IT Glossary (2017). http://www.gartner.com/it-glossary/big-data. Accessed 25 Apr 2017
8. Jewell, D., Barros, R., Diederichs, S., Duijvestijn, L., Hammersley, M., Hazra, A., Holban, C., Li, Y., Osaigbovo, O., Plach, A., Portilla, I., Saptarshi, M., Seera, H., Stahl, E., Zolotow, C.: Performance and Capacity Implications for Big Data. IBM Redbooks (2014)
9. Hammad, K., Fakharaldien, M., Zain, J., Majid, M.: Big data analysis and storage. In: International Conference on Operations Excellence and Service Engineering, Orlando, Florida, USA, pp. 648–659 (2015)
10. Khan, S., Shakil, K., Alam, M.: Cloud based big data analytics: a survey of current research and future directions. In: CSI Annual Convention, pp. 1–12 (2015)
11. Nadar, S., Gawai, N.: Unstructured big data processing: security issues and countermeasures. Int. J. Sci. Eng. Res. **6**(3), 201–204 (2015)
12. Welcome to Apache™ Hadoop®!: Hadoop.apache.org (2017). http://hadoop.apache.org/. Accessed 22 Apr 2017
13. MapReduce Tutorial: Hadoop.apache.org (2017). https://hadoop.apache.org/docs/r1.2.1/mapred_tutorial.html. Accessed 22 Apr 2017
14. Menaka, N., Jabasheela: Survey on big data processing using Hadoop, map reduce. Int. J. Innov. Res. Inf. Secur. (IJIRIS) **1**(3), 24–28 (2014)
15. Ghemawat, S., Gobioff, H., Leung, S.: The Google file system. ACM SIGOPS Oper. Syst. Rev. **37**(5), 29 (2003)
16. Deepika, P., Ananatha Raman, G.: Hadoop MapReduce - Wordcount implementation. Int. J. Sci. Dev. Res. (IJSDR) **1**(3), 130–132 (2016)
17. Hashizume, K., Rosado, D., Fernández-Medina, E., Fernandez, E.: An analysis of security issues for cloud computing. J. Internet Serv. Appl. **4**(1), 5 (2013)
18. Solanki, J., Davda, R., Jadeja, V., Patel, C.: A survey: cloud computing challenges and security issues. Int. J. Mod. Trends Eng. Res. **4**(3), 57–61 (2017)
19. Thuraisingham, B., Cadenhead, T., Kantarcioglu, M., Khadilkar, V.: Secure Data Provenance and Inference Control with Semantic Web, 1st edn., pp. 1–28. CRC Press (2014)
20. Kang, S., Lee, S., Lee, K.: Performance comparison of OpenMP, MPI, and MapReduce in practical problems. Adv. Multimedia **2015**, 1–9 (2015)
21. Tejaswi, Y., Kumar, M., Keerthi, V.: Security and privacy issues using mapreduce on clouds, Anveshana's Int. J. Res. Eng. Appl. Sci. **2**(1), 250–259 (2017)
22. Lebdaoui, I., El Hajji, S., Orhanou, G.: Managing big data integrity. In: 2016 International Conference on Engineering and MIS (ICEMIS), pp. 1–6 (2016)
23. Grolinger, K., Hayes, M., Higashino, W., L'Heureux, A., Allison, D., Capretz, M.: Challenges for MapReduce in big data. In: 2014, IEEE 10th World Congress on Services, pp. 182–189 (2014)
24. Xiao, Z., Xiao, Y.: Achieving accountable MapReduce in cloud computing. Future Gener. Comput. Syst. **30**, 1–13 (2014)
25. Glavic, B.: Big data provenance: challenges and implications for benchmarking. In: Specifying Big Data Benchmarks, pp. 72–80 (2014)

26. Satyanarayana, L.: A survey on challenges and advantages in big data. Int. J. Comput. Sci. Technol. **6**(2), 115–119 (2015)
27. Ayaydin, A., Terzi, D., Sagiroglu, S.: Security, privacy and forensics issues on big data and cloud computing. In: International Conference on Computer Science and Engineering. At Tekirdağ Namık Kemal Üniversitesi (2016)
28. Aldossary, S., Allen, W.: Data security, privacy, availability and integrity in cloud computing: issues and current solutions. Int. J. Adv. Comput. Sci. Appl. **7**(4), 485–498 (2016)
29. Jain, P., Gyanchandani, M., Khare, N.: Big data privacy: a technological perspective and review. J. Big Data **3**(1), 1–25 (2016)
30. Liu, C., Ranjan, R., Zhang, X., Yang, C., Chen, J.: A big picture of integrity verification of big data in cloud computing. In: Handbook on Data Centers, pp. 631–645 (2015)
31. Juels, A., Kaliski, B.: Pors. In: Proceedings of the 14th ACM Conference on Computer and Communications Security - CCS 2007 (2007)
32. Ateniese, G., Burns, R., Curtmola, R., Herring, J., Kissner, L., Peterson, Z., Song, D.: Provable data possession at untrusted stores. In: Proceedings of the 14th ACM Conference on Computer and Communications Security - CCS 2007, Alexandria, Virginia, USA, pp. 598–609 (2007)
33. Erway, C., Küpçü, A., Papamanthou, C., Tamassia, R.: Dynamic provable data possession. In: Proceedings of the 16th ACM Conference on Computer and Communications Security - CCS 2009, Chicago, Illinois, USA, pp. 213–222 (2009)
34. Wang, Q., Wang, C., Ren, K., Lou, W., Li, J.: Enabling public auditability and data dynamics for storage security in cloud computing. IEEE Trans. Parallel Distrib. Syst. **22**, 847–859 (2011)
35. Bowers, K., Juels, A., Oprea, A.: HAIL: a high-availability and integrity layer for cloud storage. In: CCS 2009 Proceedings of the 16th ACM Conference on Computer and Communications Security, Chicago, Illinois, USA, pp. 187–198 (2009)
36. Gentry, C.: Fully homomorphic encryption using ideal lattices. In: Proceedings of the 41st Annual ACM Symposium on Symposium on Theory of Computing - STOC 2009, pp. 169–178 (2009)
37. Miloslavskaya, N., Makhmudova, A.: Survey of big data information security. In: IEEE 4th International Conference on Future Internet of Things and Cloud Workshops (FiCloudW), Vienna, Austria, pp. 133–138 (2016)
38. Cheng, H., Rong, C., Hwang, K., Wang, W., Li, Y.: Secure big data storage and sharing scheme for cloud tenants. China Commun. **12**(6), 106–115 (2015)
39. Yang, X., Lu, R., Choo, K.K.R., Yin, F., Tang, X.: Achieving efficient and privacy-preserving cross-domain big data deduplication in cloud. IEEE Trans. Big Data (2017, in press)
40. Barbosa, M., Catalano, D., Fiore, D.: Labeled Homomorphic Encryption: Scalable and Privacy-Preserving Processing of Outsourced Data, 326 (2017)

A Scalable Semantic Resource Discovery Architecture for the Internet of Things

Rafik Mecibah(✉), Badis Djamaa(✉), Ali Yachir(✉), and Mohamed Aissani(✉)

Computer Science Department, Ecole Militaire Polytechnique (EMP), Algiers, Algeria
rafik.mecibah@gmail.com, badis.djamaa@gmail.com, ali.yachir@gmail.com, m.aissani@gmail.com

Abstract. Resource discovery in the Internet of Things (IoT) provides information about the available sensors, and actuators as well as the services and resources they provide. This task, crucial for IoT applications evolution, allows self-configuration and context awareness. When combined with semantic technologies, resource discovery increases the interoperability and backs the reasoning process. However, introducing semantics into the IoT incurs challenges regarding resources and power consumption as well as the compactness of representations and resource lookup efficiency. In this paper, we discuss such challenges and propose an architecture for including semantics in a scalable IoT system. Next, we present a mechanism aiming to increase semantic resource discovery efficiency in heterogeneous IoT systems. The proposed mechanism is based on the construction of a concept directory to accelerate semantic-based resource lookup. By using such a directory, we take benefits of semantic representations to better match user requests, without lowering discovery task efficiency. The feasibility and effectiveness of our solution was demonstrated by experimental evaluations.

Keywords: Internet of Things · Resource discovery · Semantic web
Semantic annotation

1 Introduction

By enabling everyday devices to communicate over the internet, the physical world has met the digital one to form an ecosystem known as the Internet of Things (IoT). The number of Internet connected devices such as smartphones, glasses, thermostats, bulbs, etc. has already surpassed the world population [1] and it is expected to keep growing. This remarkable growth of the IoT has among its consequences a huge number of IoT services which lead to a better control of the environment and the creation of new business opportunities. However, such an abundance of services and resources makes it difficult for applications and users to find the right service at the right time and place.

Resource discovery is a crucial task in the IoT since it allows the localization of the available resources and accelerates their retrieval in order to fulfill user and application requirements. Nevertheless, the heterogeneity of applications/users requesting

O. Demigha et al. (Eds.): CSA 2018, LNNS 50, pp. 37–47, 2019.
https://doi.org/10.1007/978-3-319-98352-3_5

sensor/actuators services along with the diversity of devices and their resource representations renders finding the right resource for a specific request more difficult.

Semantic technologies can solve the heterogeneity issue. Indeed, Web capabilities of sharing and integrating data were enhanced by adding ontology-based metadata (aka semantic annotation) to make them more understandable by other applications. These technologies, however, are considered too heavy for IoT devices, characterized by limited computation power, memory, communication and energy resources [2, 3]. These constraints push to develop new discovery solutions.

This paper tackles such an issue by proposing a new architecture for integrating sematic-rich representations in a scalable IoT resource discovery solution. The core challenges are devided and discussed in order to succeed efficient semantic resource discovery of the requested IoT services. In line with such architecture and challenges, a first mechanism allowing accelerating semantic lookup operations in an interoperable IoT system is presented. Such a mechanism is based on the construction of a Concept Directory to accelerate semantic-based resource lookups.

The remainder of this paper is organized as follows: Sect. 2 presents the recent works related to semantic resource discovery in the IoT. Section 3 introduces the proposed semantic resource discovery architecture for context-aware scalable IoT systems, along with the related challenges. This is followed in Sect. 4 by the proposition of a mechanism aiming at achieving efficient scalable semantic discovery in heterogeneous IoT systems. Section 5 is dedicated to evaluate the performance of the proposed mechanism when compared with the existing solutions. The paper is concluded in Sect. 6, where ideas of future directions are presented.

2 Related Work

Resource discovery in the IoT has been paid a big attention in both academia and industry. As a consequence, many architectures and frameworks are proposed to give solutions for its issues [4]. Such solutions are based on syntactical matchmaking between user requests and resource descriptions or considers resource and request semantics to achieve better results. A considerable number of the published syntactic-based contributions are based on Constrained Application Protocol (CoAP) [5], either as a centralized or distributed solution [6, 7]. In [8], for instance, a CoRE Resource Directory (RD) has been proposed to achieve centralized discovery of IoT resources working over CoAP. RD is an edge repository with which the devices register their resources by sending a CoRE Link Format document containing the corresponding URIs and Web linking attributes [9]. However, while the RD is well-suited for infrastructure-based environments, it suffers from the single point of failure issue. This latter can be solved by the direct solution with excessive use of multicast. Both, however, do not provide solutions addressing the complexity of IoT systems, especially concerning scalability.

In order to address the above, other contributions such as DRD (Distributed Resource Directory) [10], HRD (Hybrid Resource Discovery) [11], and P2P-based resource discovery [12] have been proposed. The latter, for instance, particularly considers the resources to be discovered by sharing the RDs following a Peer-to-Peer

(P2P) paradigm. Although experimentation and simulation results demonstrated the feasibility and the medium sized scalability of the solution, it works only with CoAP. In [13], the authors proposed another CoAP-based architecture that tackles the discovery of IoT resources at the edge (logical domain-based gateway) level, by using CoRE RD, and then used the XMHT implementation for HT to bring the RDs content to the cloud level to contribute in a P2P overlay.

Syntactic resource discovery generally lacks support for efficient interoperability between different solutions working over a multitude of protocols such as CoAP, MQTT and XMPP. To address these issues, scalable semantic resource discovery in the IoT is being actively investigated. Indeed, recent research efforts are trying to enable rich, context-aware, and interoperable discovery by proposing semantic protocol-free solutions. For instance, following the publish/subscribe (pubsub) paradigm, the authors in [14] designed an architecture that enables the interoperability between IoT application by managing the semantic annotated data at SPARQL endpoints. Desai et al. [15] proposed the concept of Semantic Gateway as Service (SGS) following the service-oriented architecture. The SGS is set between the power constrained sink node (sensors and actuators) and the IoT services (at the cloud level) and proposes a multi-protocol proxy between heterogeneous clients at the low level. It provides semantic annotation of raw sensor data using reference and domain specific ontologies following Open Geospatial Consortium (OGC) specifications [16]. The interfacing between the SGW and the external environment is done either with REST or pubsub.

Such an architecture allows for interoperability at both messaging and service levels by involving a matching mechanism between REST messages and pubsub topics at the former and using semantic technologies at the latter level. However, the semantic annotation is done after resource discovery; for instance, to answer a GET request from a REST interface; the SGW matches the id in the request with topics among the router topics and returns the most recent message then annotates the latter. This mechanism does not permit semantic-based requests for discovery and the architecture needs to contain directories where descriptions of the available resources are already annotated to better handle semantic requests. Additionally, using Semantic Web technologies directly on constrained devices adds excessive burden on these devices.

Our study of relevant solutions for resource discovery in the IoT leads us to conclude the need for a scalable solution that enables machine interpretability of resource descriptions via semantic technologies and takes into account the constrained and complex nature of IoT environments. Such an architecture is the subject of the following section.

3 The Proposed Architecture

This section presents our architecture aiming to take advantage of semantic technologies for an interoperable, scalable user- and context-aware resource discovery in the IoT. The overall architecture is presented in Subsect. 3.1. And the details of its components along with its challenges will make the subject of Subsect. 3.2.

3.1 Overall Architecture

In a scalable IoT system, the devices and their hosted resources are under the administration of different tents. The offered resources are generally provided to the outside world, for discovery and invocation, via resource directories hosting their descriptions. The descriptions might be presented in different formats and queried via a multitude of protocols. Constrained IoT devices, hosting these resources, exchange compact resource descriptions with the resource directories. Some of the resources might be observed or subscribed to. In such a case, the client (subscriber) is notified when the resource has changed. To foster interoperability between these resource directories and allow the users and applications to get the most out of the available resources, this paper presents a P2P semantic-enabled architecture depicted in Fig. 2.

Our architecture is made up of three main parts (levels), namely: the directory level; the constrained-network level; and the non-constrained network (Internet) level as can be distinguished in Fig. 1. Each part of the proposed architecture requires specific techniques and mechanisms in order to fully build an interoperable, scalable, context-aware discovery solution for the IoT.

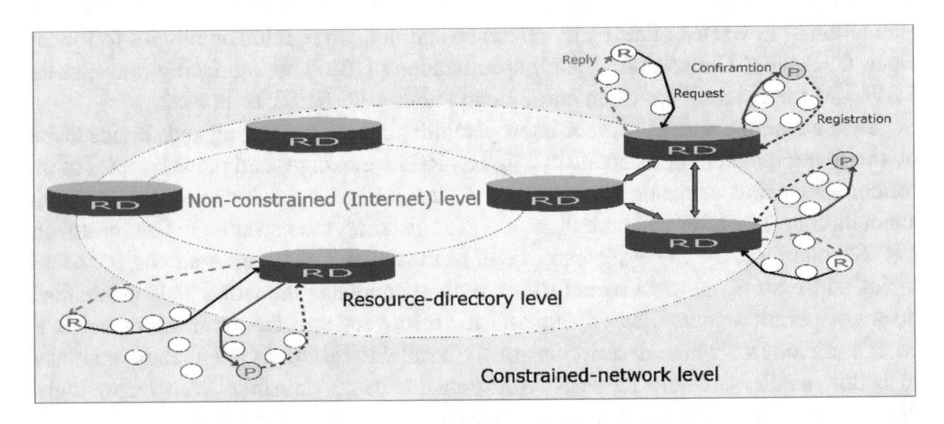

Fig. 1. Overall architecture

At the constrained-network level, for instance, special care needs to be given to compact representations of resource descriptions in order to preserve device capacities. At the same time, a context-aware solution should aim to get the most complete semantics to fulfill user/application requests with the required quality of service. Efficient communications between constrained IoT devices and the corresponding resource directories must also be considered in order not only to preserve the very-limited bandwidth available but also to address sleepy-node [17] and energy shortage requirements.

At the non-constrained-network level, although the P2P mechanism ensures dependability and provides scalability features for our architecture [13], many challenges are to be addressed in order to succeed its functions. For instance, designing efficient communication solutions between directories along with strong replication schemes that responds at the right time to the dynamicity of resources is an important challenge to be addressed. Additionally, being able to locate rapidly the directories able to answer requested services is an crucial feature for discovery success.

Finally, even with the potential features our architecture brings at the non-constrained and constrained-parts of an IoT network, the efficiency of this architecture depends on the way semantic technologies are applied at the resource directories level and the amount and relevance of the semantic information communicated within the other two parts of the architecture. Indeed, it is crucial to design efficient semantic translators that can take a user/application request expressed in a specific format and find out corresponding resources that might be represented in different formats or should be accessed through other protocols. It should be noted that the semantic-operations required by our architecture can also be physically separated from the resource directory.

Having described the overall architecture of our discovery system, the following subsection details each level and discusses the challenges that must be addressed in order to achieve scalable interoperable resource discovery in the IoT.

3.2 The Architecture in Details

This subsection details the interior composition of each part of our discovery solution with discussions of the main challenges to be addressed.

Resource-directory level, to allow heterogeneous IoT devices, applications and users to discover resources, it is necessary to dispose of a multi-proxy module as shown in Fig. 3. The latter does not only allow for translating the exchanged messages depending on devices' protocols, but also provides resource descriptions in a common format, which facilitates, thereafter, metadata extraction and adding. The outputs of this module are then annotated semantically by adding the appropriate ontologies-based concepts to them. The references and domain specific ontologies used in one RD have to be hosted by the ontology servers of the other RD, so that a semantic request can be interpreted in the same way. In [15], for instance, the authors used the OGC standard for this reason. Once the descriptions are annotated, they might be stored in a semantic resource description repository as RDF Triplestore. The latter can be seen as a result of the semantic resource discovery task at the RD level and can be used either for reasoning or for answering semantic lookup requests. Finally, the RD should contain a module to interface with the other parts of the architecture in order to allow translations between RDF and other request languages such as CORE Link Format, JSON-LD and RESTdesc.

Non-constrained (internet) level, At the Internet/non-constrained-network level, resource directories might communicate with each other over a structured P2P overlay built using distributed hash tables (DHT) [18]. In such a solution, a DHT that contains information about the locally discovered resources as well as their semantic descriptions is associated to each resource directory (Fig. 2).

The global semantic resource discovery can then be fulfilled by a P2P global lookup of DHTs contents. Such a solution, however, has the inconvenience of increasing network traffic, which might lead to congestion, especially when resource

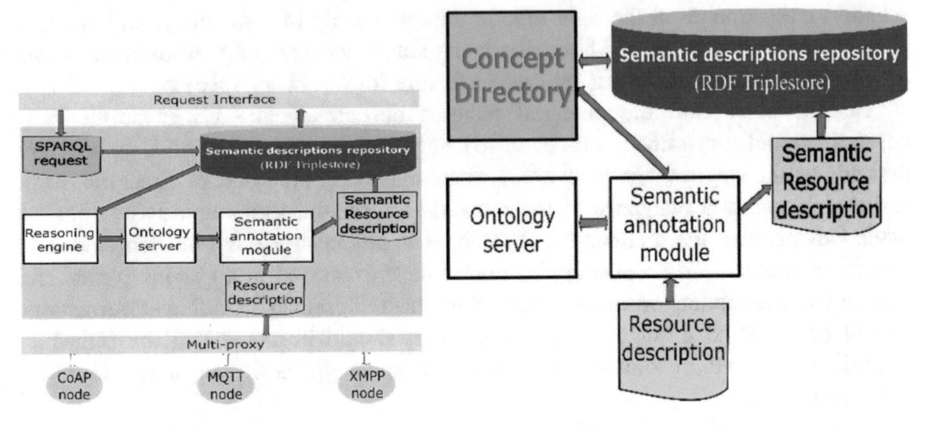

Fig. 2. The resource directory level

Fig. 3. The proposed concept directory

descriptions churn increases. In order to address these issues, new mechanisms adapted to the context of semantic discovery in the IoT should be developed.

Constrained-network level, the constrained network level of the proposed architecture is characterized by resource limited devices that have low computational capabilities, as well as short lifetime batteries, small storage memories, and limited bandwidth. These constraints have to be carefully considered when designing semantic descriptions and annotations of the provided IoT resources. Many communication protocols have been proposed to tackle the efficiency issue. Popular ones are CoAP, MQTT and XMPP [15]. Also, some semantic description enabling formats are being investigated for suitability at this level of the network. Such formats include JSON-LD, Core Link Format and RESTdesc.

Introducing semantics address interoperability issues and facilitates cross-protocol, cross-description and cross-node discovery in the IoT. However, it presents another issue regarding time-efficiency of the discovery. This might compromise the whole resource discovery process. Designing a mechanism to accelerate semantic discovery would, therefore, provide both advantages of semantics and fast request satisfaction. A technique addressing this issue will be presented in Sect. 4.

4 Delay Lowering Mechanism

Having presented our SRD architecture, detailed its components and discussed the related challenges, this section presents a contribution aiming to accelerate semantic resource discovery at the resource directories level (Sect. 3.2).

4.1 Resource-Directory Level

The proposed solution tackles the delay issue produced when retrieving matching semantic descriptions in a native way. Indeed, semantic descriptions are usually stored in RDF triple stores, which may delay information retrieval, especially when resource descriptions require sized RDF representation [19]. This section proposes a practical presentation of ontology-based descriptions via a concept directory (CD). Such directory, connected to the semantic annotation module and the semantic resource directory as shown in Fig. 3. Certainly, in a dynamic big-sized IoT environment, a huge number of resources exist and might provide similar services. Assuming that the existing resources are semantically annotated using the same ontologies; it becomes easier to find out the resources proposing the same or semantically similar services based on the concepts they refer to. This concept-based grouping might be indispensable in scenarios when a few descriptions inside a large RDF triple store match the request criteria. This observation is the drive behind our mechanism.

The CD mechanism proposes to create a directory containing the concepts available in an RD in order to allow faster semantic resource discovery. To do so, CD is constructed in a way where each entry is indexed by a concept and one instance of it and contains URIs of all the resources semantically defined, uniquely or among others, by the indexing concept as depicted in Table 1.

Table 1. An overview of the concept directory

Concept	Instance	Related resource URIs
Location	Room1	URI_1; URI_2; …; #tempsensroom1–2
	Room2	
	Corridor	
Quantity-kind	Temperature	#Room3thermometer;;#tempsensroom1–2
	Pressure	
	Humidity	
	…	

Within CD, a request is first matched with the concept directory to decide on the existence of requested resources in a given directory. This way, the CD mechanism presents a number of features allowing for a fast, scalable and reliable semantic resource discovery in the IoT. For instance, it:

- Allows for direct use of semantics to rapidly match the request criteria.
- Assembles the resources having similar ontology-based semantics.
- Avoids unnecessary exploration of the whole RDF triple store.
- Provides an ontology-based preview of the represented IoT segment for global discovery and reasoning mechanism.

The concepts chosen as indexes are those who are practically used as subject of the request. For instance, to the resource description presented in Fig. 4, only the "Quantity kind" and the "Location" will participate in CD construction as they can be retrieved in relevant requests about a specific sensing measurement or the presence of some "thing" in a specific location.

4.2 Resource-Directory Level

CD works closely with the RD to update its contents according to the dynamicity of the environment. Indeed, CD updating mechanism needs to be run: (i) every time a resource is annotated; (ii) when a resource leaves the IoT segment.

In the former (i); if a concept does not exist in any of CD entries, it is added with the corresponding instance as a secondary index and the current resource URI as a single element list for the new CD entry. Otherwise, the resource URI is added to the URIs list indexed by the concept and the current instance if this latter exists. If not, a new entry indexed by the concept and the current instance is created and inserted in the URI of the resource, as shown in Fig. 5. In the latter case (ii), all occurrences of the related URI are deleted from the CD, and if it re-joins, it is not annotated again. Therefore, CD has to retrieve its semantic description and its URI is added following the same procedure when resources are annotated.

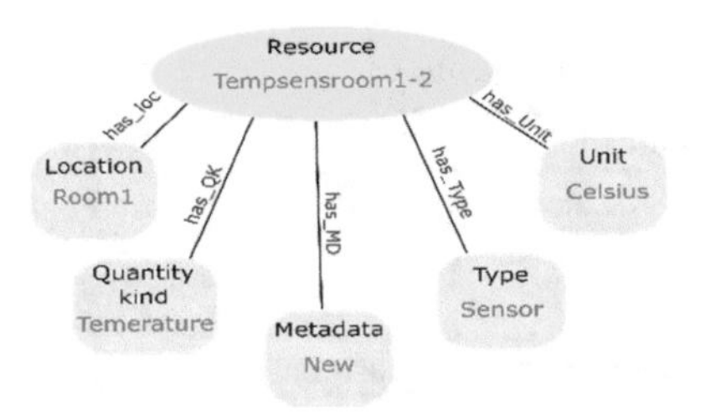

Fig. 4. Example of resource description

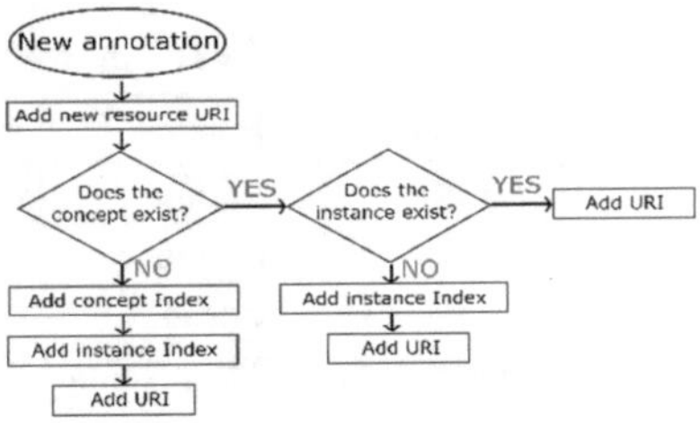

New annotation
Add new resource URI
Does the concept exist? — YES → Does the instance exist? — YES → Add URI
NO → Add concept Index → Add instance Index → Add URI
NO → Add instance Index → Add URI

Fig. 5. Organigram of CD updating

5 Performance Evaluation

This section focuses on the performance evaluation of the concept directory solution proposed in Sect. 4.

5.1 Evaluation Methodology and Metrics

The scenario considered in this evaluation uses an IoT system of one RD containing the semantic descriptions of resources provided by the constrained IoT devices constituting the constrained part of the IoT system (Sect. 3.1). This methodology is inspired from recently published work. For instance, the semantic resource descriptions repository, used in our evaluations, is constructed based on some modifications of the scheme proposed in [15]. Thus, instead of annotating the responses to external requests as in [15], we annotate the topic router, which contains sensor states and resource descriptions.

In our simulations, we calculate the average discovery time for the naïve exploration of RDF Triplestore by exploring all the semantic resource descriptions, then we compare it with that of CD. Each request is based on one concept and repeated for all the concepts involved in CD for the sake of averaging. In the first set of experiments, the average discovery time as a metric of the efficiency of the semantic discovery, was measured when varying the number of resources from a value of 100 to 5000 with a step of 100 for a fixed number of indexing concept instances (50 instances by a chosen concept). In the second set of experiments, we calculate an estimated size rate between the CD and the native approach when varying the number of resources from 1000 to 10000 with a step of 1000 for the same number of instances as in the previous scenario.

5.2 Results and Discussions

Obtained results from Matlab simulations of the above scenarios are depicted in Figs. 6 and 7.

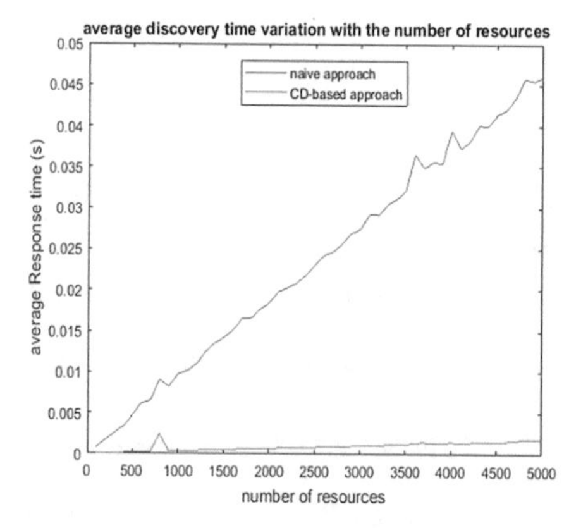

Fig. 6. Average discovery time

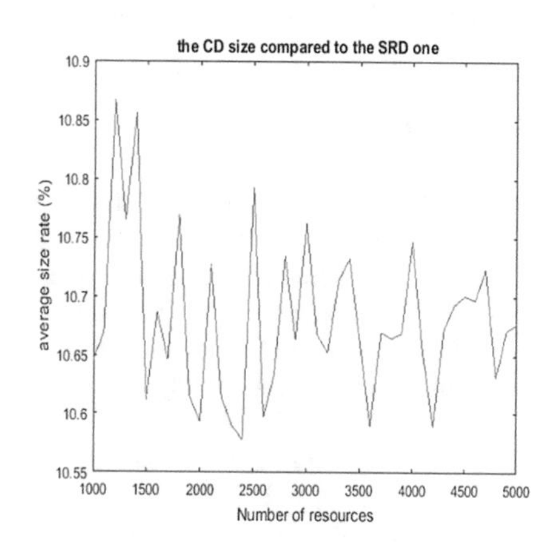

Fig. 7. Average size ratio

Figure 6 shows the average discovery time of the two evaluated solutions when varying the number of provided resources. Clearly, the discovery time in the proposed approach is, by far, shorter than the naïve approach's one, which can be explained by the fact that, unlike CD, the naïve approach checks all the resource descriptions contents which involves useless exploration for semantic resource definitions. Hence, the more the resource number increases the more the number of non-relevant descriptions increases, which delays the discovery for the naïve approach.

Figure 7 Shows the additional required storage space when implementing the concept directory, presented as a ratio between the CD size and the SRD's one. The graph presents a ratio value above 10,5% for a variable number of resources which is to be considered when choosing the hosting gateway.

As a summery, the discovery time using CD is clearly better than that of the naïve approach. This is very promising as it makes the exploitation of the semantic descriptions more efficient. However, CD requires additional memory.

6 Conclusion and Future Work

This paper deals with the semantic discovery of resources in large-scale IoT environments. We proposed an architecture that takes into account both the constrained and non-constrained network level requirements. To deal with the delay caused by the amount and nature of the semantic descriptions of resources, a CD mechanism was introduced. The obtained results show the capacity and advantages of the proposed semantic discovery mechanism when compared to the naïve exploration of RDF triple stores. Ongoing work considers larger scale simulations along with experiments on publically available IoT testbeds. Ideas on how to enhance the CD mechanism are also being investigated. Future work will tackle some of the challenges facing the proposed architecture at the constrained and non-constrained parts of the network.

References

1. Evans, D.: The Internet of Things - how the next evolution of the internet is changing everything. CISCO White Paper, pp. 1–11, April 2011
2. RFC 7228: Terminology for Constrained Node Networks, pp. 1–3 (2014)
3. Howitt, I., Gutierrez, J.A.: IEEE 802.15.4: Low-Rate Wireless Personal Area Networks (LR WPANs), vol. 2011, September 2011
4. Chikkamannur, A.A.: Study of resource discovery trends in Internet of Things (IoT). Int. J. **3089**, 3084–3089 (2016)
5. Shelby, Z., Hartke, K., Bormann, C.: Constrained Application Protocol (CoAP). CoRE Work. Gr., pp. 1–118 (2013)
6. Djamaa, B., Yachir, A.: A proactive trickle-based mechanism for discovering CoRE resource directories. Procedia Comput. Sci. **83**(Ant), 115–122 (2016)
7. Liu, M., Leppanen, T., Harjula, E., Ou, Z., Ylianttila, M., Ojala, T.: Distributed resource discovery in the machine-to-machine applications. In: Proceedings - IEEE 10th International Conference on Mobile Ad Hoc and Sensor Systems, MASS 2013, pp. 411–412 (2013)
8. Shelby, Z., Bormann, C.: CoRE Resource Directory. IETF draft-ietf-core-resource-directory-02 (2014)
9. Shelby, Z., Vial, M.: CoRE Interfaces. IETF draft-ietf-core-interfaces-02 (2014)
10. Liu, M., et al.: Distributed resource directory architecture in Machine-to-Machine communications. In: International Conference on Wireless and Mobile Computing, Networking and Communications, pp. 319–324 (2013)
11. Cheshire, S., Krochmal, M.: RFC 6763: DNS-based service discovery. Internet Engineering Task Force, pp. 1–49 (2013)
12. Cirani, S., et al.: 3A scalable and self-configuring architecture for service discovery in the Internet of Things. IEEE Internet Things J. **1**(5), 508–521 (2014)
13. Tanganelli, G., Vallati, C., Mingozzi, E.: Edge-centric distributed discovery and access in the Internet of Things. IEEE Internet Things J. **4662**(c) (2017)

14. Roffia, L., et al.: A semantic publish-subscribe architecture for the Internet of Things. IEEE Internet Things J. **3**(6), 1274–1296 (2016)
15. Desai, P., Sheth, A., Anantharam, P.: Semantic gateway as a service architecture IoT interoperability. In: Proceedings of the IEEE 3rd International Conference on Mobile Services, MS 2015, pp. 313–319 (2015)
16. Borges, K.a.V.: Open geospatial consortium. Measurement **2007**(5), 1–17 (2011)
17. Shelby, Z., Chakrabarti, S., Nordmark, E., Bormann, C.: Neighbor discovery optimization for IPv6 over low-power wireless personal area networks (6LoWPANs). Stand. Track, pp. 1–55 (2012)
18. Dabek, F.: A distributed hash table. Science **80**(2000), 2–134 (2005)
19. Khodadadi, F., Sinnott, R.O.: A semantic-aware framework for service definition and discovery in the Internet of Things using CoAP. Procedia Comput. Sci. **113**, 146–153 (2017)

Dynamic Clustering for IoT Key Management in Hostile Application Area

Soumaya Souaidi[1](✉), Tayeb Kenaza[1], Badis Djamaa[1], and Monther Aldwairi[2]

[1] Information Security Laboratory, Ecole Militaire Polytechnique, Bordj El Bahri, Algeria
souidi.sou@gmail.com, ken.tayeb@gmail.com, djamaa.badis@gmail.com

[2] College of Technological Innovation, Zayed University, P.O. Box 144534, Abu Dhabi, United Arab Emirates
monther.aldwairi@zu.ac.ae

Abstract. The IoT development area has drawn the attention of nowadays researchers, some of them made assumptions regarding the use of clustering in their key management schemes. For example, in CL-EKM *(Certificateless Effective Key Management)* protocol, cluster-heads are assumed to be with high-processing capabilities and deployed within a grid topology. In fact, this is only possible in a controlled environment. In a hostile environment, such as battlefields, this assumption cannot be satisfied. In this work, an enhancement of the CL-EKM scheme has been proposed by introducing a distributed clustering algorithm. The performance of the implemented and enhanced system proved our assumptions.

Keywords: Clustering · Certificate-less public key cryptography
Key management · Internet of Things (IOT) security
Elliptic curve cryptography (ECC)
Certificate-Less Effective Key Management (CL-EKM)
Dynamic networks · Mobility

1 Introduction

Due to recent technological developments, the IoT has grown, and it raises questions concerning the security of people and properties. This security can be ensured by securing the communications and the management of mobility in the network, since the mobility of objects is the main feature of the IoT [1]. Since ECC *(Elliptic curve cryptography)* is more competent in terms of computation and ensures high security with only short key [2], several approaches have been proposed based on it. Such as Chatterjee et al. in [3], where they proposed an ID-PKC *(Identity-based Public Key Cryptography)* based key management scheme. However, the pairing operations in this scheme are very expensive in terms of energy and calculation time compared to standard operations like ECC point multiplication. There is also Alagheband et al. in

O. Demigha et al. (Eds.): CSA 2018, LNNS 50, pp. 48–56, 2019.
https://doi.org/10.1007/978-3-319-98352-3_6

[4] who proposed a ECC-based signcryption scheme for key management. Unfortunately, their proposition was insecure against message forgery.

CL-EKM *(Certificate-Less Effective Key Management)* is a certificate-less key management scheme proposed by Seo et al. in [5], which supports the creation, distribution and revocation of four (04) key types. They assume that the cluster heads are of different types from the ordinary nodes, with higher computation power and backup, their deployment is deterministic and uniform and their role is predefined. This assumption renders the deployment very difficult in real world because it is only possible in a controlled environment. In hostile environments such as battlefields or forest fire monitoring, this assumption cannot be satisfied.

This fact encouraged us to propose an improved scheme of the CL-EKM protocol where it allows the creation, distribution and revocation of four keys. It is designed to protect a dynamic network against various attacks such as compromised node attack, cloning and impersonation attacks, and to ensure forward and backward secrecy. After the analysis of [6, 7], the clustering phase was improved by proposing the use of a distributed clustering algorithm. Cluster heads should be changed periodically to guarantee a balancing between nodes' energy consumption. All these operations regarding a distributed clustering algorithm has been implemented and evaluated to highlight the effect of the proposed improvements.

The rest of this paper is organized as follows; Sect. 2 gives an overview of the proposed improvements. In Sect. 3 detailed implementation was presented. The performance analysis results are discussed in Sect. 4, followed by a conclusion.

2 Overview

Seo et al. [5] suppose that the cluster heads are of a different nature from the ordinary nodes with higher performances in terms of processing and storage. Besides, the cluster heads (CH) are static and their role is predefined. In addition, their deployment is deterministic and they are uniformly scattered in the network according to a grid. This assumption is unrealistic especially in hostile environments. Our solution aims to consider the general case of a dynamic distributed network. The following subsections explain the general scheme based on dynamic clustering.

2.1 Key Types

Like the CL-EKM [5], the modified scheme requires the use of four (04) types of keys:

- ***Partial Public/Private Key:*** The KGC (*Key Generation Center*) in the base station (BS) generates a unique pair of private/public keys for each node.
- ***Individual Node Key:*** Each node in the network shares with the BS a unique individual key used to encrypt the exchanged messages between them.
- ***Pairwise Key:*** Each node shares a different key with each of its neighbors. This key is used to encrypt the exchanged messages between the node and its neighbor.
- ***Cluster Key:*** In a cluster, all the members share a unique key called the cluster key or group key. This key is used to encrypt the messages broadcasted in the cluster.

2.2 Protocol Scheme

The scheme is composed of six (06) phases namely: (1) the system setup phase where the system parameters are fixed, it takes place before the deployment of the network; (2) The pairwise key generation phase in which all nodes establish their pairwise keys; (3) The cluster forming phase where the clusters are formed by executing the LEACH *(Low Energy Adaptive Clustering Hierarchy)* clustering algorithm, then the group keys are established; (4) The key updating phase which is executed periodically to update the encryption and the group keys; (5) The node movement phase where the nodes can move within the network; (6) Last and not least, the node revocation phase where some nodes can be considered malicious and excluded from the network with all its keys. The general diagram of the proposed scheme is illustrated in Fig. 1 where the colored zones represent the phases we have intervened. After going into the steady state, the nodes use the AES-128 to encrypt the exchanged messages.

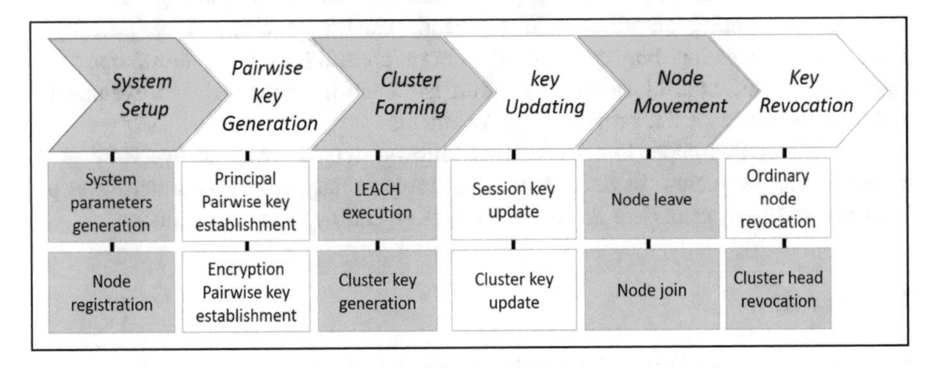

Fig. 1. Protocol scheme.

2.3 Implemented Improvements

CL-EKM is a certificate-less key management scheme provided by [5], which supports the creation, backup, and revocation of the four (04) key types listed above. It is designed to protect a dynamic network against various attacks such as replication attack and compromised key attack. Our improvements can be summarized in:

1. Key management: here a key pre-distribution method was implemented, where all the nodes retrieve their individual and partial keys from the KGC in the system setup phase before the deployment.
2. Unlike CL-EKM [5] where the CH are powerful, predefined and immobile, the improved scheme chooses the CH periodically in the cluster forming step. They are mobile and can move declining their cluster heading role while joining another CH.
3. In the cluster forming phase we aim to consider the general case of a dynamic distributed network where all the nodes are of identical features and randomly deployed. Thus, we integrated a distributed clustering algorithm namely LEACH, which is periodically launched to allow all nodes in the network to become CH, in

order to calibrate the power consumption and expend network lifetime. The proposed solution is based on the following assumptions:

- The base station (BS) is powerful and cannot be compromised;
- The KGC is at the base station;
- Random deployment; where the neighbors of each node are not known before deployment.
- All nodes are identical in terms of energy, computing capacity and storage.

3 Implementation

3.1 System Setup

A key pre-distribution mechanism was adopted in our proposed scheme to obtain the system parameters and the nodes' secret information before deployment. In order to create the individual keys, partial private/public keys and the primary private/public key, we implemented a firmware on the KGC that calculates the listed keys in the following order:

- Generate the primary public/private key;
- For each node, generate the partial public/private key;
- For each node, compute the individual key using ECDH (Elliptic Curve Diffie-Hellman) [8].

3.2 Pairwise Key Generation

The output of this step is the pairwise master key P_{MK} that will be used in the establishment of the second key which is the encryption (or session) key P_{EK}, used later in encrypting the exchanged messages.

This step is triggered by a node *A* broadcasting its identifier and public key. On the reception of this advertisement each neighbor *B* of this node starts the encapsulation process to generate an encapsulation set noted Φ_B and the pairwise master key of the two nodes *A* and *B* noted P_{MK}.

After that, the node *B* should send the obtained encapsulation set Φ_B alongside with its identifier and public key to the node *A* (the advertising node). Once the message is received, the node *A* starts the decapsulation process to obtain P_{MK}. Afterward, the node *A* calculates the pairwise encryption key P_{EK} using a random number *r*. Then it sends *r* to node *B* that calculates P_{EK}.

3.3 Cluster Forming

After establishment of the pairwise keys, the clustering phase is executed to divide the network into subgroups named clusters. This phase was run periodically to allow all nodes to become cluster heads to calibrate the energy consumption and extend the network lifetime. This phase uses the clustering algorithm LEACH [9] and is executed on two (02) stages:

a. ***Cluster head designation***
   Each node calculates its probability of becoming a cluster head and compares it to a random number, if it is greater than this number this node will take the role of cluster head. Later, the cluster head will invite its neighbors to join it. Ordinary nodes may receive more than one invitation message. Therefore, they wait for some random time to ensure the reception of the whole advertisement, then choose the cluster head with the highest power of RSSI (*Radio Signal Strength Indication*), as it gives sign of the neighbors' position (near or far) and their residual energy. Later the ordinary nodes send a request (NODE-JOIN) to join the selected CH. After receiving requests by neighbors, a CH responds by a confirmation message (YOU-CAN-JOIN). In the end of this stage, each CH sends the list of its members to the BS which checks all the network CHs lists for possible redundancies; a node that tries to join more than one CH is considered as malicious and will be revoked by the BS that also notifies all the CHs.
b. ***Group key generation***
   After forming the clusters, each cluster head "CH_j" calculates its group key GK_j, encrypts it using the session key of each node i of the members then transmits it to the node i. It repeats this process for each member of the cluster. Once the node i receives the message from the CH, it decrypts it to restore GK_j.

3.4 Key Updating

To be protected against cryptanalysis, a frequent encryption keys updating is crucial. Thus, we used a periodic update of the encryption keys namely the session keys and the group keys.

To update the session key both nodes should use the pairwise master key to generate a new session key. Only the cluster head can trigger the group key update process, so any node attempts to update the group key is considered malicious and will be revoked. To update this key the CH calculates the new key *GKj*', encrypts it using the old group key then broadcasts the encryption in the cluster. After receiving the encrypted key, each node decrypts it to get the new group key.

3.5 Node Movement

Because this work considers a dynamic network, all the nodes regardless of its nature can move physically among the network area. A moving node can leave a cluster and join another.

When an ordinary node intends to move between clusters, it informs the cluster heads to correctly manage their group keys and inform the BS of the new state. Otherwise, if a cluster head decides to move it informs the cluster members and become an ordinary node. Then, the members dedicate one of them to be a cluster head in case they cannot join any nearby cluster.

3.6 Node Revocation

A node is considered compromised or malicious if:

- It tries to modify the group key while it is still an ordinary node;
- It disappeared for enough time to an attacker to change its behavior, and reappears in the network later;
- It tries to join more than one cluster at the same time.

A compromised node can be an ordinary node as well as a cluster head. The revocation of the two is slightly different:

- In case of the compromised node (noted C) is an ordinary node, the BS updates its revocation list and informs the network CH of node C status, so that they inform their members to delete any shared keys with it. Finally, the CH of the revoked node's cluster updates the group key with the remaining members.
- If the compromised node is a CH (denoted by CH_j), the BS updates its revocation list and informs all the members in the cluster j of their leader's status as well as the rest of the CH in the network, so that they delete their shared keys. As for the compromised CH member nodes, they delete all the keys established with it, then each of them tries to join a new cluster. If a node fails to find a neighbor CH it decides to become a CH itself and informs its neighborhood.

4 Performance Analysis

In the simulation phase, an EXP5438 sensor is used as network nodes. It uses a 16 bits CPU (MSP430F5438A) with a clock rate varying from 8 MHz to 25 MHz, and memory of 16 kb. The BS is simulated using Cooja-Mote, which is a powerful virtual sensor with 32 bits CPU.

After an empirical study of the parameters of the elliptic curve, we chose the curve "SECP192R1" which represents the most consuming curve in terms of storage capacity and computation time, since it is the curve that ensures the highest level of security compared to SECP160R1 and SECP128R2. SECP192R1 requires 1.5 times more time than SECP160R1 [5], and the SECP128R2 requires almost 4% less time than SECP160R1. Simulation parameters are given in Table 1.

Table 1. Simulation parameters

Parameter	Value
Sensor	EXP5438 (clock rate 8 MHz)
CPU	MSP430F5438A
Simulation area	$(100 \times 100)\ m^2$
Number of nodes	up to 100
Communication range	25 m
Routing protocol	RPL
Number of maximum hops	50
Cryptography	ECC (SECP192R1), AES-128, SHA2-256

a. ***Necessary time for key generation***

To calculate the necessary time to generate each of the keys, two nodes are used to exchange the messages needed in calculating those keys.

As it is illustrated in Table 2, compared to other processes, encapsulation and decapsulation are the most consuming, this is caused by the high number of point multiplications needed; four (04) multiplications in the encapsulation and six (06) in the decapsulation. Noting that the master key P_{MK} is established (using encapsulation and decapsulation) only once in the network lifetime and used to derive the session key. Once two nodes establish the pairwise keys, they do not require further ECC operations. Partial Keys use about a quarter (0.29) of the time needed for the encapsulation. The required time for session and cluster key is negligible, this is due to no ECC multiplications are used.

Table 2. Keys generation time

Key	Time (ms)
Partial key	4664
Encapsulation	15667.5
Decapsulation	17550
Session key	23
Cluster key	62

b. ***Network density***

A network is called dense if each node has at least eight (08) neighbors. According to Fig. 2 the network will be denser by containing sixty (60) nodes going from just one neighbor to fourteen (14).

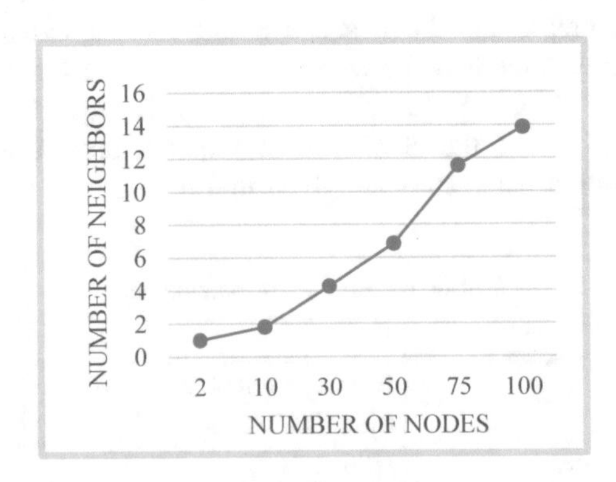

Fig. 2. Network density.

c. ***Necessary time for the whole protocol execution***

The objective here is to examine the impact of the network's density on the required time to each phase of the protocol, namely; system setup, key generation, and clustering. As shown in Fig. 3, the clustering phase just added few seconds (about 8 s) to the global protocol time. Most of the time elapsed in this phase is a random waiting time up to 5 s to avoid collisions. Moreover, the clustering is not affected by the network density because of the distributed execution of the algorithm where all clusters can be built in parallel. We conclude that even for a high dense network (more than 14 neighbors) running the modified protocol, it can be ready and secure in less than 50 s.

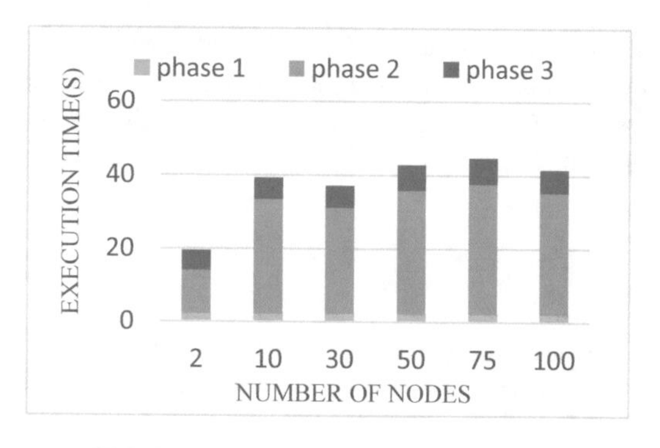

Fig. 3. Execution time for the 3 phases.

d. ***Energy consumption***

As shown in Fig. 4, energy consumption in LMP (Low Power Mode) is insignificant compared to the other modes. In general, CPU energy consumption is negligible compared to the transmission energy consumption. However, in this case, a high-energy consumption of the CPU is caused by the computing complexity of the ECC operations. Nevertheless, this huge CPU energy is consumed only once after deployment, by each node during the entire network lifetime.

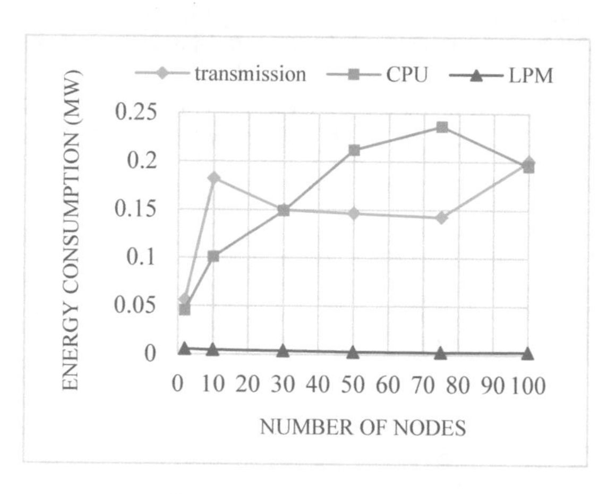

Fig. 4. Energy consumption.

5 Conclusion and Future Works

This work helped the examination of the certificate-less key management scheme, based on elliptic curve cryptography characterized by a reduced key size, which makes it suitable for the context of miniaturized networks. The proposed scheme focused on managing keys in a dynamic hierarchical network to ensure security and minimize power consumption.

The implemented propositions have been tested and the performances have been evaluated. The obtained results showed the effectiveness of the proposed improvements in terms of computing capacity and low energy consumption while keeping the high level of security.

Nevertheless, a large-scale simulation is required. Besides, as the used clustering algorithm is based on random calculations, it can affect the performance, hence the use and test of different clustering algorithm is desirable to get the best compromise between security and clustering.

References

1. Moskowitz, R.: HIP diet exchange (DEX). Internet draft, IETF (2011)
2. Gura, N., Patel, A., Wander, A., Eberle, H., Shantz, S.C.: Comparing Elliptic Curve Cryptography and RSA on 8-bit CPUs (2004)
3. Chatterjee, K., De, A., Gupta, D.: An improved ID-based key management scheme in wireless sensor network. In: Proceedings of 3rd International Conference on ICSI, vol. 7332, pp. 351–359 (2012)
4. Alagheband, M.R., Aref, M.R.: Dynamic and secure key management model for hierarchical heterogeneous sensor networks. IET Inf. Secur. **6**(4), 271–280 (2012)
5. Seo, S.H., Won, J., Sultana, S., Bertino, E.: Effective key management in dynamic wireless sensor networks. 3rd IEEE Trans. Inf. Forensics Secur. **10**(2) (2015)
6. Du, X., Xiao, Y., Guizani, M., Chen, H.-H.: An effective key management scheme for heterogeneous sensor networks. Ad Hoc Netw. **5**(1), 24–34 (2007)
7. Zhu, S., Setia, S., Jajodia, S.: LEAP: efficient security mechanisms for large-scale distributed sensor networks. In: Proceedings of the 10th ACM Conference on Computer and Communications Security (CCS 2003), Washington, DC, USA, pp. 62–72, October 2003
8. Hankerson, D., Vanstone, S., Menezes, A.J.: Guide to Elliptic Curve Cryptography. Springer (2004)
9. Arora, V.K., Sharma, V., Sachdeva, M.: A survey on LEACH and others routing protocols in wireless sensor network. Optik – Int. J. Light Electron Optics **127**(16), 6590–6600 (2016)

Towards a New Framework for Service Composition in the Internet of Things

Samir Berrani(✉), Ali Yachir, and Mohamed Aissani

Artificial Intelligence Laboratory, Military Polytechnic School (EMP), PO BOX 17, 16111 Bordj-El-Bahri, Algiers, Algeria
samir.berrani@yahoo.fr, a_yachir@yahoo.fr, maissani@gmail.com

Abstract. Internet of Things' (IoT) services represent a great interest research topic for both academic and industrial communities due to their large application domains. New dynamic and automatic development approaches are created in order to improve their effectiveness based on service composition process. Indeed, they allow the aggregation of smart object's services to meet complex requirements from various application areas. This paper provides an architecture that describes a framework for service composition in IoT. Moreover, it proposes a rule-based reasoner where several facts and rules (knowledge base) are inferred to satisfy user queries. This approach is designed and implemented using SysML and Prolog platform respectively. The use-cases scenario and extensive tests show clearly the interest, feasibility, and suitability of the rule-based system for service composition in the IoT.

Keywords: Internet of Things · Service composition
Rule-based system · Semantic web

1 Introduction

Internet of Things' (IoT) services are developed to meet new consumer needs on real-time observation and control of particular phenomena at a specific environment, which can be a space, or an object. IoT devices are characterized by mobility, availability, autonomy, and adaptability to environment changes, which make the development of their applications an arduous task [1]. In addition, these solutions should be low cost, intuitive to use, robust, easy to maintain and achievable in a competitive time.

To address these challenges, it is crucial to create new development approaches that are dynamic and automatic. This active research topic is known as service composition. It is a process that allows the creation of new services, which provide non-existing functions, based on existing ones [2]. This idea, when applied to IoT, promises to provide added-value services that none of IoT smart objects could provide individually. In fact, service composition allows the aggregation of smart objects' services to meet complex requirements from various application domains [3].

O. Demigha et al. (Eds.): CSA 2018, LNNS 50, pp. 57–66, 2019.
https://doi.org/10.1007/978-3-319-98352-3_7

In this paper, we provide a global framework and a rule-based reasoner module for service composition in IoT. In such a module, several facts (services) and rules (solvers) are inferred to find a sequence of facts that stands for composite service which fulfills the user query (request). The remainder of this paper is organized as follows. Section 2 gives a background and discusses some existing works on IoT service composition. Section 3 presents an overview of the proposed IoT service system (IoT framework). Section 4 describes an e-health use case scenario. Section 5 details the reasoner modules. Section 6 shows the performance evaluation of the proposed solution. Section 7 gives a conclusion and ideas for future works.

2 Related Work

Several frameworks have been proposed to enable service composition in the IoT environments. Due to space limitations, this section summarizes the most significant approaches. In [4], a framework based on service selection and matching is presented. It takes into account the QoS preferences using fuzzy cognitive map model. To improve the reliability of service composition, a K-Means framework [5] is proposed to classify candidate services into clusters representing QoS levels. In [6], a service-oriented, user-centered and event-aware framework for service discovery and selection in IoT is proposed. To enhance the complexity of service composition process, a new service selection approach based on reputation criteria [7] is proposed. In [8], a semantic approach is given for robot interaction with humans, agents, and systems using natural language. To enable the integration of IoT applications into the Web, a service provisioning architecture for smart objects with semantic annotation [9] is introduced. Besides, an adaptive service composition framework [10] for IoT-based Smart Cities is proposed. In [11], a publish-subscribe architecture, based on a generic SPARQL endpoint, is designed for interoperability in IoT. To deal with self-configurable service composition in resource-constrained environments using CoAP, a semantic description of IoT devices [12], based on the RESTdesc format is designed. In [13], a CoAP extension with lightweight semantic-rich information is proposed by defining appropriate CoRE link format attributes describing both IoT resources and user requests in IoT. Moreover, a hybrid centralized/distributed architecture [14] is developed for resource discovery in IoT based on CoAP. In [15], things are represented as nodes in a social network which is created by analyzing the proximity of two users' environment in the IoT.

The works discussed above show that various frameworks and semantic models are proposed to describe things in IoT. However, they deal only with some kind of things such as sensors and actuators without addressing the more general notion of thing in IoT including the users, the system itself as well as the physical (devices) and the digital (services) worlds. Moreover, the relationships between such entities should also be precisely described. Especially, it should be specified how users interact with the framework and how their requests are handled based on the available devices and services. To overcome the aforementioned shortcomings of the existing works, we propose in this paper, a new IoT framework that supports many IoT entities including devices, services, and users. These entities and their relationships are well described using the different SysML diagrams. Specifically, we describe the core module of the proposed framework which handles user requests in detail using rule-based service composition.

3 Proposed IoT Framework Overview

The IoT service ecosystem is an extra complicated domain because it contains several actors of different natures. We describe the IoT service system using SysML language[1] in order to design a model that includes the interactions between their parts. The proposed IoT service model is described through functional, structural, and behavioral aspects. This conception style enables a better comprehension of the system of interest. To simplify this task, we define the structure of the whole system through the SysML Block Definition Diagrams (BDD). As shown in Fig. 1, the IoT service system is composed of IoT user, IoT target, IoT device and IoT service platform block. Each entity among the mentioned elements represent an essential component of the studied system. The IoT service platform component provides user-friendly software for operating the IoT facilities. In this section, we introduce the mentioned parts of IoT service system, one by one, in order to give a clear idea of the whole proposed solution.

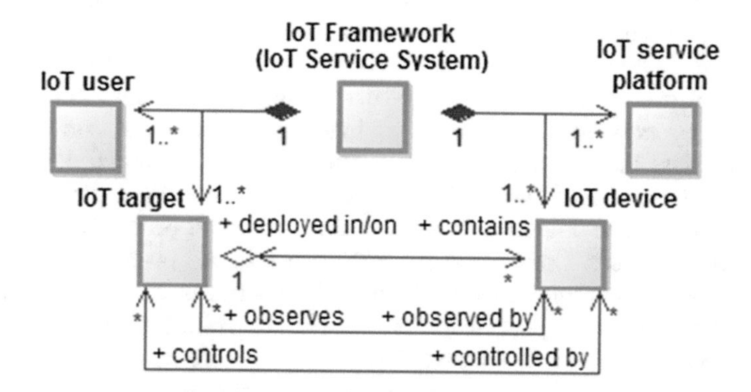

Fig. 1. IoT Framework (IoT Service System) block definition diagram

3.1 IoT Target Block Definition Diagram

As represented in BBD:01 of Fig. 2, IoT target BDD stands for a space or an object. The space entity can contain sub-spaces and/or objects. The object entity can be mobile or contextual, fixed or portable. This target model allows us to describe any target under surveillance including its relations with the other defined targets. Hence, it is possible to deduce new (direct/indirect) relations from the existing ones between the available IoT targets.

[1] http://www.omgsysml.org/.

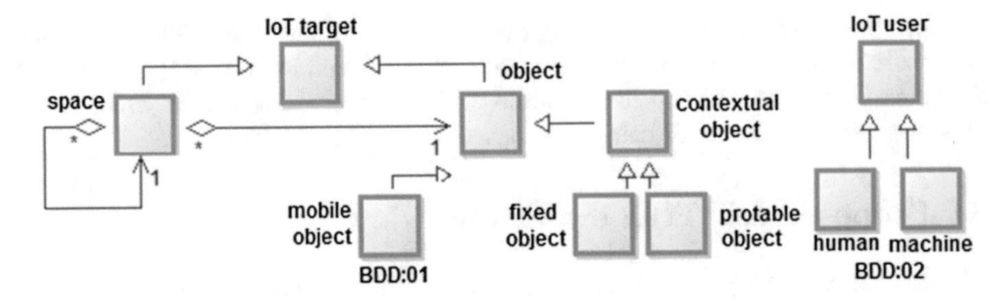

Fig. 2. IoT target and IoT user block definition diagram

3.2 IoT User Block Definition Diagram

For the user block, it represents a client or an operator of the IoT service system. In BDD:02 of Fig. 2, IoT user can be a human or a machine thanks to the semantic aspect included in the proposed solution. The user of the IoT system deploys IoT device in/on target in order to observe or control it.

3.3 IoT Device Block Definition Diagram

To realize an observation and/or a control of targets, appropriate devices for each IoT application domain are used. These smart IP-objects provide some special physical and/or digital functions published as services. As shown in BDD:01 of Fig. 3, an IoT device can be an actuator that applies effects in/on targets, a sensor that measures some metrics in the surrounding environment and/or detects events, or a processing board which enables the treatment of context data or calculates orders (commands) to handle other devices. A virtual controller is a software/hardware component (digital sensor) that provides a direct control of an actuator device to the user. It can be a driver which provides commands to the user as a service. This digital sensor catches direct order of user and formulates its related command for the appropriate actuator devices. The actuator devices provide actuating services that enable controlling targets. In order to carry out treatments of raw context data, the processor boards enable processing services. The sensor devices supply the acquisition of context data through trigger services. Also, the digital sensors allow the acquisition of user orders.

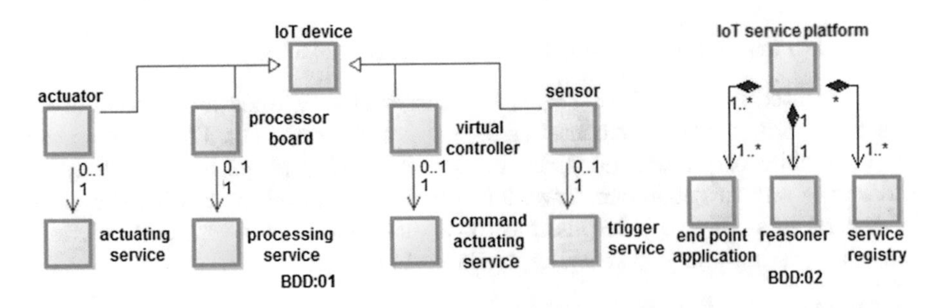

Fig. 3. IoT device and IoT service platform block definition diagram

3.4 IoT Service Platform Block Definition Diagram

Regarding the IoT service platform shown in BDD:02 of Fig. 3, it is composed of three main parts, namely: endpoint application, service registry, and reasoner entity. First, the endpoint application allows the providers to describe and publish their services. Also, it assists the user to specify their requests correctly. This latter will be decomposed to sub-requests which are less complex. Next, they will be sent to the reasoner in order to obtain responses for each one. In the end, these responses will be composed together in order to reconstitute a single response to the former user request. Secondly, the service registry lists all the available services. Finally, the reasoner gets the list of the available services from their registries and calculates responses for the user's request. In this work, we focus on the reasoner system. However, the endpoint application and the service registry will be the subject of other future works.

4 Proposed IoT Framework: An e-Health Use Case Scenario

To illustrate some concepts of our proposed IoT framework, we introduce an e-health application, called "Medic Eye: ME", that can be deployed and used, for example, at a hospital in order to observe or to control the evolution of a patient's medication. Some potential IoT targets in ME application are patients, nurses, doctors, etc. Moreover, Table 1 illustrates some IoT devices and services that can be used in that application.

Let us suppose that "Rafik" is a diabetic old man, who needs to keep his blood sugar under control. Therefore, he is considered as an IoT target and identified by "Rafik-01" in our Framework. In this scenario, "Rafik-01" wears, in his right upper arm, an actuating device, called insulin pump. Such a device provides the list of services defined at line 02 and 03 in Table 1. In addition, the proposed example contains a Raspberry device that provides a processing service which converts temperature values from Celsius to Fahrenheit. Some other devices like blood glucose meter and body temperature sensor can be worn by "Rafik-01" to measure respectively his blood glucose (BG) and body temperature (BT) via

Table 1. Illustration of IoT devices and their provided services

Actuator device	act-01: Insulin Pump Implant-IPI
Actuating service	act-sce01: getInslinInjection (get inject of insulin)
Processing device	proc-01: Raspberry "Pi" (processing board)
Processing service	proc-sce01: convertC2F (convert temp. in "C" to "F")
Trigger device	trig-01: blood glucose meter implant
Trigger service	trig-sce-01 getBGLevel (get level of the BG)
Trigger device	trig-02: body temperature sensor implant
Trigger service	trig-sce-02 getBTemp (get body temperature)
Virtual controller	vclt-cmd-01: bracelet contains button to order insulin injection
Cmd. Act. Service	cmd-act-sce-01 orderInsulinDose (format a user order to a command)

the following trigger services: getBGLevel and getBTemp. The last example of devices is the virtual controller that intercepts user orders to be transformed on commands to the appropriate actuating services.

5 Detailed Description of the Reasoner Module

In this section, we propose a rule-based reasoner for service composition in IoT. We believe that the rule-based system is suitable for service composition due to many similarities, between facts and services, queries and requests, and rules and composition process. This mapping is illustrated along this section, in which we describe the proposed reasoner through functional, structural and dynamic SysML diagrams.

5.1 Functional Description of the Proposed Reasoner

Based on SysML, the functional description includes the requirements and use case diagrams. The reasoner requirements, as shown in Fig. 4, consist in the calculation of the user's request responses which must be performed in a finite time. The response to the users' request can be negative (empty response), or positive (a simple or composite service that satisfies the request). The reasoner must avoid starvation or deadlock of user's requests in a case of concurrency access.

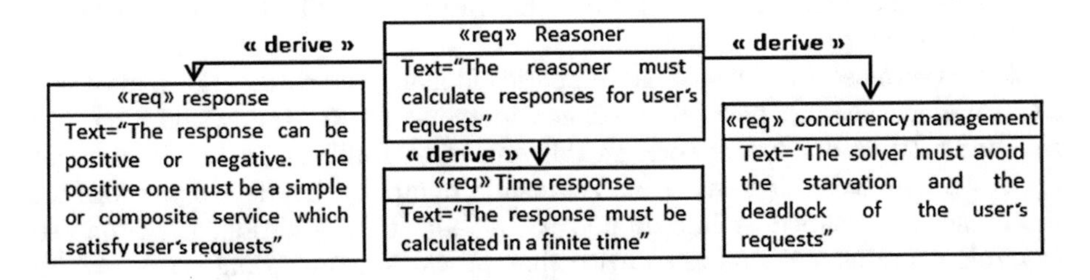

Fig. 4. Reasoner requirement diagram

To complete this description, we propose the reasoner use case diagram shown in Fig. 5. As can be seen from this figure, the system calculates responses for three kinds of requests namely: actuating, processing and trigger.

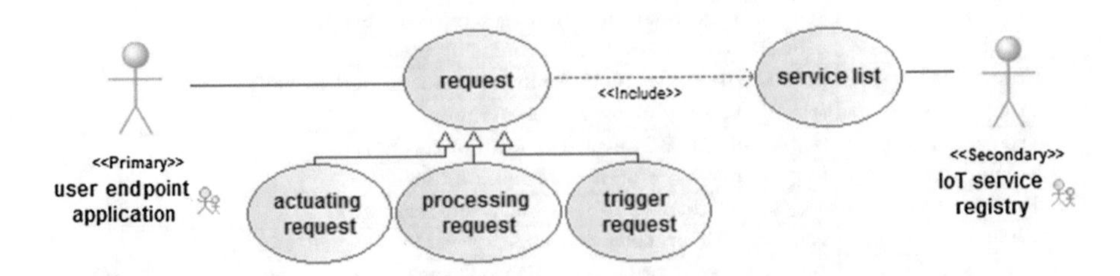

Fig. 5. Reasoner use case diagram

5.2 Structural Description of the Proposed Reasoner

In this work, we adopt the service definition given by [16], in which the authors introduce concrete and abstract services. The concrete service is a service that performs concretely specific functionalities. However, the abstract service defines a class of concrete services which are functionally equivalent. In Table 2, we illustrate the temperature abstract service at the first line and its relative concrete services at the remained ones in Prolog format. We assume that the temperature concept of the "ME" ontology contains two sub-concepts namely: body and ambient temperature. Each service is defined as a fact through the following parameters: input, output, category, id, subject, and QoS. The input shows the required messages or effects. The output represents the provided messages or effects. The category depicts the service category, trigger, processing or actuating. The id is the service identifier. The subject determines the observed/controlled target. The QoS represents the number of concrete services for abstract services. In the following examples, # represents a physical message obtained from a target.

Table 2. Illustration of temperature service descriptions

getTemp: ["#",["Temp.","abstract","abstract"],"trigger","trig-sce-06","Rafik-01","02"]
getAmb.Temp: ["#",["Amb. temp.", "double", "C"], "trigger","trig-sce-07","Room-01","0"]
getB.Temp: ["#",["Body temp.", "double", "F"],"trigger","trig-sce-08","Rafik-01","0"]

The proposed reasoner is composed of a request "query", a solver "rules" and a service "facts", as shown in BDD:01 of Fig. 6. The services (facts) symbolize the services of IoT systems. A service is featured by an input message, an output message, a category, a number of contained concrete services and an identifier. The service can be an actuating service that acts directly on a target, a processing service that provides treated data, a trigger service which supplies context data, or a virtual controller that provides commands for actuating services. BDD:02 of Fig. 6 shows the proposed service types introduced above. The request (query) represents the user request. It is described by an output message and a category. The output message denotes the expected user result, and the category is introduced to reduce the request complexity. Three types of categories can be accounted, namely: a trigger, a processing and an actuating request. Each one among these categories will be satisfied by a simple or special composite service (sequence of facts). BDD:03 of Fig. 6 depicts user request types. Based on these descriptions, we have adopted the same definition format of a service for a request. As an illustration, let us suppose the following trigger request(query) related to the body temperature of an observed target "Rafik-01". So, it is defined as follow: bodyTempRequest1: (Nil,(BodyTemperature, double, Celsius), trigger, id-req, "Rafik-01",0). Finally, the solver (composer) is based on rules. As represented in BDD:04 of Fig. 6, the solver contains three kinds of sub-solvers, namely: trigger, processing and actuating. Each is dedicated to performing sequences of facts that satisfy user queries. Together, rules and facts are considered for the reasoner as knowledge base.

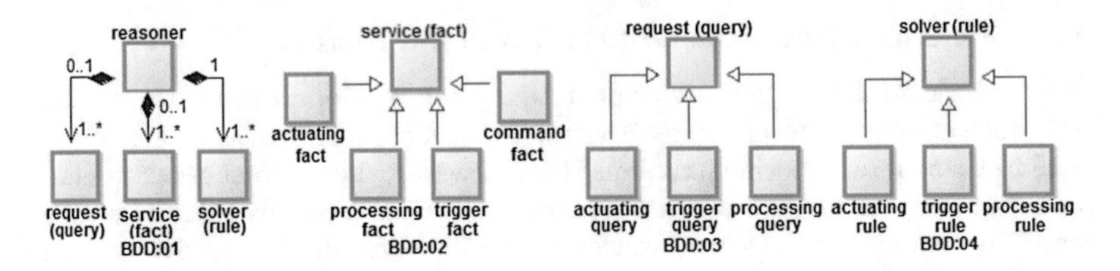

Fig. 6. Reasoner system block definition diagrams

5.3 Dynamic Description of the Proposed Reasoner

The composition process is initiated by the user request, which is described mainly by an expected result, an observed target, and a category. The trigger, processing, and actuating requests are performed respectively by a trigger, processing or actuating solver. Figure 7 shows the dynamic behavior of the reasoner. Each state depicts which kind of facts are involved in the composition process. In addition to the request, three states can be distinguished, namely: trigger, processing, and actuating. Each transition is featured by a condition that must be validated to cross it. Meeting such condition means, also, that the aggregation of their facts is allowed.

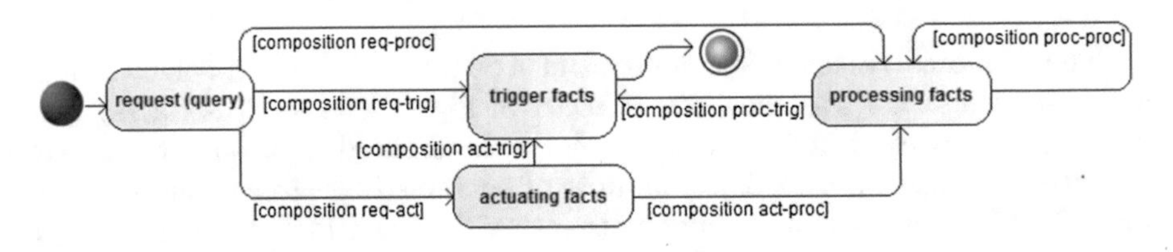

Fig. 7. Reasoner state machine diagram

6 Performance Evaluation

To evaluate the performance of our proposed rule-based reasoner, we have adopted the test scenario given by [12]. In such an evaluation, tests were carried out in a specific environment constituted of 64, 256 and 1024 services. Each time, we varied the composite service size, respectively to 3, 16 and 32, and measured its needed time to be calculated. We have used a laptop with windows 7 professional, Intel(R) Core(TM) i7-4610M CPU 3.00 GHz, 4Go RAM, Oracle Java 1.8.0_112 and SWI-Prolog version 7.5.15-18-g1a41862 to run programs. Each experiment is repeated at least 20 times.

As shown in Graph: A of Fig. 8, a medium execution plan length of ten (10) services needs approximately one second to be calculated in an environment of 256 services. As well, for 1024 services, it takes 1.5, 8 and 80 s respectively to calculate an execution plan length of 3, 16 and 32 services. Comparing these results

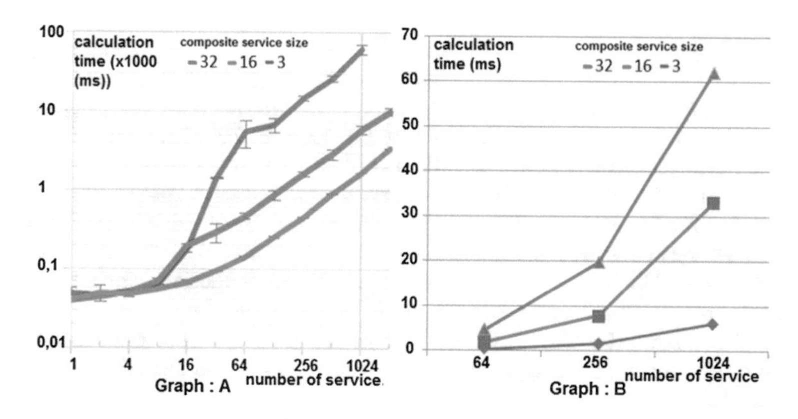

Fig. 8. Variation of reasoning time according to the number of services and the length of the execution plan, composite service size. Graph: A represents the results obtained by [12], and Graph: B depicts the rule-based reasoner results.

to the obtained ones by the proposed rule-based reasoner, shown in Graph: B of Fig. 8, we notice that the needed time to perform a composite service with a length equal to 32 in an environment that contains 1024 services is roughly 0.1 s. So, it seems that our proposed system is more performer than the one proposed in [12]. We explain these results by the fact that we have subdivided services and requests respectively into four and three categories as was mentioned above. By making these distinctions, the number of services involved in the composition process is reduced and, indeed, the computing time is improved.

7 Conclusion

In this work, we presented an architecture that describes a framework for service composition in IoT. We also proposed a rule-based reasoner where several facts and rules (knowledge base) are inferred to satisfy user queries. In order to assess our proposed solution, we illustrated each part of the given model by an example from "ME" application. The ME use-cases scenario and extensive tests show clearly the interest, feasibility, and suitability of the knowledge base system for service composition in the IoT. As a future work, we plan to develop an endpoint application and a service registry to confirm the obtained results by a realistic IoT scenario.

References

1. Razzaque, M.A., Milojevic-Jevric, M., Palade, A., Clarke, S.: Middleware for internet of things: a survey. IEEE Internet of Things J. **3**, 70–95 (2016)
2. Lemos, A.L., Daniel, F., Benatallah, B.: Web service composition: a survey of techniques and tools. ACM Comput. Surv. **48**, 33:1–33:41 (2015)
3. Chen, I., Guo, J., Bao, F.: Trust management for SOA-based IoT and its application to service composition. IEEE Trans. Serv. Comput. **9**, 482–495 (2016)
4. Li, Z., Fang, H., Xia, L.: Increasing mapping based hidden Markov model for dynamic process monitoring and diagnosis. Expert Syst. Appl. **41**, 744–751 (2014)
5. Dou, W., Zhang, X., Liu, J., Chen, J.: Towards privacy-aware cross-cloud service composition for big data applications. IEEE Trans. Parallel Distrib. Syst. **26**, 455–466 (2015)
6. Yachir, A., Amirat, Y., Chibani, A., Badache, N.: Event-aware framework for dynamic services discovery and selection in the context of ambient intelligence and internet of things. IEEE Trans. Autom. Sci. Eng. **13**, 85–102 (2016)
7. Bossi, L., Braghin, S., Trombetta, A.: Multidimensional reputation network for service composition in the internet of things. In: 2014 IEEE International Conference on Services Computing, pp. 685–692 (2014)
8. Ayari, N., Chibani, A., Amirat, Y., Matson, E.T.: A novel approach based on commonsense knowledge representation and reasoning in open world for intelligent ambient assisted living services. In: 2015 IEEE/RSJ International Conference on Intelligent Robots and Systems, pp. 6007–6013 (2015)
9. Han, S.N., Crespi, N.: Semantic service provisioning for smart objects: integrating IoT applications into the web. Future Gener. Comput. Syst. **76**, 180–197 (2017)
10. Urbieta, A., Gonzalez-Beltran, A., Mokhtar, S.B., Hossain, M.A., Capra, L.: Adaptive and context-aware service composition for IoT-based smart cities. Future Gener. Comput. Syst. **76**, 262–274 (2017)
11. Roffia, L., Morandi, F., Kiljander, J., D'Elia, A., Vergari, F., Viola, F., Bononi, L., Cinotti, T.S.: A semantic publish-subscribe architecture for the internet of things. IEEE Internet Things J. **3**, 1274–1296 (2016)
12. Kovatsch, M., Hassan, Y.N., Mayer, S.: Practical semantics for the internet of things: physical states, device mashups, and open questions. In: 2015 5th International Conference on the Internet of Things, pp. 54–61 (2015)
13. Yachir, A., Djamaa, B., Zeghouani, K., Bellal, M., Boudali, M.: Semantic resource discovery with CoAP in the Internet of Things. In: WINSYS 2017, pp. 75–82 (2017)
14. Djamaa, B., Yachir, A., Richardson, M.: Hybrid CoAP-based resource discovery for the Internet of Things. J. Ambient Intell. Humaniz. Comput. **8**, 357–372 (2017)
15. Xu, W., Hu, Z., Gong, T., Zhao, Z.: Towards a dynamic social-network-based approach for service composition in the Internet of Things, vol. 8350 (2012)
16. Yachir, A., Amirat, Y., Chibani, A., Badache, N.: Towards an event-aware approach for ubiquitous computing based on automatic service composition and selection. Ann. Telecommun. - annales des télécommunications **67**, 341–353 (2012)

A Preamble Sampling Scheme Based MAC Protocol for Energy Harvesting Wireless Sensor Networks

Abdelmalek Bengheni[1(✉)], Fedoua Didi[1], Ilyas Bambrik[1], and Adda Boualem[2]

[1] Computer Science Department, Laboratory of Research in Informatics of Tlemcen (LRIT), Tlemcen University, Tlemcen 13000, Algeria
baghanimalik@yahoo.fr, fedouadidi@yahoo.fr, ilyas9111@yahoo.fr
[2] High National School of Computer Science, ESI, Algiers 16000, Algeria
ab_boualem@esi.dz

Abstract. The operation of the wireless sensor networks (WSNs) with limited energy resources is an important challenge for researchers. In fact, many schemes have been proposed to save the wasted energy in a WSN or to equip its sensor nodes with an energy harvesting system. In this paper, we propose a preamble sampling scheme based MAC protocol for energy harvesting WSNs (PS-EHWSN). PS-EHWSN leverages the advantage of transmitter-initiated schemes and uses the low power listening (LPL) technique with short preamble messages where each sensor node in the network can determine its next sleep period to reduce the duty-cycle by the use of its residual energy, which can increase over time. Moreover, our proposition promotes even energy consumption of the sensor nodes with the energy harvesting capability. PS-EHWSN protocol was simulated using OMNeT++/MiXiM. The simulation results show that our proposed protocol exceeds some existing MAC protocols such as the BMAC.

Keywords: PS-EHWSN · Wireless sensor networks (WSNs) Energy harvesting (EH) · OMNeT++ · MiXiM

1 Introduction

A wireless sensor network (WSN) is a type of Ad-Hoc network, composed of small devices called «sensors nodes» forming a network ranging from dozens of elements to hundreds [1], sometimes more. Generally, sensors communicate wirelessly in a small coverage area and can be deployed very densely in diverse environments such as houses, shops, warehouses, cities, or hostile environments such as military fields, volcanos, or even in a forest. Each node in a WSN can sense its environment and process the data collected locally or sends them to one or more collection points called «sinknode or base station» [2]. Typically, components of the WSN have a limited energy source which is difficult or impossible to replace/replenish. This implies that the sensors must conserve maximum of their energy in order to operate. Thus, currently, the energy conservation in

O. Demigha et al. (Eds.): CSA 2018, LNNS 50, pp. 67–78, 2019.
https://doi.org/10.1007/978-3-319-98352-3_8

WSNs plays an important role in prolonging the network lifetime, which became the predominant performance criterion [3] in this type of network. In fact, many techniques have been proposed to conserve or save the wasted energy of this latter. «Duty-cycling» [4] is one of these techniques used for conserving the energy in WSN where each node in the network can switch between the active state and the sleep state; this is provided by the duty-cycle MAC protocols that can be categorized into two classes: synchronous and asynchronous MAC protocols [3]. In synchronous MAC protocols such as S-MAC [5], the nodes share the schedule information that specifies their cycle of active and sleep periods through control packets. Asynchronous protocols, on the other hand do not exchange the synchronization information to send or receive data. Instead, they employ preamble sampling or through sending beacons to do that, such as in B-MAC [6], X-MAC [7] and RI-MAC [8]. Nevertheless, the network lifetime stays always bounded. Another solution which newly appeared is to incorporate an energy harvesting system [9] in each sensor node. This latter harnesses energy from the environment or other energy sources (foot strike, finger strokes… etc.) and converts it to electrical energy directly useable in order to prolong the network lifetime and improve its performance.

Therefore, in order to control the energy consumption and how it is collected by the nodes in WSN, the majority of the duty-cycle is expended in the low power sleep state where the nodes can harvest more energy, which results in reducing the time of node's activity; thus minimizing the duty-cycle [10]. In case where multi-hop transmitting for a packet is wanted, a high node's activity causes a great latency and also leads to packet collisions when multiple packets are transmitted. Therefore, transmissions are deferred by the concurrent transmitters. Since most of the ambient energy sources have diverse features that depend on the different environmental situations, our main objective is reducing the duty-cycle of the nodes based on their energy levels, in order to improve the WSN performances.

In A Preamble Sampling Scheme MAC Protocol for Energy Harvesting WSNs (PS-EHWSN) that we propose in this paper, all sensor nodes are supplied with an energy harvesting system and the communication between these nodes is initiated by sending several short preambles after checking the channel where the sender node plays an active role in communication. The initiator adopts advanced features of the transmitter-initiated mechanisms devised in previous works [6, 7]. To further increase the efficiency of PS-EHWSN, an estimation method of the duration of the sleep period of nodes is also adopted from previous works [11], and it is used to adjust their duty-cycles based on the future-presented harvested energy; the computation of this next sleep period duration is based on the residual energy. The main objective of PS-EHWSN is to increase the packet delivery ratio, whilst reducing the mean latency and improving the throughput in the WSN using the harvested energy.

The remainder of the paper is structured as follows: Sect. 2 describes the operations of PS-EHWSN protocol; Sect. 3 describes the simulation tools and the parameters used. Section 4 present the performance evaluation of PS-EHWSN and its comparison with BMAC protocol. Finally, Sect. 5 concludes the paper and presents future work.

2 PS-EHWSN: A Preamble Sampling Scheme MAC Protocol for Energy Harvesting WSNs

In this section, A Preamble Sampling Scheme MAC Protocol for Energy Harvesting WSNs (PS-EHWSN) is presented. PS-EHWSN uses the strobed preamble method by broadcasting a sequence of short preamble packets. Lesser breaks between preamble packets enable the board receiver sending an acknowledgment that stops the sequence of preamble packets. It operates with dynamic duty-cycle scheduling scheme, which exploits the energy harvested by the nodes.

2.1 Communication Scheme in PS-EHWSN Protocol

PS-EHWSN uses a CSMA/CA protocol, in which the sender uses carrier wave to make the channel busy and the receiver listens to the wireless channel to avoid collision. In PS-EHWSN, all nodes that have queued packets that need to be forwarded to the sink, are listening to the channel waiting for a suitable acknowledgement (ACK) after each sending of the short preamble. Upon receiving the corresponding ACK, the data packet transmission is initiated.

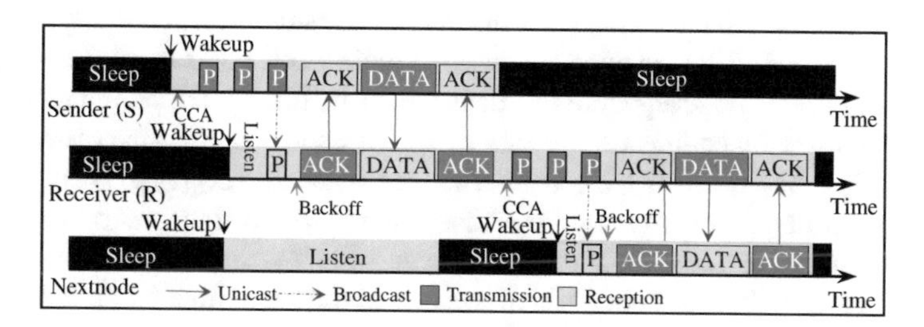

Fig. 1. Illustration of the operation of our protocol.

Our proposed protocol is based on duty-cycling, which interweaves active and sleep periods, delivers a respectable chance for energy harvesters to recharge the energy source before going into an unloading state where the deposited energy is consumed by the sensor node. PS-EHWSN is based on the following principle of communication as shown in the Fig. 1:

1- The nodes in our protocol wakeup periodically to carry out the transmission operation if there is data to send, and switch to sleep state when there is no traffic in the network after a listening time or after each successive data transmission;

2- All the nodes switch to sleep state at startup and then switch to listen state based on their own schedule to listen for the traffic in the network;

3- If a sender node S wishes to transmit data, it initiates a preamble session by sending on broadcast several short preamble packets P to its neighboring nodes after checking the medium availability by using the CCA (Clear Channel Assessment), and it pauses after each transmission of short preamble to listen the carrier. An awake receiver node R responds to the sender by sending an ACK (acknowledgement preamble) packet in unicast mode after receiving a short preamble if their back off is expired;

4- The Sender S stops sending the remaining P short preambles after the reception of the acknowledgment ACK and starts sending the data packet to the intended recipient if the channel is free, which will be acknowledged by R with ACK; otherwise, it switches to sleep state after a predefined duration of time allocated to sending several short preambles P;

5- The sender S switches to sleep state after the reception of the acknowledgment of the data packet ACK;

Subsequently, receiving node immediately begins transmitting his data to the next node by executing steps 3, 4 and 5.

2.2 The Proposed Mechanism for Determining the Next Sleep Period

We consider one type of periodic energy harvesting, where each node i is composed of a set of units: the environment detection unit, an Analog to Digital Converter (ADC), processing unit, multiple types of memory (program, data and flash memories), communication unit (RF transceiver), a power unit of maximum capacity (E_{MAX}) and an energy harvesting system. Additionally, an energy threshold (E_{th}) is used by each node in order to ensuring a balance between the energy consumption and energy harvesting ability. With PS-EHWSN, the duty-cycling of a node's activity is fixed at startup where the active time and the sleep time are determined in full by the administrator and it becomes dynamic depending on the rate at which energy can be harvested once the residual energy $E_{rs_i}(t_{check})$ *at* t_{check} instant is less than E_{th} ($E_{rs_i}(t_{check}) < E_{th}$). Repeatedly, nodes with better energy supply wakeup more, with a short sleep time, and thus receive and transmit more packets. Moreover, this mechanism reduces collisions resulting from neighboring nodes which awaken at the same time and compete for the channel. Through PS-EHWSN, the nodes sleep times are chosen deterministically as a function of nodes' energy availability which cannot be probably identical for a set of nodes. Consequently, a node with a lesser energy level selects a long sleep time to harvest as much energy as possible in order to stock an assured energy level. Trough PS-EHWSN, the sleep time for node $i(t_{sleep_i})$ is determined by the following procedure:

a. In the wakeup state, each node i checks its residual energy just before it turns to sleep state.
b. In sleep state, if $(E_{rs_i}(t_{check}) < E_{th})$, then node i computes the sleep time elapsed (t_{Ste_i}), until its residual energy reaches threshold E_{th}, by using the following equation:

$$T_{Ste_i} = \min(T_{Ste_i}, 1) \tag{1}$$

c. In the next wakeup states, while $E_{rs_i}(t_{check}) \geq E_{th}$, node i computes its sleep time (t_{sleep_i}) by using Eq. (2); Otherwise, node i executes steps (a), (b) then (c) respectively.

$$t_{sleep_i} = \left(E_{MAX} - E_{rs_i(t_{check})} / E_{MAX} - E_{th}\right) \times T_{Ste_i} \tag{2}$$

3 Simulation

3.1 Simulation with OMNeT++/MiXiM

We have used for the implementation of our proposed protocol PS-EHWSN the OMNeT ++simulator with MiXiM simulation framework [12]. The main implementation of PS-EHWSN is PSEHWSN Layer, which is a module of the MAC layer that is integrated in a new network interface NicPSEHWSN (based on Wireless NicBattery [13]; which is composed of a set of specific modules such as: the BatteryStats that computes the energy consumption for each module, the network interface WirelessNic Battery and the power supply SimpleBattery). Figure 2 shows a finite state machine of the transmitter and receiver nodes. It illustrates all self-messages and the network packets that are used in the PSEHWSN Layer to change the state of a HostPSEHWSN node.

After initialization in INIT, the node goes to SLEEP state. First, we explain the finite state machine from the sender standpoint. If the latter has a packet to send, then it will first switch to CCA state from SLEEP state and permit variable value set to true. If no communication is detected during CCA (CCA_timeout is reached), then it goes to SEND_PREAMBLE and starts sending short-preamble packets. After each preamble transmission, it goes to WAIT_ACKK state if time to send preamble is not reached (StopPreambles); otherwise, it goes back to SLEEP.

The node will move to SEND_DATA after that it receives an ACK preamble (ACKK). Once the sender sends DATA packet to the receiver node (if the channel is free), it goes automatically to WAIT_ACK state immediately. Finally, during WAIT_ACK, if the node receives (a data acknowledgement) or (ACK_timeout is reached and TxAttemps > MaxAttemps), then it goes back automatically to SLEEP state immediately; otherwise if (ACK_timeout is reached and TxAttemps < MaxAttemps), then it goes back to SEND_PREAMBLE.

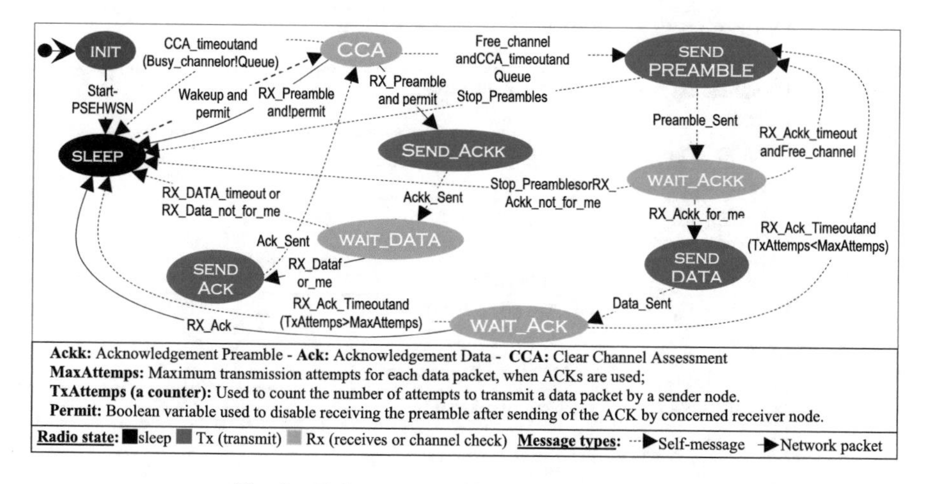

Fig. 2. Finite state machine of HostPSEHWSN.

On the receiver node side, it wakes up to detect the channel for periodic listening by switching from SLEEP state to CCA state. If the node receives a preamble from his neighbors' and the permit value equal true during CCA, it goes to SEND_ACKK state to rely to the corresponding sender node when their back off is expired and goes automatically in WAIT_DATA state immediately; otherwise, it goes back to SLEEP state. If no data packet is received from corresponding sender node during WAIT_DATA, the node goes to SLEEP state immediately; otherwise, it goes back to SEND_ACK state and permit variable value set to false. Once the receiver sends ACK to sender node, it goes automatically to CCA state immediately to send the received DATA packet to the next node.

3.2 Simulation Parameters

We form a WSN in OMNeT++ 4.6 version and the MiXiM-2.3 framework to evaluate our PS-EHWSN protocol. The various parameters used in the simulation are summarized in Table 1, which are specified in omnetpp.ini file. Moreover, physical layer parameters

Table 1. Simulation parameters at different layers

Layer	The parameter	Value
General	Number of nodes	10, 25, 50, 75, 100
	Playground Size (x, y)	400 m × 400 m, 600 m × 600 m, 800 m × 800 m, 950 m × 950 m, 1100 m × 1100 m.
	Mobility Type	ConstSpeedMobility
	Mobility Speed	2 mps
	Initial residual energy E_{rs_i}	50 J
	Energy threshold E_{th}	30 J
App Layer	Application Type	Sensor App Layer
	Traffic Type	Periodic
	Traffic Param	0.1 s, 0.5 s
	Number of Packets sent	500, 1000
Network Layer	Network Type	BaseNetwLayer
	Destination Node	9, 24, 49, 74, 99
Mac Layer	MAC Type	PSEHWSNLayer–BMACLayer
	Queue Size	2 pkts (packets)
	MaxAttempts	2
Phy Layer	Transmission power	0.05 W
	Carrier frequency	2.412e + 9 Hz

including analog models and decision type are declared in config.xml and MAC Type is specified in the Nic.ned file. Here, the Traffic Param is the maximum time interval between two packets transmission expressed in seconds. Many scenarios are implemented and tested, from 10 deployed mobile sensor nodes in a 400 m $\times$ 400 m area to 100 deployed mobile sensor nodes in 1100 m $\times$ 1100 m area where the distance between two neighbors is 100 meters for 500 s of simulated time. EH rate is periodic with harvesting param (is the maximum time interval between two harvesting energy moments expressed in seconds) is equal to 0.1 s for each sensor node randomly set between [0 mW… 50 mW]. The initial energy of each mobile sensor node is set to 50 J and the maximum energy value is 100 J with the value of energy threshold E_{th} set to 30 J.

4 Performance Evaluation

Through extensive simulations using the OMNeT++ simulator with MiXiM framework, we compare our PS-EHWSN protocol against to B-MAC [14] protocol. The performance metrics used to evaluate the two protocols are:

Packet Delivery Ratio: The ratio of the data packets delivered to the destination to those generated by the sources. This should be maximized.

Mean Packet Latency: The time used by the packet to reach the destination node is averaged for all the packets. This should be minimized.

Throughput: The total rate of transfer (received bytes/simulation time) which should be maximized.

Balance Factor (BF) of the Energy Consumption: It is used to examine the extent of the equilibrium of energy consumption through sensor nodes [15] and with k nodes it is defined by the following equation:

$$\left(\sum_{j=1}^{k} E_{rs_j}\right)^2 \div k \sum_{j=1}^{k} E_{rs_j}^2 \tag{3}$$

Figure 3 presents the simulation results for the delivery ratio metric with various TrafficParam values (0.1 s and 0.5 s) by different number of sent packets (500 and 1000) where the number of nodes is between 10 and 100. The results show that th packet delivery ratio decreases with the increase of the nodes number. Furthermore, the results show that PS-EHWSN is able to give a high packet delivery ratio than BMAC for various EH rates thanks to adjusting the duty-cycles dynamically based on the residual energy, which can increase over time. It is also due to the use of the strobed preamble method. On the other hand, BMAC provides a low packet deliver ratio with the increase in the traffic load, with the same parameters used by PS-EHWSN, due to the use of the static duty-cycle; additionally, the transmission of a long continuous preamble sequence before transmitting the data packet hinders its performance further and causes to packet drop due to collision. For example, in the case where the number of packets sent is set to 500 and when the number of nodes increases from 10 to 50 as shown in Fig. 3(a), the PS-EHWSN outperforms BMAC in terms of the packet delivery ratio ranging approximately between 95.74% and 97.36% when the TrafficParm is set to 0.1 s and between 84.53% and 89.12% when the TrafficParm is set to 0.5 s. For PS-EHWSN, Fig. 3(a) shows a gain in terms of higher delivery ratio of about 97.28% when TrafficParam = 0.1 s and 87.85% when TrafficParam = 0.5 s in comparison with BMAC when the number of nodes is equal to 100 and the number of packets sent = 500. Therefore, it can be seen that significant performance gains in the delivery ratio are obtained with PS-EHWSN over BMAC.

Figure 4 presents the results obtained for the Mean Latency. We observe that the Mean Latency increases when the number of nodes increases. The results show that PS-EHWSN outperforms BMAC moderately when the number of nodes increases. Also, it's perceivable that the mean latency of PS-EHWSN is lower than that of BMAC. The two reasons are: (1) our PS-EHWSN protocol leverages nodes with high energy level to be active more time thanks their low sleep period duration which is computed based on equation, and (2) use of the strobed preamble method. Both the reasons are explained by the fact that data packets are not buffered, which greatly reduces the mean latency compared to the transmission of a long continuous preamble sequence used by BMAC. On the other hand, our proposed mechanism uses an approximation method of sleep period duration of nodes to adjust their duty-cycle based on future-presented harvested energy. For PS-EHWSN, Fig. 4(a) shows a gain in terms of lower mean latency of about 34.93% when TrafficParam = 0.1 s and 45.27% when TrafficParam = 0.5 s in comparison with BMAC when the number of nodes is equal to 100 and the sent packets = 500.

Figure 5 presents the throughput value measured in bits per second (bps) that is generated by varying nodes number. The result indicates that the throughput increases with the increase of the number of nodes. Furthermore, the results show that PS-EHWSN is able to give a higher throughput than BMAC thanks to the use of the strobed preamble method and the dynamic decrease in active period of sensor nodes. This latter depends on the amount of the harvested energy and allows dynamic adjustment of the duty-cycles. This reduces the number of collisions and the maximum number of retransmissions because the more collisions take place, the more time is needed for a successful transmission that causes packet drop. On the other hand, BMAC provides a low packet throughput because of the transmission of a long continuous preamble sequence before transmitting the data packet which causes packet drop due to collision. For example, in Fig. 5(a) when the number of nodes increases from 50 to 100, the throughput increases by 17.46% for PS-EHWSN and 22.98% for BMAC of which TrafficParmis set to 0.1 s and the number of packets sent is set to 500. Also, it can be noticed from this figure that significant performance gains (approximately 85.69% in throughput) was obtained from PS-EHWSN over BMAC, when the number of nodes = 100 and with TrafficParm set to 0.5 s.

We carry out another simulation with the number of nodes k = 100 and the number of packets sent = 500 where the trafficparam set to 0.5 s for 5 different simulation times: 50, 100, 150, 200, 250 and 300 s. Figure 6 shows that BMAC protocol has a great reduction in BF over time, as compared to PS-EHWSN. Initially, PS-EHWSN and BMAC protocols have both (BF) equals to 1 at the first time of the simulation because the sensor nodes are tuned to have the same primary energy. After that and as the simulation time goes, the results show that PS-EHWSN has a better BF than BMAC, since they favors nodes having a high energy level to be frequently active thanks their low sleep period duration which is computed based on Eq. (2). However, BMAC protocol allows all sensor nodes to use a fixed sleep time which is defined by administrator. Thus, PS-EHWSN can contribute to balancing energy consumption among sensor nodes in WSN in comparison with BMAC.

As a result, in order to increase the performance of a WSN in terms of delivery ratio, the mean latency and throughput in the area of EH, the reduction of duty-cycle is important to avoid energy consumption caused by the collision and high period of node's activity.

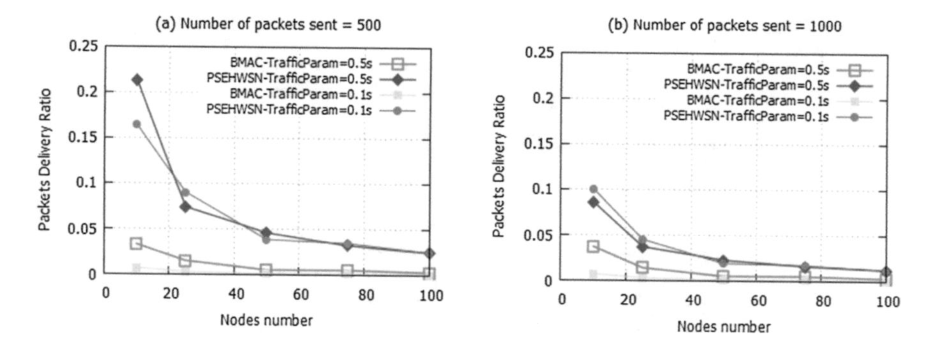

Fig. 3. Packet delivery ratio with various numbers of the sent packets

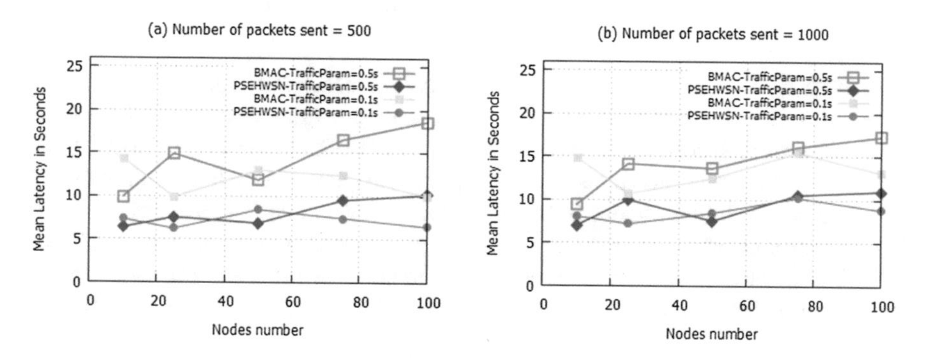

Fig. 4. Mean latency in seconds with various numbers of the sent packets

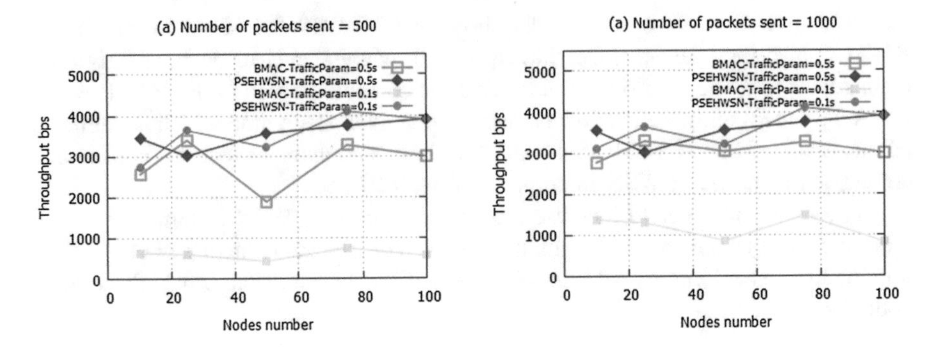

Fig. 5. Throughput with various numbers of the sent packets

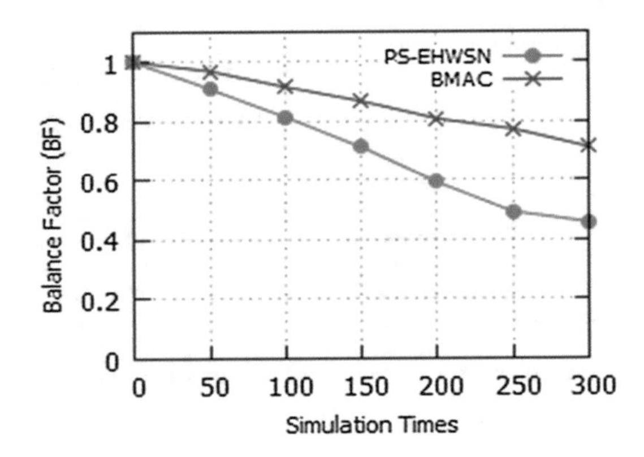

Fig. 6. Balance Factor of the energy consumption

Thanks to this variable duty-cycle reduction that makes it dynamic where the sleep times are selected consciously and differently among neighbor nodes (which reduces the number of collisions and retransmissions) PS-EHWSN outperforms BMAC which uses a static duty-cycle and transmission of preamble packets for a longer duration. The receiving nodes sleep durations in BMAC are increased which causes a drop in packets due to collisions, and makes the receiver nodes stay awake until the end of this long preamble. Overall, we can observe a significant performance improvements in throughput, mean latency and packet delivery ratio in PS-EHWSN over BMAC.

5 Conclusion

Mobile wireless sensor networks are characterized by their absence of infrastructure, their dynamicity and with restricted resources such as processing power; storage and more importantly energy resources because they are usually powered by batteries. Recharging the batteries of the sensor nodes is sometimes impossible due to their unreachable locations. This paper describes PS-EHWSN protocol: a new approach to low-power communication in WSNs based on EH which is suitable to mobile WSN. PSEHWSN employs the strobed preamble approach by broadcasting a series of short preamble packets, and it operates with a dynamic duty-cycle scheduling scheme. The main contribution of this work is that our protocol defines the next sleep period durations of the nodes using their residual energy level. The computation of this duration is based on the residual energy where each node can adjust its duty-cycle dynamically. This concept improves the performance of the network when compared with BMAC protocol in various ways. By comparing the performance of PS-EHWSN with that of BMAC protocol, PS-EHWSN is able to manage the energy harvested of a sensor node used for detection, transmission, reception, processing as well as increasing the energy efficiently; it decreases the mean latency, has a higher packet delivery ratio and has a higher throughput because the duty-cycle of each node in the network depends on the value of the energy harvest rate. Using OMNeT++/MiXiM simulations, PS-EHWSN is evaluated with many test cases of variable traffic flows. In all experiments, PS-EHWSN outperforms BMAC protocol in terms of packet delivery ratio, mean latency and throughput. It also contributes to balancing energy consumption among sensor nodes in WSN. The future work includes more thorough evaluation of our protocol PS-EHWSN with other energy harvesting protocols, taking into account the impact of the mobility of nodes on its performance and on the duty-cycle reducing.

References

1. Akyildiz, I.F., Su, W., Sankarasubramaniam, Y., Cayirci, E.: Wireless sensor networks: a survey. Comput. Netw. **38**(4), 393–422 (2002)
2. Vaidya, N.H., Bahl, P., Gupta, S.: Distributed fair scheduling in wireless LAN. In: 6th Annual International Conference on Mobile Computing and Networking, Boston, USA, Aout, pp. 67–78 (2000)
3. Lee, P., Han, M., Tan, H.P., Valera, A.: An empirical study of harvesting-aware duty cycling in environmentally-powered wireless sensor networks. In: 2010 IEEE International Conference on Communication Systems (ICCS) (2010)
4. Maheswar, R., Jayarajan, P., Sheriff, F.N.: A survey on duty cycling schemes for wireless sensor networks. Int. J. Comput. Netw. Wireless Commun. **3**(1), 37–40 (2013)
5. Ye, W., Heidemann, J., Estrin, D.: An energy-efficient MAC protocol for wireless sensor networks. In: Proceedings of the Twenty-First Annual Joint Conferences of the IEEE Computer and Communications Societies, INFOCOM 2002. IEEE (2002)
6. Polastre, J., Hill, J., Culler, D.: Versatile low power media access for wireless sensor networks. In: SenSys 2004: Proceedings of the 2nd International Conference on Embedded Networked Sensor Systems, pp. 95–107. ACM, New York (2004)
7. Buettner, M., Yee, G.V., Anderson, E., Han, R.: X-MAC: a short preamble MAC protocol for duty-cycled wireless sensor networks. In: SenSys 2006: Proceedings of the 4th International Conference on Embedded Networked Sensor Systems, pp. 307–320. ACM, New York (2006)
8. Sun, Y., Gurewitz, O., Johnson, D.B.: RI-MAC: a receiver initiated asynchronous duty-cycle MAC protocol for dynamic traffic loads in wireless sensor networks. In: SenSys 2008: Proceedings of the 6th ACM Conference on Embedded Networked Sensor Systems (2008)
9. Sudevalayam, S., Kulkarni. P.: Energy harvesting sensor nodes: survey and implications. Commun. Surv. Tutorials, 1–19 (2010)

10. Jeličić, V.: Power management in Wirless sensor networks with high-consuming sensors. Qualifying Doctoral Examination, January 2011
11. Bengheni, A., Didi, F., Bambrik, I.: Energy-harvested management mechanism for wireless sensor networks. In: 2017 5th International Conference on Electrical Engineering - Boumerdes (ICEE-B), Boumerdes, Algeria, pp. 1–4 (2017)
12. MiXiM Documentation, December 2017. http://mixim.sourceforge.net/
13. Nguyen, V.T., Gautier, M., Berder, O.: Implementation of an adaptive energy-efficient MAC protocol in OMNeT++/MiXiM" 1st OMNeT++ Community Summit, France, pp. 1–4, September 2014
14. Förster, Implementation of the B-MAC Protocol for WSN in MiXiM, April 2015. http://www.omnetworkshop.org/2011/uploads/slides/OMNeT_WS2011_S5_C1_Foerster.pdf
15. Yoo, H., Shim, M., Kim, D.: Dynamic duty cycle scheduling schemes for energy-harvesting wireless sensor networks. IEEE Commun. Lett. **16**(2), 202–204 (2012)

A New Key Agreement Method for Symmetric Encryption Using Elliptic Curves

Nissa Mehibel(✉), M'hamed Hamadouche, and Amina Selma Haichour

LIMOSE Laboratory, Faculty of Science, University M'hamed Bougara of Boumerdes, Independence Avenue, 35000 Boumerdès, Algeria
n.mehibel@univ-boumerdes.dz,
hamadouche-mhamed@hotmail.com,
selma.haichour@gmail.com

Abstract. The Elliptic Curve Diffie-Hellman (ECDH) is the basic protocol used for key agreement on Elliptic Curves (EC) and it is analogue to the standard Diffie-Hellman key exchange. The security of ECDH relies on the difficulty of the Elliptic Curve Discrete Logarithm Problem (ECDLP), however this protocol is vulnerable to man in the middle attack. In this paper, we first analyze the Ahirwal and Ahke encryption scheme which is based on ECDH key exchange and then we propose a new key agreement method to secure it from man in the middle attack.

Keywords: Elliptic Curve Cryptography · Elliptic Curve Discrete Logarithm Problem · Elliptic Curve Diffie-Hellman · Man in the middle attack

1 Introduction

Miller [1] and Koblitz [2] introduced independently Elliptic Curves in Cryptography (ECC) in the mid 80s, and Lenstra [3] showed the factorization of integers using elliptic curves. Since then, elliptic curves have played an increasingly important role in many cryptographic situations. ECC is based on the arithmetic on elliptic curves and security of the hardness of the Elliptic Curve Discrete Logarithm Problem (ECDLP) [4]. The advantages of EC cryptosystem over other public key cryptosystems are: robust security, less and faster computation, less storage space required and shorter keys [5]. The Elliptic Curve Diffie-Hellman (ECDH) is the basic protocol used to key agreement on Elliptic Curves (EC) and it is analogue to the standard Diffie-Hellman key exchange [6]. The security of ECDH relies on the difficulty of ECDLP, however this protocol is vulnerable to man in the middle attack.

Ahirwal and Ahke [7] proposed two symmetric cryptosystems to encrypt and decrypt messages by using ECDH key agreement, making their encryption susceptible to man in the middle attack.

Two basic methods have been developed and used to solve the Diffie-Hellman key exchange authentication problem. The first method is to add a digital signature to the exchange [8–10]. The second one is to modify the Diffie-Hellman key exchange [11, 12]

O. Demigha et al. (Eds.): CSA 2018, LNNS 50, pp. 79–88, 2019.
https://doi.org/10.1007/978-3-319-98352-3_9

by introducing a specific public key for each communicator, such as digital authentication is integrated in the exchange [13–15].

In this paper, we propose a new key agreement method based on the introduction of a specific public key [14, 15] in order to reduce the probability of exposure in Ahirwal and Ahke encryption, to man in the middle attack. The rest of the paper is organized as follows: Sect. 2 summarizes Elliptic Curve Cryptography, Sect. 3 presents the Ahirwal and Ahke encryption scheme, Sect. 4 describes the proposed method and Sect. 5 concludes the paper.

2 Elliptic Curve Cryptography

2.1 Elliptic Curve

The elliptic curve E over a finite field k, is an algebraic nonsingular curve which can be represented by the following generalized Weierstrass equation:

$$E : \{(x, y)|y^2 + \alpha_1 xy + \alpha_3 y - x^3 - \alpha_2 x^2 - \alpha_4 x - \alpha_6 = 0\} \cup \{o\} \tag{1}$$

where α_1, α_2, α_3, α_4 and $\alpha_6 \in k$ and 0 is the point at infinity.

In the present paper it suffices to limit the study of elliptic curve of third degree over a finite field $k = F_p$ having the form:

$$y^2 = x^3 + ax + b \tag{2}$$

where $a, b \in k$ and $4a^3 + 27b^2 \neq 0$. Together with an extra point 0, called the point at infinity.

2.2 Geometric Addition

Let $P(x_1, y_1)$ and $Q(x_2, y_2)$ be two points on the elliptic curve $E_p(a, b)$

- *Point adding*: $P \neq Q$, the group operator will allow us to calculate a third point $P + Q = R(x_3, y_3) \in E_p(a, b)$ where:

$$x_3 = (y_2 - y_1/x_2 - x_1)^2 - x_1 - x_2 \quad \text{and} \quad y_3 = (y_2 - y_1/x_2 - x_1)(x_1 - x_3) - y_1.$$

- *Point doubling*: $P = Q$, the group operator will allow us to calculate a third point $P + P = 2P = R(x_3, y_3) \in E_p(a, b)$ where:

$$x_3 = (3x_1^2 - a/2y_1)^2 - 2x_1 \text{ and } y_3 = (3x_1^2 - a/2y_1)^2 - (x_1 - x_3) - y_1.$$

- *Adding vertical point:* $P = -P$, the group operator will give us the point at infinity $P + (-P) = 0$.

2.3 Point Multiplication (Scalar Multiplication)

Let $P(x, y)$ be any point on the elliptic curve $E_p(a, b)$ and k is a large integer. Scalar multiplication consists of computing the value $k*P$ by doing a series of point doublings and additions until the product point is reached. $kP = P + P + P + \ldots + P$, k times.

2.4 Elliptic Curve Discrete Logarithm Problem (ECDLP)

Elliptic Curve Cryptography (ECC) is based on the difficulty of Elliptic Curve Discrete Logarithmic Problem (ECDLP). The difficulty of this problem is to determine the value of k of the equation $Q = k*P$ for the known points P and Q on the elliptic curve $E_p(a, b)$, where k is a Large random number less than p.

2.5 Elliptic Curve Diffie-Hellman (ECDH)

Figure 1 shows the secret key sharing based on the Diffie-Hellman protocol. In the latter, Alice and Bob agree on an elliptic curve $E_p(a, b)$ and a generator $G \in E_p(a, b)$, then they share the secret key K as follows:

- The private key for each of Alice and Bob is generated by selecting random number a and b respectively.
- The public key for each of Alice and Bob is calculated as follows: $a*G$ and $b*G$ respectively and each sends it to opposite side.
- Alice and Bob both compute the shared secret key $K = a*(b*G) = b*(a*G) = abG$.

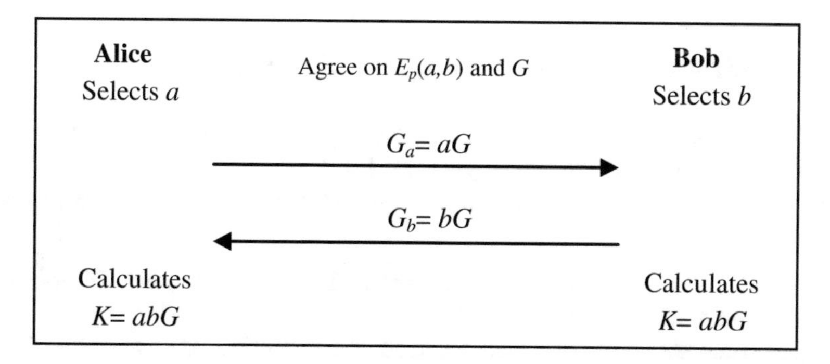

Fig. 1. Elliptic Curve Diffie-Hellman key exchange.

The above key exchange technique is secured by the difficulty of the discrete logarithm problem on an elliptic curve. However, this technique remains susceptible to man in the middle attack. The latter is therefore the basic weakness of the Diffie-Hellman key exchange protocol [14, 15].

This attack allows a third party (attacker) to intercept the communication between two endpoints (Alice and Bob), to create with the first one a common key and to do the same with the second one. Thus, Alice and Bob think to communicate directly whereas in reality, they communicate with the attacker as in Fig. 2.

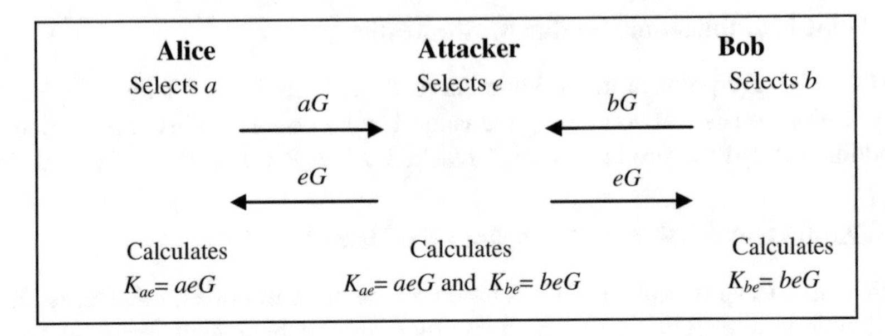

Fig. 2. Man in the middle attack during ECDH key exchange.

2.6 Elliptic Curve Encryption and Decryption

The Elliptic curve encryption/decryption of messages using the secret key shared by the Diffie-Hellman protocol is a symmetric procedure that consists of the following steps [14]:

- Alice and Bob compute the secret key $K = abG$ (Using ECDH).
- Message M to be encrypted is coded on to an elliptic curve point $M \in E$.
- Alice/Bob computes the encrypted message $C = K + M$ and sends it to Bob/Alice.
- Bob/Alice receives the encrypted message C and decrypts it by computing: $M = C - K$
  $= K + M - K \cdot$

3 The Ahirwal and Ahke Encryption Scheme

Ahirwal and Ahke proposed two methods to encrypt and decrypt messages using the protocol of Diffie-Hellman for key agreement [7].

In the first method, to encrypt and decrypt a message, Alice and Bob proceed as follows:

1. Alice and Bob compute the secret key shared using Elliptic Curve Diffie-Hellman (ECDH) protocol, $K = abG = (x, y)$.
2. So that Alice sends an encrypted message $M \in E$ to Bob, she calculates the multiplication between the coordinates of the secret key $S = (x * y) \bmod N$.
3. Then she multiplies the result S by the message M which is a point of the elliptic curve $C = S * M$, and sends the encrypted message C to Bob.
4. Bob receives the encrypted message C and to decrypt it, he calculates the multiplication between the coordinates of the secret key $S = (x * y) \bmod N$.
5. He calculates its inverse $S^{-1} \bmod N$, where $N = \#E$ (number of points in the elliptic curve $E_p(a, b)$).
6. Then, he multiplies S^{-1} by the encrypted message in order to find the message $M = S^{-1} * C = S^{-1} * S * M$.

In the second method, to encrypt and decrypt a message, Alice and Bob proceed as follows:

1. Alice and Bob compute the secret key shared using Elliptic Curve Diffie-Hellman (ECDH) protocol, $K = abG = (x, y)$.
2. So that Alice sends an encrypted message $M \in E$ to Bob, she calculates the exponentiation function between the coordinates of the secret key $S = (x^y) \bmod N$.
3. Then she multiplies the result by the message M which is a point of the elliptic curve $C = S * M$ and sends the encrypted message C to Bob.
4. Bob receives the encrypted message C and to decrypt it, he calculates the exponentiation function between the coordinates of the secret key $S = (x^y) \bmod N$.
5. He calculates its inverse for the exponentiation function $S^{-1} \bmod N$, where $N = \#E$ (number of points in the elliptic curve $E_p(a, b)$).
6. Then he multiplies S^{-1} by the encrypted message in order to find the message $M = S^{-1} * C = S^{-1} * S * M$.

The authors claim that the second method is more secured than the first one, because the sender of an encrypted message calculates the exponentiation function between the coordinates of the key (use a fast exponentiation method), and the receiver calculates the inverse of the exponentiation function between the coordinates of the key in order to decrypt a message [7]. Although this method is more secured, it remains susceptible to man in the middle attack as it uses the Diffie-Hellman key exchange protocol.

4 The Proposed Method

4.1 The Proposed Algorithm

We propose a new key agreement method to secure Ahirwal and Ahke encryption (second method) from the man in the middle attack. We propose to introduce into our key exchange protocol a specific public key for each communicating party [14, 15]. This key can only be used by the communicating party to which is intended the public key. Each communicating party calculates the specific public key for the other by multiplying its private key by the sum of its public key and the public key of the other concerned communicating party. This allows us to ensure the authenticity of the shared secret key (Fig. 3).

Firstly, Alice and Bob calculate the shared secret key, for this they proceed as follows:

1. Alice and Bob agree upon to use an elliptic curve $E_p(a, b)$ where p is a prime number and a generator G of $E_p(a, b)$.
2. Alice and Bob respectively select their private keys, a large random number $a <$ ord (G) and a point A on the elliptic curve for Alice's private keys, and a large random number $b <$ ord (G) and a point B on the elliptic curve for Bob's private keys.
3. Alice and Bob respectively calculate their public key, $G_a = aG$ Alice's public key, and $G_b = bG$ Bob's public key.

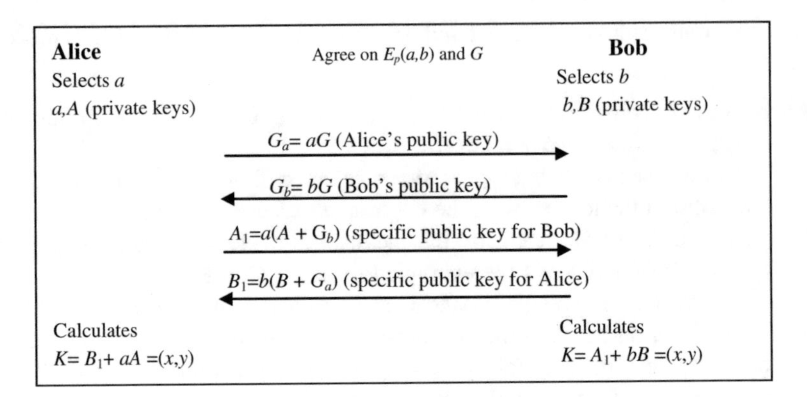

Fig. 3. Proposed method for key agreement.

4. Alice and Bob respectively calculate the specific public key for each other, $A_1 = a\ (A + G_b)$ Alice's specific public key for Bob and $B_1 = b\ (B + G_a)$ Bob's specific public key for Alice.
5. Now Alice and Bob can calculate the shared secret key as follows:
   - Alice calculates $K = B_1 + aA = b\ (B + G_a) + aA = bB + baG + aA = (x, y)$.
   - Bob calculates $K = A_1 + bB = a\ (A + G_b) + bB = aA + abG + bB = (x, y)$.

Secondly, so that Alice sends an encrypted message M to Bob, she must convert all the characters of the message into points of the elliptic curve by using a code table, which is agreed upon by the two entities Alice and Bob. Then, each point is encrypted as follows:

1. Alice calculates the exponentiation function between the coordinates of the secret key $S = (x^y) \bmod N$.
2. Then she multiplies the result by each character of the message C which is a point of the elliptic curve $M' = \{C'_1 = S * C_1, C'_2 = S * C_2, C'_3 = S * C_3 \ldots\}$ and sends the encrypted message M' to Bob.
3. Bob receives the encrypted message M' and to decrypt it, he calculates the exponentiation function between the coordinates of the secret key $S = (x^y) \bmod N$.
4. He calculates its inverse for the exponentiation function $S^{-1} \bmod N$, where $N = \#E$ (number of points in the elliptic curve $E_p(a, b)$).
5. Then he multiplies S^{-1} by each encrypted character in order to find the decrypted message $M = \{C_1 = S^{-1} * C'_1, C_2 = S^{-1} * C'_2, C_3 = S^{-1} * C'_3 \ldots\}$

4.2 Security Considerations

Our main contribution is to introduce the specific public key in the key exchange protocol [14, 15] to guarantee the authenticity of the shared secret key used in the Ahirwal and Ahke encryption, which will reduce its probability of exposure risk to man in the middle attack. Our scheme is based on the robustness of the Elliptic Curve Discrete Logarithm Problem (ECDLP). Moreover, each communicating party publishes a specific public key that is destined to a specific communicator, which ensures the authenticity of the shared secret key, because each communicating party calculates its specific public key with the public key of the specific communicator, therefore the resulting specific public key can only be used by the specific communicator only. In addition, by constructing the secret key shared by the communicator's private key and the specific public key for him only, this ensures that the shared secret key is implicitly signed. Therefore, the shared secret key achieves the qualities of confidentiality, authentication and non-repudiation.

4.3 Example

Let E: $y^2 = x^3 - 5x + 21$ an elliptic curve represented by the graph in Fig. 4.

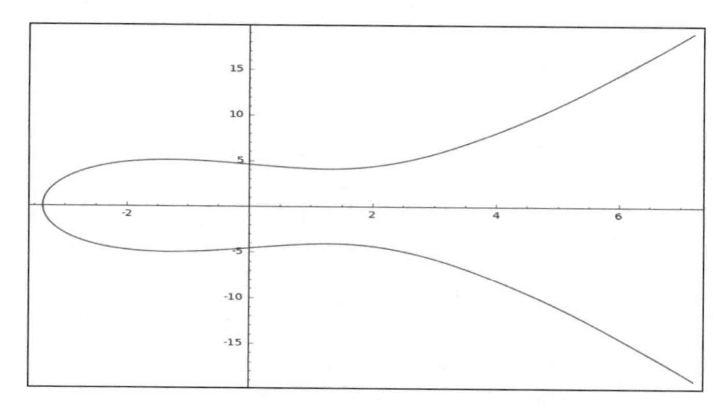

Fig. 4. Elliptic curve $y^2 = x^3 - 5x + 21$.

E is defined in the finite field F_{37}, where $4 * (-5)^3 + 27 * (21)^2 \bmod 37 = 16 \neq 0$ and $N = 43$.

The points on the elliptic curve $E_{37}(-5,21)$ are:

{∞, (0, 13), (0, 24), (3, 12), (3, 25), (4, 18), (4, 19), (5, 11), (5, 26), (7, 12), (7, 25), (8, 7), (8, 30), (10, 3), (10, 34), (13, 9), (13, 28), (15, 18), (15, 19), (16, 2), (16, 35), (18, 18), (18, 19), (20, 15), (20, 22), (21, 1), (21, 36), (23, 14), (23, 23), (26, 15), (26, 22), (27, 12), (27, 25), (28, 15), (28, 22), (29, 17), (29, 20), (30, 3), (30, 34), (34, 3), (34, 34), (36, 5), (36, 32)}.

The points on the elliptic curve $E_{37}(-5,21)$ are shown in the Fig. 5.

Firstly, Alice and Bob calculate the shared secret key, let G = (30, 3) is the point generator.

- Alice and Bob respectively select their private keys:
  - Alice's private keys: $a = 11$ and $A = (5, 26)$,
  - Bob's private keys: $b = 31$ and $B = (13, 9)$.
- Alice and Bob respectively calculate their public key:
  - Alice's public key: $G_a = aG = 11* (30, 3) = (21, 1)$,
  - Bob's public key: $G_b = bG = 31* (30, 3) = (13, 28)$.

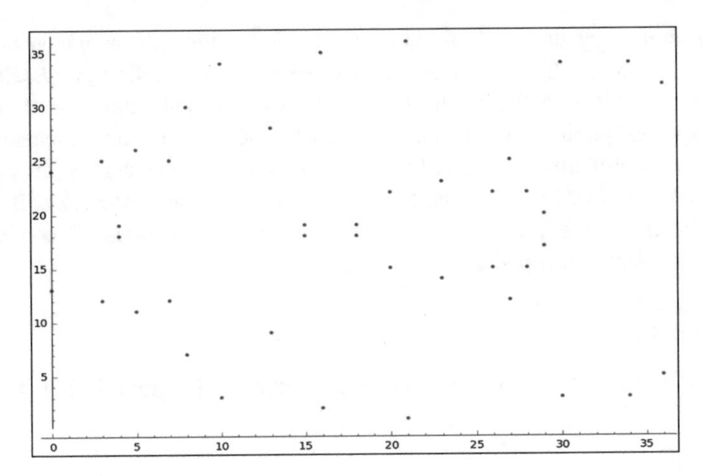

Fig. 5. Elliptic curve $y^2 = x^3{-}5x + 21$ over finite field F_{37}.

- Alice and Bob respectively calculate the specific public key for each other:
  - Alice's specific public key for Bob $A_1 = a\ (A + G_b) = (8, 30)$,
  - Bob's specific public key for Alice $B_1 = b\ (B + G_a) = (27, 25)$.
- Now Alice and Bob can calculate the shared secret key as follows:
  - Alice calculates $K = B_1 + aA = (20, 25) + 11{*}(5, 26) = (27, 12)$,
  - Bob calculates $K = A_1 + bB =$.(16, 2) + 31*(13, 9) = (27, 12).

Secondly, so that Alice sends an encrypted message M = "CSA18" to Bob, she must convert all the characters of the message to the points of the elliptic curve $E_{37}(-5,21)$ by using the code Table 1, which is agreed upon by Alice and Bob.

Table 1. Corresponding characters to the co-ordinate points.

	A	**B**	**C**	**D**	**E**
∞	(0,13)	(0,24)	(3,12)	(3,25)	(4,18)
F	**G**	**H**	**I**	**J**	**K**
(4,19)	(5,11)	(5,26)	(7,12)	(7,25)	(8,7)
L	**M**	**N**	**O**	**P**	**Q**
(8,30)	(10,3)	(10,34)	(13,9)	(13,28)	(15,18)
R	**S**	**T**	**U**	**V**	**W**
(15,19)	(16,2)	(16,35)	(18,18)	(18,19)	(20,15)
X	**Y**	**Z**	**0**	**1**	**2**
(20,22)	(21,1)	(21,36)	(23,14)	(23,23)	(26,15)
3	**4**	**5**	**6**	**7**	**8**
(26,22)	(27,12)	(27,25)	(28,15)	(28,22)	(29,17)
9	**+**	**–**	*	**?**	**!**
(29,20)	(30,3)	(30,34)	(34,3)	(34,34)	(36,5)
%					
(36,32)					

CSA18 = {(3, 12), (16, 2), (0, 13), (23, 23), (29, 19)}.

- Alice calculates the exponentiation $S = (27^{12}) \bmod 43 = 21$.
- She encrypts the message as follows:
  - $C'_1 = S * C_1 = 21 * (3,\ 12) = (34,\ 34) \rightarrow ?$
  - $C'_2 = S * C_2, = 21 * (16, 2) = (10,\ 3) \rightarrow \mathrm{M}$
  - $C'_3 = S * C_3 = 21 * (0, 13) = (29,\ 20) \rightarrow 9$
  - $C'_4 = S * C_4 = 21 * (23, 23) = (15,\ 18) \rightarrow \mathrm{Q}$
  - $C'_5 = S * C_5 = 21 * (29, 17) = (23,\ 23) \rightarrow 1$

  Alice sends the encrypted message M' = {(34, 34), (10, 3), (29, 20), (15, 18), (23, 23)} = "?M9Q1" to Bob.
- Bob receives the encrypted message M' and to decrypt it, he calculates:
  - $S = (27^{12}) \bmod 43 = 21$.
  - $S^{-1} \bmod N = 21^{-1} \bmod 43 = 41$.
- He multiplies S^{-1} by each encrypted character in order to find the decrypted message M as follows:
  - $C_1 = S^{-1} * C'_1 = 41 * (34,\ 34) = (3,\ 12) \rightarrow \mathrm{C}$
  - $C_2 = S^{-1} * C'_2 = 41 * (10,\ 3) = (16, 2) \rightarrow \mathrm{S}$
  - $C_3 = S^{-1} * C'_3 = 41 * (29,\ 20) = (0, 13) \rightarrow \mathrm{A}$
  - $C_4 = S^{-1} * C'_4 = 41 * (15,\ 18) = (23, 23) \rightarrow 1$
  - $C_5 = S^{-1} * C'_5 = 41 * (23,\ 23) = (29, 17) \rightarrow 8$

5 Conclusion

In this paper, we analyzed the Ahirwal and Ahke encryption scheme and we proposed a new key agreement method to secure it by guaranteeing the authenticity of the shared secret key. Thus, our proposed method reduces the probability of exposure risk, in Ahirwal and Ahke encryption scheme, to the man in the middle attack. In addition, the security of our method relies on the difficulty of the discrete logarithm problem in elliptic curve (ECDLP). Therefore, the shared secret key achieves the qualities of confidentiality, authentication and non-repudiation.

References

1. Koblitz, N.: Elliptic curve cryptosystems. Math. Comput. **48**(177), 203–209 (1987)
2. Miller, V.S.: Uses of elliptic curves in cryptography. In: Advances in Cryptology, CRYPTO 1985. Lecture Notes in Computer Science, vol. 218, pp. 417–428. Springer (1986). 5. N
3. Trappe, W., Washington, L.C.: Introduction to Cryptography with Coding Theory, 2nd edn. Pearson-Prentice-Hall, Upper Saddle River (2006)
4. Haichour, A.S., Hamadouche, M., Khouas, A.: Hardware design and implementation of ElGamal elliptic curve cryptosystem. Wulfenia J. **23**(2), 62–85 (2016)
5. Chande, M.K., Lee, C.C.: An improvement of a elliptic curve digital signature algorithm. Int. J. Int. Technol. Secur. Trans. **6**(3), 219–230 (2016)

6. Diffie, W., Hellman, M.E.: New directions in cryptography. IEEE Trans. Inform. Theory **IT-22**, 644–654 (1976)
7. Ahirwal, R.R., Ahke, M.: Elliptic curve Diffie-Hellman key exchange algorithm for securing hypertext information on wide area network. Int. J. Comput. Sci. Inf. Technol. **4**(2), 363–368 (2013)
8. Elgamal, T.: A public key cryptosystem and a signature scheme based on discrete logarithms. IEEE Trans. Inform. Theory, **IT-31**(4) (1985)
9. Biswas, G.P.: Establishment of authenticated secret session keys using digital signature standard. Inf. Secur. J. Glob. Perspect. **20**(1), 9–16 (2011)
10. Stallings, W.: Digital signature algorithms. Cryptologia **37**, 311–327 (2013)
11. Vasudeva Reddy, P., Padmavathamma, M.: An authenticated key exchange protocol in elliptic curve cryptography. J. Discrete Math. Sci. Crypt. **10**(5), 697–705 (2007)
12. Suneetha, C., Kumar, D.S., Chandrasekhar, A.: Secure key transport in symmetric cryptographic protocols using some elliptic curves over finite fields. Int. J. Comput. Appl. **36**(1) (2011)
13. Kumar, D.S., Suneetha, C.H., Chandrasekhar, A.: Authentic key transport in symmetric cryptographic protocols using some elliptic curves over finite fields. Int. J. Math. Arch. **3**(1), 137–142 (2012)
14. Mehibel, N., Hamadouche, M.: A new approach of elliptic curve Diffie-Hellman key exchange. In: The 5th International Conference on Electrical Engineering, Boumerdes, Algeria, 29–31 October 2017 (2017)
15. Mehibel, N., Hamadouche, M.: A public key data encryption based on elliptic curves. Int. J. Math. Comput. Methods **2**, 393–401 (2017)

Information Systems and Software Engineering

Data Lakes: New Generation of Information Systems?

Omar Boussaid(✉)

Université de Lyon 2, Lyon, France
omar.boussaid@univ-lyon2.fr

Abstract. Information systems are strongly impacted by new information technologies. New challenges emerge in the scientific and technological research caused by the craze around data and more particularly around big data. The data poses many challenges related to its exploitation. Its scope goes beyond traditional structures, such as databases or data warehouses. Furthermore, it is mostly unstructured which results in a need for new approaches to explore it. The different existing processes need to be redesigned. Their evolution poses new scientific obstacles as soon as they are projected in a big data framework. The Data Lakes represent today an emerging concept how to organize around the data to rethink innovation cycles within companies. On the other hand, this concept opens up new issues of investigation and promises real challenges, which will allow information technologies to evolve towards new perspectives in the professional world, and calls on new skills to develop and to capitalize on future users. What is it really about?

O. Demigha et al. (Eds.): CSA 2018, LNNS 50, p. 91, 2019.
https://doi.org/10.1007/978-3-319-98352-3_10

Using Fast String Search for Quran Text Auto-completion on Android

Djalel Chefrour(✉) and Abdallah Amirat(✉)

Mathematics and Computer Science Department, Souk Ahras University, 41000 Souk Ahras, Algeria
djalel.chefrour@univ-soukahras.dz,
amirat_abdallah@hotmail.fr

Abstract. Text auto-completion speeds up user input on the desktop PC by proposing a list of alternative words to select from after typing few characters. This useful feature is even more important on mobile devices because their human-computer-interface is restricted. We find it very helpful to extend auto-completion to the Holy Quran text, by allowing the user to select from alternative Ayat (آيات) where the characters he typed appear, instead of typing the Aya (آية) he wants to cite manually or copying it from another source. To achieve this goal, we have implemented Quran text auto-completion on the Android platform by extending an open-source software keyboard application. As this feature needs to search quickly and repetitively the whole Quran text, we tested a number of string search methods, including the Boyer Moore algorithm, to determine the fastest search solution to use. The result is an application that proposes a fast and smooth user experience that is available freely in the play store under the name (لوح مفاتيح القرآن الكريم).

Keywords: Holy Quran text auto-completion · Fast string search
Android software keyboard

1 Introduction

Typing Quran text manually on mobile devices is an error prone task. First, one needs to remember the Quran text accurately and also remember its Rasm (الرسم), whether it is Imla (إملائي) or Uthmani (عثماني), with or without Shakl (الشكل). A more difficult task is to memorize where each Aya (آية) starts and ends, in which Surah (سورة) or Suwar (سور) it is present and what is its number and recall similar Ayat (المتماثلات). Few people in the Muslim world are capable of such achievements without external help. We demonstrate that such help can be provided by implementing auto-completion of Quran text at the Aya level. So when the user types few characters the device can quickly find the Ayat where they appear, and display them as alternatives, then the user is able to select the one he wants to cite. Second, even if one remembers exactly the Aya he wants to type, it is not easy to do so on mobile devices because of their tiny keyboards, which are very often soft keyboards displayed on limited size screens, or - for few device models - a set of hard keys disposed on a small surface. This makes it hard to select the correct Arabic keys at a decent typing pace for most users. As text auto-completion is generally

O. Demigha et al. (Eds.): CSA 2018, LNNS 50, pp. 92–101, 2019.
https://doi.org/10.1007/978-3-319-98352-3_11

limited to one word or few words (i.e., the rest of a sentence), and because of the mentioned screen limits, our feature does not include the completion of multiple consecutive Ayat or the completion of a whole Surah. Rather, these two use-cases require a full screen application.

Because the auto-completion feature needs to refresh the list of alternative Ayat proposed to the user with each newly typed character, it has to perform a repetitive fast string search overall the Quran text. For that purpose, to find a fast search solution, we benchmarked four alternative search methods implementations, which are the Brute force algorithm, Java's *String.indexOf* method, Java's regular expressions library and the Boyer Moore algorithm that is considered as the fastest string search algorithm in the literature [1, 2].

The rest of this paper is organized as follows: Sect. 2 explains how the Boyer Moore algorithm works. Section 3 describes the open source soft keyboard application that we extended with the fast search solution to add the Quran text auto-completion feature. Section 4 details the new structure of the new application and the main modifications we brought to it. Section 5 analyses the benchmark results of the alternative search methods we considered during development and Sect. 6 concludes this work by emphasizing the advantages of the proposed solution and outlining possible future improvements.

2 The Boyer Moore Search Algorithm

The Boyer Moore algorithm [1, 2] compares the search text to the pattern from right to left. Hence, if there is a mismatch, it can skip the rest of the characters in the pattern and try to shift it (to the right, relative to the text) the furthest possible. Shifting amplitude is determined as the maximum of two functions called the bad character shift and the good suffix shift, which are precomputed from the pattern and the text alphabet. If n is the length of the text and m is the length of the pattern, then the maximum possible shift amplitude in Boyer Moore is m. This gives it a best case time complexity of $O(n/m)$, which is an absolute minima among all search algorithms. Its worst case time complexity is $O(n * m)$, which is similar to the Brute Force algorithm. Whereas its average time complexity is sub-linear.

As shown by Fig. 1 from [2], if a bad character $\boldsymbol{b}$ is found in the text, the pattern can be shifted either fully past this bad character if it is not present in it, or to align the bad character with its right most occurrence in the pattern. In addition, if the already matched suffix $\boldsymbol{u}$ re-occurs elsewhere in the pattern, preceded by a character $\boldsymbol{c}$ different from the bad one $\boldsymbol{b}$, then this new occurrence is aligned with the matched text to shift of the pattern by hopefully more than one position. Otherwise, if only a portion $\boldsymbol{v}$ of the matched suffix, the longest right most one, re-occurs in the pattern, then it is aligned with the matched text to perform the shift.

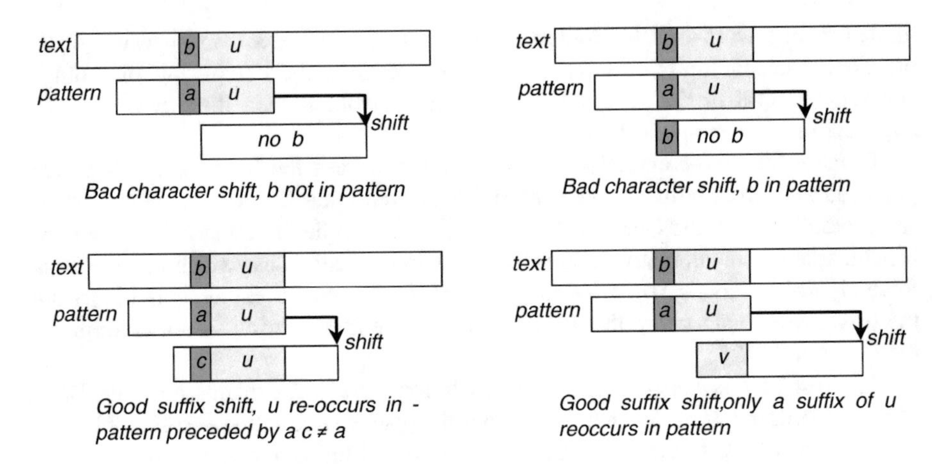

Fig. 1. Examples of Boyer Moore shifts [2].

3 Android Software Keyboard

Soft keyboards on the Android platform support text auto-completion by displaying a list of candidate words in a horizontal bar above the keyboard itself. These words are usually fetch from a dictionary corresponding the language selected by the user for input. Adding the auto-completion feature to the keyboard application, that is where text is typed, make its reusable in all the applications that which to employ it. In addition, if an application wants to provide an alternative list of completion words, other than the one provided by the keyboard itself, it can do so by sending to the keyboard which will display it in its candidates bar.

To develop Quran text auto-completion, we started from the simple soft keyboard published by the Android Open Source Project [3]. This one contains a QWERTY layout and a candidate bar, which just displays the exact same text entered by the user, as it has no dictionary to choose the alternative words from and no search method. The *SoftKeyboard* application is made of the main classes explained in Table 1.

A soft keyboard, or an *Input Method Editor (IME)* in Android parlance, is an interface element that allows user to enter text. Following the Android developer reference guide, the IME main class must extend the framework class *InputMethodService*, which has the life cycle depicted in Fig. 2. The modifications we brought to this cycle are shown in gray boxes.

The Android framework calls the method *onCreate* when an application requests the keyboard. This is where the keyboard initializes its resources. In our modifications to this class, we load the whole Quran text here once, from secondary storage to main memory, to speed up access to this text later and avoid a latency with each new search.

The method *onCreateInputView* is called by the when the keyboard layout is displayed. It selects the type of the keyboard to show (QWERTY, symbols, etc.). Similarly, *onCreateCandidatesView* is called when the candidates bar is displayed. It instantiates the *CandidateView* class. Then when the text field in the application that needs user input is displayed the method *onStartInput* is called. It allows the IME to

Table 1. List of classes of the open source SoftKeyboard application.

Class	Role
CandidateView	Handle the display of the candidate words separated vertical bars. When the user selects one of them, it sends it to the main class (cf. below)
ImePreferences	Handles user preferences for this keyboard. In this case, only the choice between American and British English
LatinKeyboard	Handles the keyboard as list of Latin letters, symbols, numerical and special keys (e.g. shift, enter, etc.) and manages the switch between them and between supported languages
LatinKeyboardView	Handles the display of the keyboard layout itself, in the form of horizontal rows containing a set of keys each. The layout is defined in XML
SoftKeyboard	This is the application main class, which is extended from the android framework *InputMethodService* class. It handles the keyboard logic and passes the typed characters to the application that uses the keyboard for input. During typing, this class is responsible for passing the list of alternative words to the candidate bar so the user can choose from them

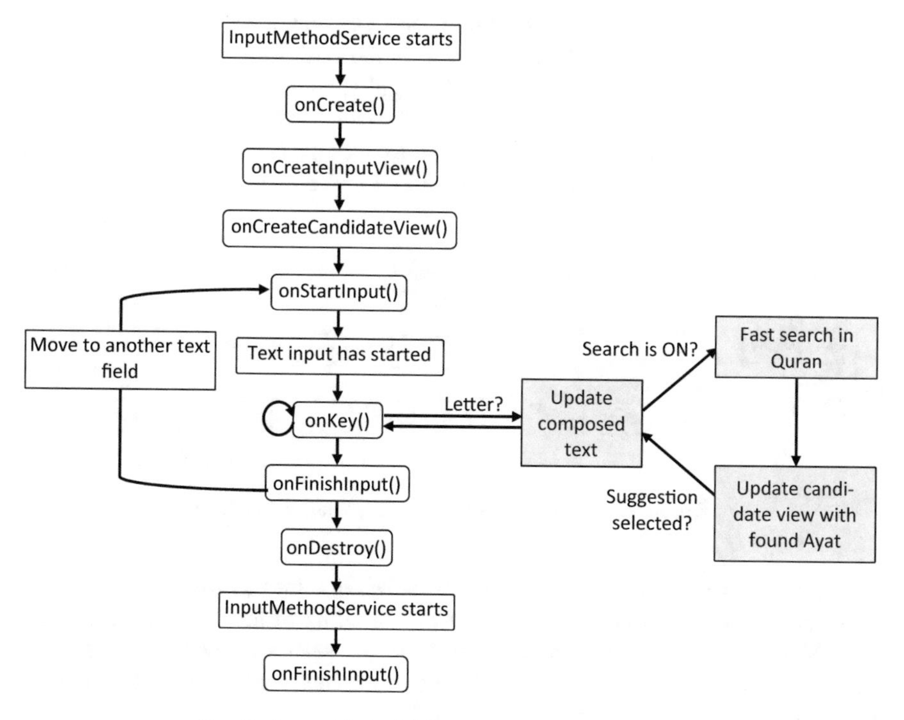

Fig. 2. Input Method Life cycle (modified from [4]).

clear the content of the candidates bar and reset its fields, especially the one used to save the text composed by key strokes.

Afterwards, with each new typed key, the method *onKey* is called to handle it. If it is a character, it will be appended to the composed text. In our modifications to the code, we send this text to the fast search function as a pattern for look up in the Quran text. That function returns the matching Ayat, which will be fed to the candidates bar using the method *updateCandidates*. This method will call the *setSuggestions* method to perform this task on the *CandidateView* object. If the user selects one of the suggested Ayat, this object will send its index back to the IME via a call to its *pickSuggestionManually* method. This method will replace the pattern characters of the composed text with the Aya text. When the user finishes typing the method *onFinishInput* is called and the composed text is passed to the application before the keyboard is removed from the screen and its *onDestroy* method is called to free the allocated resources.

4 Holy Quran Keyboard

To develop the modifications to the existing application, mentioned in the previous section, we designed a new application with structure shown in Fig. 3. The old Latin specific classes have been replaced by their Arabic counter parts as the QWERTY keyboard layout left room for a classic Arabic one. One important change we did to this layout is to replace the caps lock key (i.e. shift in the QWERTY view) with a new one that shows a Mushaf (مصحف) and that can be toggled on and off to enable and disable the search in the Quran function. Therefore, the new keyboard can be used to type normal Arabic text and it is only when the user toggles the Mushaf key, that the Quran Ayat matching the newly typed characters will be looked up.

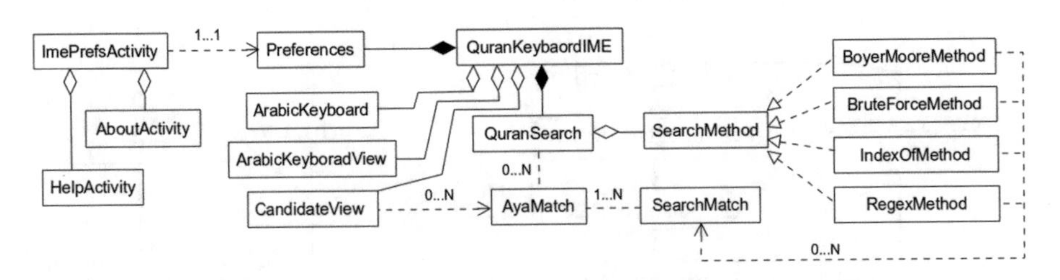

Fig. 3. Structure of Holy Quran Keyboard application.

In addition, we renamed the old *SoftKeyboard* class into *QuranKeyboardIME*. Its fields contain a copy of the whole Quran text, the string of the text composed by the user, which is the search pattern, a reference to the search method instance used to lookup the pattern quickly and the list of matching, which are shown in the candidates bar as suggestions for the user to pick from.

Following the Model View Controller design pattern, we have isolated the logic of "fast search in the Quran text" in a set of classes that are separate from the Android framework. Hence, these classes (*QuranSearch*, *SearchMatch* and *AyaMatch*) can be reused in a new application on any other platform. Furthermore, inspired by the Strategy design pattern, we separated the search function from the Quran text specific details by introducing a text search interface called *SearchMethod*. We then added four alternative implementations to this interface to determine the fastest one that will go into the final application. These are the *BoyerMooreMethod*, which implements the fast search algorithm of the same name. We expect it, at the design stage, to be the fastest. The *BruteForceMethod*, which implements a straightforward brute force search algorithm. The *IndexOfMethod*, which uses Java's *String.indexOf*. The *RegexMethod*, which uses the Matcher class of Java's regular expressions library. The full Quran is saved locally in a UTF-8 text file. Each line of this file contains one Aya prefixed with its Surah number and the Aya number in that Surah. So the matcher uses the pattern in formula (1).

$$([0-9]+)\backslash|([0-9]+)\backslash|(.*(\mathrm{userinput}).*) \tag{1}$$

The *SearchMethod* interface has one method that takes references to the whole Quran text and the search pattern string and returns a list of corresponding matches. The Quran text Rasm used for search is Imla without Shakl. The whole Quran string is preloaded from secondary memory to RAM when the keyboard application starts to speed up consecutive search operations. The *SearchMatch* class models a match by saving: its index in the Quran string, the beginning and the end of its line, and the index of the word containing the matched pattern. It has also an extra time stamp used to measure the execution time.

Once the *QuranSearch* object gets back a list of matches from *SearchMethod* instance, it needs to format the corresponding Ayat based on some user preferences. This might include using a Rasm different from the simple one used for search, like Imla with Shakl or Uthmani Rasm. In addition, the user might want to append the Aya number and its Surah name to its text. As he might want to truncate the beginning of the Aya to keep the portion starting with the word of containing the search pattern. All these processing is handled in the class *AyaMatch*. If a pattern is matched multiple times in one Aya (e.g. الرحمن in البسملة), then one *AyaMatch* will group the information obtained from many *SearchMatch* objects. In the end, the list of matches passed as suggestions to the *CandidateView* object is in fact a list of *AyaMatch* objects.

Besides, the keyboard itself (with its candidate view), the application user interface contains three separate screens (i.e. activities in Android jargon) which can be accessed from the system setting. These are handled by *ImePresActivity*, which manages the user preferences explained above, *AboutActivity*, which displays some info about the application and *HelpActivity*, which gives some help to the user on how to use the search feature and configure the preferences.

5 Benchmark Results Analysis and Discussion

Following the design detailed in the previous section, we wrote our application, Holy Quran Keyboard (لوح مفاتيح القرآن الكريم), using Android SDK version 23 [5] and Android Studio version 2.3 [6]. Its code is available online under the open source Apache 2.0 license [7]. Then we tested it on a Samsung S5 mini smart phone, which runs the same SDK version. The goal of the test is to determine how fast a Java implementation of the Boyer Moore algorithm can be, compared to a Brute Force one and to the other two search solutions offered by the language library (i.e. *indexOf* and *regex*).

An additional goal is to see how the speed of these four methods may vary with the search patterns lengths and positions. We used small and long patterns, located at the beginning, at the middle and at the end of the Holy Quran text as shown in Table 2.

Table 2. Benchmark results

Search patterns		Search times in micro-seconds			
Pattern string	Index of 1st match	Brute Force	IndexOf	Boyer Moore	RegEx
مالك	24	190	36	197	1855
ذلك الكتاب لا ريب فيه هدى	78	136	7	131	807
ذي القرنين	207576	111012	688	28213	217479
إن الذين آمنوا وعملوا الصالحات	209251	110505	721	12425	203683
مسد	416608	240374	2760	154757	426981
قل هو الله أحد	416999	234838	1931	43787	434529
Quran text total length	417316				

The above benchmarks show clearly that the Java's *indexOf* method is the fastest in all scenarios. The explanation of this result that goes against our initial expectation is found in the Java source *indexOf* [8]. It looks for the first character of the pattern very quickly, by using the native C function *fastIndexOf* [9]. The latter is not interpreted by the Java Virtual Machine but it is compiled to run directly on the device hardware, which makes up to 100 times faster compared to the Brute Force algorithm, written in Java.

The second fastest method is the Boyer Moore algorithm, which performs much better than Brute Force when both are written in the same language as expected. The only exception to this rule is with the first occurrence of the pattern (مالك) where the latter is slightly faster. This is explained by the time spent in the preprocessing phase of the Boyer Moore algorithm, which computes the bad character shift and good suffix shift tables, and which becomes negligible only for the patterns that are not located at the beginning of the Quran text.

The slowest search method in our test is the *Matcher* class pf the regular expression library, which even takes double the time used by Brute force, in some cases. This because the regular expressions engine first converts the pattern into non-deterministic finite automata (NFA) [10], then advances through it and through the search text as long as there is a character match. However, when there is no match, the engine backtracks to a previous position where it can take a different path in the NFA, which makes the search exponential in time.

The only advantage from using regular expressions is to support search in the Quran text with the * meta-character in the middle of the user input, in order to match the Ayat containing specific letters or words with anything in between. Although this method's worst search times are below half a second, as shown in Table 2, this is not a typical use case. Therefore, it is not included in the current version of the application.

Another peculiar finding in Table 2 is the search time of the second pattern (الكتاب لا ريب فيه ذلك) that is lower than the first one (مالك), for all four methods, even though it is located after it in the Quran. This is explained by the cache effect as the test patterns are run in a loop so the first one takes the penalty of loading the code of the four methods into memory.

Table 2 results for the third and fourth patterns (الصالحات كانتإن الذين آمنوا وعلموا and ذي القرنين), and the fifth and the sixth (قل هو الله أحد and مسد), confirm the fact that Boyer Moore is much faster at finding long patterns than short ones in the same region of the searched text. This is due to the nature of the shifting function of this algorithm, which jumps through the text as explained earlier. As such a function do not exist in the other search methods, the length of the search pattern has no influence on their speed.

Last but not least, the search time for all tested methods increases with the first occurrence of the pattern in the Quran text. This is a normal behavior as all these methods perform a full text search and not an index based one. The use of an index based search technique, as in web search engines, has a tangible advantage with large text corpus (in the order of Terabytes) that can still grow with time and where the words are often matched in a relaxed way (i.e. by searching all terms derived from a particular root). This not required in our application due to the nature of the searched text, which is below 1 MB in dynamic memory, which will never evolve in time and which is searched by exact string matching for an auto-completion purpose only.

6 Conclusion

This paper presented a new Android application that offers Holy Quran text auto-completion. Although, auto-completion is a well-known feature, its application to Quran Ayat is a novel idea. This idea exploits fast string search to allow users to cite the Holy Quran in smooth and fast way that avoids typing mistakes and human memory errors. Figure 4 shows screen captures of an example scenario where, after typing the letters "إن هذا القـ", the users sees the matching Ayat in the candidate view above the keyboard, and then scrolls horizontally to select the one he wants to cite. The application is available freely from the Google Play Store [11]. Its code source is fully open to allow contributions from the community.

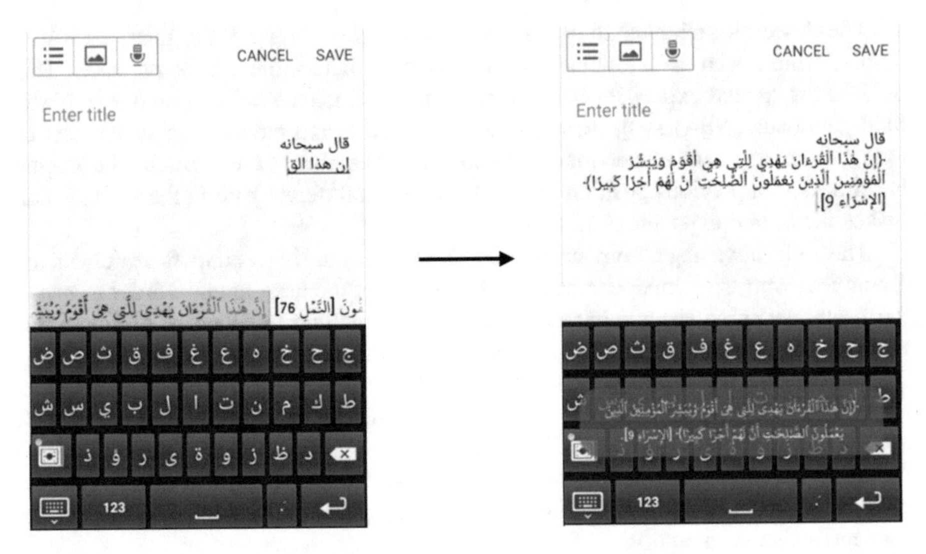

Fig. 4. Holy Quran Keyboard usage example.

Finally, this application can be improved in the future by addressing the next points:

- Implement Boyer Moore algorithm as a native function to improve its speed. The main difficulty here is to allow access to the long Quran string in the JVM from C runtime without copying it. A possible solution would to mimic the existing *fastIndexOf* function.
- Port the Quran text auto-completion feature to other platforms such as iOS and the desktop PC. It would a useful addition to an office suit like LibreOffice and to its word processor.
- Avoid repeating the search overall the Quran, when the user adds new characters to the previous pattern. Currently, the search restarts with each new typed character. This has no visible consequence, as the worst search times with *IndexOfMethod* are in the order of milliseconds. Nevertheless, it would be an optimization of resource usage to repeat such a search only over the already matched Ayat.

References

1. Boyer, R.S., Moore, J.S.: A fast string searching algorithm. Commun. Assoc. Comput. Mach. **20**(10), 762–772 (1977)
2. Charras, C., Lecroq, T.: Handbook of Exact String-Matching Algorithms. King's College, London (2004)
3. SoftKeyboard. https://android.googlesource.com/platform/development/+/master/samples/SoftKeyboard. Accessed 29 Dec 2017
4. Android API guide: Creating an Input Method. https://developer.android.com/guide/topics/text/creating-input-method.html. Accessed 29 Dec 2017

5. Android SDK version 23. https://developer.android.com/about/versions/marshmallow/android-6.0.html. Accessed 29 Dec 2017
6. Android Studio version 2.3. https://developer.android.com/studio/releases/index.html. Accessed 29 Dec 2017
7. QuranKeyboard source repository. https://github.com/cdjalel/QuranKeyboard. Accessed 29 Dec 2017
8. Android indexOf Java source code. https://github.com/AndroidSDKSources/android-sdk-sources-for-api-level-23/blob/master/java/lang/String.java#L702. Accessed 29 Dec 2017
9. Android fastIndexOf C source code. https://android.googlesource.com/platform/art/+/android-6.0.1_r79/runtime/native/java_lang_String.cc#66. Accessed 29 Dec 2017
10. Android Regex Reference: The Pattern class. https://developer.android.com/reference/java/util/regex/Pattern.html. Accessed 29 Dec 2017
11. Quran Keyboard application. https://play.google.com/store/apps/details?id=com.djalel.android.qurankeyboard. Accessed 29 Dec 2017

Software Implementation of Pairing Based Cryptography on FPGA

Azzouzi Oussama[1,2(✉)], Anane Mohamed[1], and Haddam Nassim[1]

[1] Laboratoire des Méthodes de Conception des Systèmes, Ecole nationale Supérieur d'Informatique ESI, BP 68M, 16309 Oued-Smar, Alger, Algérie
{m_anane,n_haddam}@esi.dz

[2] Centre de Développement des Technologies Avancées, CDTA Baba Hassen, Alger, Algérie
oazzzouzi@cdta.dz
http://www.esi.dz

Abstract. This paper presents the software implementation of *Weil*, *Tate*, *Ate* and *Optimal Ate* pairings in *Jacobean* coordinates, over *Barreto-Naehrig* curve, on Virtex-5 using the *MicroBlaze* software processor and the ZedBoard Zynq-7000 platform using *ARM* hardcore processor. The most pairing functions are constructed on the same model, one execution of the *Miller*'s algorithm plus a final exponentiation, which can be programed with addition chain method. Our flexible system can be performed for any curve parameters.

Keywords: Pairing · *Weil* · *Tate* · *Ate* · *Optimal Ate* · *Barreto-Naehrig* curve *Jacobean* coordinates · Addition chain · *MicroBlaze* · *ARM*

1 Introduction

A pairing is a bilinear function that at two points given of an elliptic curve associates an element of an extension field F_{p^k}. The pairings on elliptic curves was introduced for the first time in cryptography for the purpose of cryptanalysis. The property of bilinearity made it possible to transfer the discrete logarithm problem (DLP) from an elliptic curve to a finite field F_p. This introduces the MOV (Menezes, Okamoto and Vanstone) attack and the Frey-Rück attack. However, the constrictive use of pairings in cryptography appears from the 2000s. Joux proposed the tripartite key exchange scheme for Diffie-Hellman. After that, the identity-based encryption (IBE) proposed by Boneh and Franklin is without doubt the most remarkable application for pairings.

To execute the pairings in an efficient way, several curves have been discovered for providing better pairing computation as well as for achieving better security. We refer to Freeman et al. [4] for a taxonomy of pairing-friendly curves. In fact, *Barreto-Naehrig* (BN) curve is the most popular pairing-friendly curve in current days.

Generally, the construction of pairing functions is based on the *Miller* algorithm [5], dating from 1986, whose performances depend on the arithmetic used in the basic field F_p and its extensions F_{p^k}. Many articles propose protocols using pairings [1, 6] others improve the computation of pairings [9, 10] and few articles propose hardware implementations to compute pairings [14].

O. Demigha et al. (Eds.): CSA 2018, LNNS 50, pp. 102–112, 2019.
https://doi.org/10.1007/978-3-319-98352-3_12

In this paper, we propose two fully software PSoC (Programmable System on Chip) implementations of pairing functions such as *Weil* [6], *Tate* [7], *Ate* [8] and *OptimalAte* pairing [9] in *Jacobean* coordinate, over BN curve using two Boards based on Virtex-5 and Zynq-7000 Xilinx FPGA circuits. The main objective of such an implementation is to provide a flexible system which can be reconfigured for any curve parameters.

The paper is organized as follows: Sect. 2 gives a brief overview of pairing functions over BN curves with specific parameter selection. Section 3 presents the *Miller* algorithm and the step of doubling and addition lines in *Jacobean* coordinates. Section 4 describes the addition chain method to calculate the final exponentiation. Section 5 shows the software implementation of pairings function in Virtex-5and Zynq-7000 platforms. Finally, Sect. 6 concludes the paper.

2 Cryptographic Pairing Over Barreto-Naehrig Curves

A pairing is a bilinear and non-degenerate function $e : (G_1, +) \times (G_2, +) \rightarrow (G_3, \times)$, where G_1 and G_2 are additive groups and G_3 is a multiplicative group. The most useful property derived from the bilinearity is: for $P \in G_1$, $Q \in G_2$, we have:

$$\forall j \in \mathbb{N} : e([j]P, Q) = e(P, Q)^j = e(P, [j]Q) \quad (1)$$

Pairings are generally constructed on elliptic curves E defined over finite field F_p. In 2005, *Barreto* and *Naehrig* [2] described a method to construct pairing friendly elliptic curves with embedding degree $k = 12$. These curves are defined by the following equation:

$$E : y^2 = x^3 + b \quad (2)$$

Where $b \neq 0$ is neither a square nor a cube in F_p. The finite field, order and trace of *Frobenius* of the curve are defined by the following polynomial equations:

$$\begin{aligned} p(t) &= 36t^4 + 36t^3 + 24t^2 + 6t + 1 \\ r(t) &= 36t^4 + 36t^3 + 18t^2 + 6t + 1 \\ t_r(t) &= 6t^2 + 1, \ t \in \mathbb{Z} \end{aligned} \quad (3)$$

The parameter t is chosen such that both p and r are prime numbers. The selection of parameters has an important impact on the security and the performance of the pairing. For a security level equivalent to 128 bits in Advanced Encryption Standard (AES), we should select t such that $log_2(r(t)) \geq 256$ and $3000 \leq k.log_2(p(x)) \leq 5000$, which corresponds to the recommendations of National Institute of Standards and Technology (NIST) [3].

The BN curves admit a sextic twist $E'/F_{q^2} : y^2 = x^3 + b/\xi$ where ξ is neither a square nor a cube in F_{p^2}, and that has to be carefully selected such that $r|\#E'(F_{q^2})$ holds [10]. This allows us to map points in the sextic twist $E'(F_{q^2})$ to points in $E(F_{q^{12}})$.

The main objective to use a twisted curve is to reduce the computation in extension fields and to avoid the denominator evaluation in *Miller* algorithm.

Popular pairings such as *Weil, Tate, Ate* and *OptimalAte* pairing choose G_1 and G_2 to be cyclic subgroups of $E(F_{q^k})$, and G_3 to be a subgroup of $F^*_{q^k}$. Let $E[r]$ denotes the r-torsion subgroup of E and π_p be the *Frobenius* endomorphism $\pi_p : E \to E$ given by $\pi_p(x, y) = (x^p, y^p)$. We define $G_1 = E(F_p)[r]$, $G_2 \subset E(F_{q^{12}})[r]$, and $G_3 \subset F^*_{q^{12}}$ (*i.e.* the group of r-th roots of unity). The *Weil* pairing is defined for $P \in G_1$ and $Q \in G_2$. We note it e_w, such that:

$$\begin{aligned} e_w : G_1 \times G_2 &\to G_3 \\ (P, Q) &\mapsto \frac{f_{r,P}(Q)}{f_{r,Q}(P)} \end{aligned}$$

For computing $f_{r,P}(Q)$, we use the *Lite Miller* and the evaluation of $f_{r,Q}(P)$ deals with *Full Miller*. The computation cost of *FullMiller* is more than the one of *Lite Miller*. This is due to the fact that the majority of operations in the computation steps (doubling and addition) are in F_{q^k} for *FullMiller* and F_p for *Lite Miller*.

The *Tate, Ate* and *Optimal Ate* pairings are constructed on the same model, one execution of the *Miller's* algorithm plus a final exponentiation.

The *Tate* pairing, denoted by e_t, is defined by:

$$\begin{aligned} e_t : G_1 \times G_2 &\to G_3 \\ (P, Q) &\mapsto f_{r,P}(Q)^{\frac{p^{12}-1}{r}} \end{aligned}$$

The main idea of *Tate* pairing is to optimize the *Weil* pairing by replacing the *Full-Miller* execution with a final exponentiation.

The *Ate* pairing is given by:

$$\begin{aligned} e_{Ate} : G_2 \times G_1 &\to G_3 \\ (Q, P) &\mapsto f_{T,Q}(P)^{\frac{p^{12}-1}{r}} \end{aligned}$$

Where $T = t_r - 1$. The *Ate* pairing has an iteration number for the *Miller* loop reduced compared to the *Tate* pairing. However, we do not perform the *Ate* pairing with *LiteMiller* but we use *Full Miller*.

By the same way, the *Optimal Ate* pairing is defined by:

$$\begin{aligned} e_{Opt} &: G_2 \times G_1 \to G_3 \\ (Q, P) &\mapsto (f_{s,Q}(P).f_{[s]Q, \pi_p(Q)}(P).f_{[s]Q + \pi_p(Q), -\pi_p^2(Q)}(P))^{\frac{p^{12}-1}{r}} \end{aligned}$$

Where $s = 6t + 2$. The *Optimal Ate* pairing minimizes the number of iterations for *Miller* algorithm, but it is defined for a small group of curves.

3 Miller Loop

The most famous pairings such as *Weil*, *Tate*, *Ate* and *Optimal Ate* pairing are based on *Miller* algorithm [5]. This latter builds a rational function $f_{r,P}(Q)$ associated to the point P and evaluated at the point Q, with an iterative process using the *double&add* method. The *Miller* algorithm is given as follows:

Algorithm 1. $Miller\ (P, Q, r)$

Data: $r = (r_n \dots r_0)$, $P \in G_1(\subset E(F_p))$ and $Q \in G_2(\subset E(F_{p^k}))$
Result: $f_{r,P}(Q) \in G_3(\subset E(F^*_{p^k}))$
1.$T \leftarrow P$
2.$f_1 \leftarrow 1$
3.$f_2 \leftarrow 1$
for $i = n - 1$ to 0 then
  4.$T \leftarrow [2]T$
  5.$f_1 \leftarrow f_1^2 \times l_{T,T}(Q)$,the tangent at point$T$
  6.$f_2 \leftarrow f_2^2 \times v_{T,T}(Q)$,the vertical line at $[2]T$
  if$r_i = 1$ then
    7.$T \leftarrow T + P$
    8.$f_1 \leftarrow f_1 \times l_{T,P}(Q)$,the straight line (PT)
    9.$f_2 \leftarrow f_2^2 \times v_{T,P}(Q)$,the vertical line at $P + T$
  end
end
return$\frac{f_1}{f_2}$

There are several types of system coordinates used for computing line 4 and 7 in Algorithm 1, namely the affine, *Projective* and *Jacobian* coordinates. The affine coordinate needs modular division operation which is a complex operation to implement in FPGA. The *Jacobian* one generally has fewer multiplications than the *Projective* coordinates but more doubling operations. It is generally beneficial to use *Jacobian* coordinate because there are more optimized algorithms for doubling operations. In what follows, all equations are given in *Jacobian* coordinate.

3.1 Doubling and Tangent Line Equations

The formulas for $T = 2Q = (X_T, Y_T, Z_T)$ in *Jacobian* coordinates are defined as follows:

$$\begin{aligned} X_R &= 9X_T^4 - 8Y_T Y_T^2 \\ Y_R &= 3X_T^2(4X_T Y_T - X_R) - 8Y_T^4 \\ Z_R &= 4X_T Y_T \end{aligned} \quad (4)$$

Let $P \in E(F_p)$ be given in affine coordinates as $P = (x_p, y_p)$. The tangent equation line at T evaluated at P can be calculated as follows:

$$l_{T,T}(P) = (4Z_R Z_T^2 y_P) - (6X_T^2 Z_T^2 x_p)w + (6X_T^3 - 4Y_T^2)w^2 \in F_p^{12} \tag{5}$$

3.2 Addition and Line Equations

The formulas for addition $R = T + Q = (X_R, Y_R, Z_R)$ are defined as follows:

$$\begin{aligned} X_R &= (2Y_Q Z_T^3 - 2Y_T)^2 - 4(X_Q Z_T^2 - X_T)^3 - 8(X_Q Z_T^2 - X_T)^2 X_T \\ Y_R &= (2Y_Q Z_T^3 - 2Y_T)\left(4(X_Q Z_T^2 - X_T)^2 X_T - X_R\right) - 8Y_T(X_Q Z_T^2 - X_T)^3 \\ Z_R &= 2Z_T(X_Q Z_T^2 - X_T) \end{aligned} \tag{6}$$

The line through T and Q evaluated at the point P is given by:

$$\begin{aligned} l_{T,Q}(P) = & (4Z_T(X_Q Z_T^2 - X_T)y_p) - (4x_p(Y_Q Z_T^3 + Y_T))w + (4X_Q(Y_Q Z_T^2 X_Q - Y_T) \\ & - 4Y_Q Z_T(X_Q Z_T^2 - X_T))w^2 \in F_p^{12} \end{aligned}$$

After computing the *Miller* loop, we must carry out an extra step which is called the final exponentiation, where the *Miller* loop result must be raised to the power $\frac{p^k - 1}{r}$.

4 Final Exponentiation

Here we exploit the structure of BN curves to improve the computation of the final exponentiation. This latter is the last step to be performed in *Tate*, *Ate* and *Optimal Ate*. We can calculate it in several ways. The most traditional way is to use the "*square and multiply*" method. This method takes a lot of time because the exponent $e = \frac{p^{12} - 1}{r}$ is too large. This exponent can be divided as:

$$e = (p^6 - 1).(p^2 + 1).\frac{p^4 - p^2 + 1}{r} \tag{7}$$

To calculate the first part of $f = f_1^{(p^6 - 1)(p^2 + 1)} \in F_{p^{12}}$ which is the easy part, we must rise f_1 to the power p^6 and p^2 which are simple *Frobenius* operations. Many methods are developed to calculate the hard part of the final exponentiation. In 2009 Scott et al. [11] proposed a new approach based on addition chain for computing the hard part. Indeed, powering f to the $p^6 - 1$ makes the result unitary [12]. By this way, during the hard part, all the elements involved are unitary. This simplifies computations, as squaring of non-unitary elements is significantly cheaper than squaring of unitary elements, and any future inversion can be implemented as a *Frobenius* operator, more precisely $f^{-1} = f^{p^6}$ which is just a simple conjugation [12].

The addition chain method takes the advantage of the fact that p and m have a polynomial representation in t in order to effectively break down the hard part. This method describes a smart procedure that requires the calculation of ten temporary values, namely:

$$f^t, f^{t^2}, f^{t^3}, f^p, f^{p^2}, f^{p^3}, f^{(tp)}, f^{\left(t^2p\right)}, f^{\left(t^3p\right)}, f^{\left(t^2p^2\right)}$$

These elements are the necessary elements for the construction of a multiplication chain whose evaluation produces the final exponentiation f^e

$$\left[f^p.f^{p^2}.f^{p^3}\right].[1/f]^2.\left[\left(f^{t^2}\right)^{p^2}\right]^6.[1/(f^t)^p]^6.\left[1/\left(f^t.\left(f^{t^2}\right)^p\right)\right]^{18}.\left[1/f^{t^2}\right]^{30}.\left[1/\left(f^{t^3}.f^{t^3}\right)^p\right)\right]^{36}$$

This chain can be calculated by performing 13 multiplication operations and four doubling operations in $F_{p^{12}}$. To raise an element to the power p, we can compute it by the application of *Frobenius* operation. Also, to raise an element to the power t, which takes a lot of time, we can use the "square and multiply" method.

5 Implementation Results

As $k = 12 = 2^2.3$, we can construct the arithmetic in F_{q^k}, step by step in smaller extensions fields, with the polynomial irreducible $X^k - \beta$ using what is called Friendly Fields. Therefore, $F_{q^k} = F_p[X]/\left(X^k - \beta\right)$, it can be considered as a tower of extensions of degree 2 (quadratic extension) and 3 (cubic extension). We can choose β as a small value of F_p, then the multiplications by β can be reduced to few additions, and its cost can be neglected.

In fact, we represent $F_{q^{12}}$ using the same tower extension of [13], namely, we construct a quadratic extension, which is followed by a cubic extension and then by a quadratic one, using the following irreducible binomials:

$$F_{q^{12}} \rightarrow^2 F_{q^6} \rightarrow^3 F_{q^2} \rightarrow^2 F_q$$

$$\begin{cases} F_{q^2} = F_q[u]/(u^2 - \beta), \textit{where } \beta = -5 \\ F_{q^6} = F_{q^2}[v]/(v^3 - \xi), \textit{where } \xi = u \\ F_{q^{12}} = F_{q^6}[w]/(w^2 - v) \end{cases} \tag{8}$$

An element $f \in F_{q^{12}}$ can be represented as $f = g + hw$, with $g, h \in F_{q^6}$, such that, $g = g_0 + g_1v + g_2v^2$ and $h = h_0 + h_1v + h_2v^2$ where $g_i, h_i \in F_{q^2}$ for $i = 0, 1, 2$. This construction helps us to speed up the arithmetic computation.

We have implemented our software solution on Virtex-5 "xc5vlx50t" using the *Microblaze* processor which is available as IP (Intellectual Property) core in the EDK library (Embedded Development Kit), and on the ZedBoard Zynq-7000"xc7z020clg484-1" platform. The Zynq platform contains two *ARM* Cortex-A9 hardcore processors whose maximum frequency is 1 GHz.

The architecture of our proposed system on Virtex-5 is shown in Fig. 1.

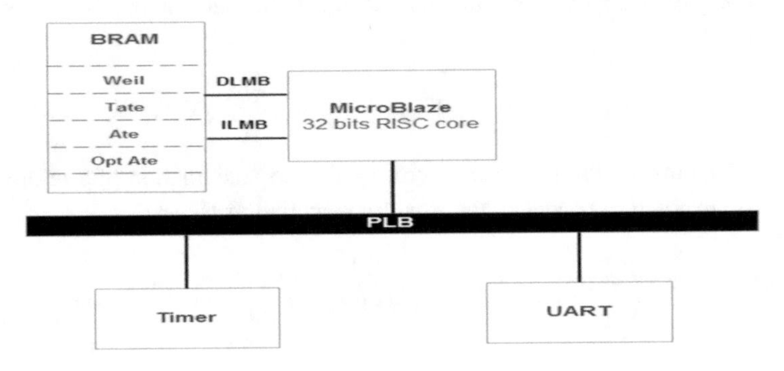

Fig. 1. The architecture of proposed system on Virtex-5.

Similarly, Fig. 2 shows the architecture of our proposed system on Zynq-7000.

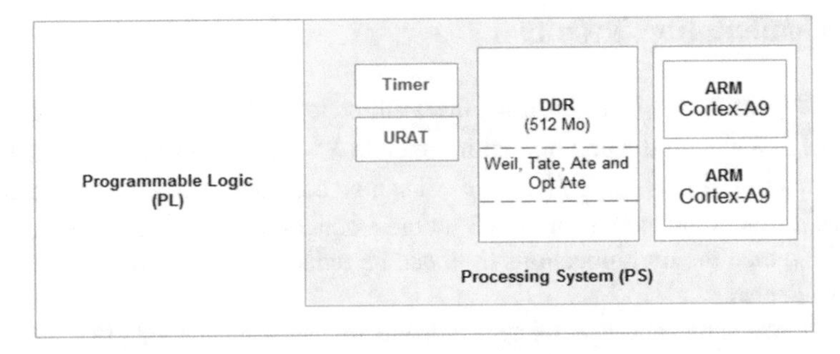

Fig. 2. The architecture of proposed system on Zynq-7000.

The implementation of *Weil*, *Tate*, *Ate* and *Optimal Ate* algorithms on the two proposed systems are fully software where all the computation are made on *MicroBlaze* processor for the system on the Virtex-5 platform and *ARM*Cortex-A9 for the Zynq-7000 platform. The computations are organized as follows. In the field F_p, we use *Montgomery* algorithm to multiply two elements in the same field. Modular inversion can be computed via «*square and multiply*» method by using Fermat's theorem, as: $a^{-1} \equiv a^{p-2} (mod\, p)$. In F_{p^2}, we use *Karatsuba* multiplication and the complex method for squaring. Inversion can be found from the identity, $(a_0 + a_1 u)^{-1} = (a_0 - a_1 u)/(a_0^2 - \beta a_1^2)$. Same methods are programmed to compute addition, multiplication, squaring and inversion in F_{q^6} and $F_{q^{12}}$.

Table 1 presents the implementation results of different operations in F_q, F_{q^2}, F_{q^6} and $F_{q^{12}}$ using Xilinx FPGA boards.

Table 1. Clock cycles and execution time of operations in F_q, F_{q^2}, F_{q^6} and $F_{q^{12}}$.

Operations	Virtex-5 (125 MHz)		Zynq-7000 (666.66 MHz)	
	Clock cycles	Execution time	Clock cycles	Execution time
Add/Sub F_q	366	2.92 us	240	0.36 us
Add/Sub F_{q^2}	696	5.56 us	470	0.71 us
Add/Sub F_{q^6}	2018	16.14 us	1118	1.68 us
Add/Sub $F_{q^{12}}$	4170	33.36 us	2820	4.23 us
Mult F_q	12968	103.74 us	2522	3.79 us
Mult F_{q^2}	53942	431.53 us	10966	16.47 us
Mult F_{q^6}	359562	2.87 ms	69654	104.59 us
Mult $F_{q^{12}}$	1104127	8.83 ms	214512	322.09 us
Sqar F_q	12968	103.74 us	2522	3.79 us
Sqar F_{q^2}	53934	431.47 us	9988	15.00 us
Sqar F_{q^6}	302357	2.41 ms	56878	85.40 us
Sqar $F_{q^{12}}$	757880	6.06 ms	147136	220.92 us
Inv F_q	4906983	39.25 ms	833114	1.25 ms
Inv F_{q^2}	4944293	39.55 ms	844630	1.26 ms
Inv F_{q^6}	5647073	45.17 ms	971050	1.45 ms
Inv $F_{q^{12}}$	6989503	55.91 ms	1226470	1.84 ms
Frob (p)	133109	1.06 ms	23668	35.54 us
Frob (p^2)	131013	1.04 ms	22064	33.13 us
Frob (p^3)	133082	1.06 ms	23244	34.90 us
Exp (t)	48645507	0.38 s	9499350	14.26 ms

We remark that the squaring operation is faster than multiplication in all fields F_{p^k}. This result is predictable because the squaring methods are more optimized than those used for multiplication. These methods are called the complex method for squaring.

Doubling and addition lines results in *Jacobean* coordinate are shown in Table 2.

Table 2. Doubling and addition lines in *Jacobean* coordinates.

Operations	Virtex-5 (125 MHz)		Zynq-7000 (666.66 MHz)	
	Clock cycles	Execution time	Clock cycles	Execution time
PtDbTang (Tate)	201726	1.61 ms	38736	58.16 us
PtAddLine (Tate)	254056	2.03 ms	47180	70.84 us
PtDbTang (Ate)	653724	5.22 ms	125026	187.73 us
PtAddLine (Ate)	866698	6.93 ms	162690	244.28 us

We note that the doubling point and the tangent evaluation take less time than the addition point and the line evaluation. This is due to the complexity of each method.

Table 3 shows the clock cycles and the execution times for both *Lite Miller* and *Full Miller* in *Jacobean* coordinates, and for the final exponentiation.

Table 3. *Lite Miller, Full Miller* in *Jacobean* coordinates and final exponentiation results.

Operations	Virtex-5 (125 MHz)		Zynq-7000 (666.66 MHz)	
	Clock cycles	Execution time	Clock cycles	Execution time
Lite Miller (Tate)	631656491	5.05 s	122217374	183.50 ms
Full Miller (Ate)	345695225	2.76 s	66849082	100.37 ms
Full Miller (optAte)	171428706	1.37 s	33055428	49.63 ms
Final exp	173341106	1.38 s	33696800	50.59 ms

Table 4 presents the results of our implementation of pairings functions such as *Weil, Tate, Ate* and *Optimal Ate* in *Jacobean* coordinates on Virtex-5 and Zynq-7000platforms.

Table 4. *Weil, Tate, Ate* and *Optimal Ate* functions.

	Virtex-5 (125 MHz)		Zynq-7000 (666.66 MHz)	
	Clock cycles	Execution time	Clock cycles	Execution time
Weil	1442572362	11.54 s	278252922	417.79 ms
Tate	804397188	6.43 s	156028110	234.27 ms
Ate	525076635	4.2 s	100389804	150.73 ms
Opt Ate	348793411	2.79 s	67551738	101.42 ms

We remark that *Ate* has a significant improvement in execution time compared to *Weil* and *Tate*. Finally, *Optimal Ate* pairing is the culmination of efforts to reduce the *Miller*'s loop. As a result, this pairing is the most efficient one.

Table 5 presents a comparison of our results with the related implementations achieving 128 bits security in different platforms. [10, 15] present software implementations on Intel Pentium 4 and Intel Core i7 with a maximum frequency of 3 GHz and 2.8 GHz, respectively. In [14], the authors present a hardware implementation of *Tate, Ate* and *Optimal Ate* on Virtex-6 with 23 k slices. In comparison with our results, we note that a software system on FPGA generally takes more time than another one on PC because this latter can achieve a high level of frequency. On the other hand, our software system is a flexible design which can suitable to several curve parameters.

Table 5. Results comparison with existing designs

Ref.	Platform Freq.	Design	Type	Cycles	Times
Our	Virtex-5 125 MHz	SW	Tate	804397188	6.43 s
			Ate	525076635	4.2 s
			optAte	348793411	2.79 s
	Zynq-7000 666.66 MHz	SW	Tate	156028110	234.27 ms
			Ate	100389804	150.73 ms
			optAte	67551738	101.42 ms
[15]	Intel P IV 3 GHz	SW	Tate	103833600	33.8 ms
			Ate	71270400	23.2 ms
[10]	Intel Core i7 2.8 GHz	SW	optAte	2330000	0.83 ms
[14]	Virtex-6 145 MHz	HW	Tate	1730000	11.93 ms
			Ate	1206000	8.32 ms
			optAte	821000	5.66 ms

6 Conclusion

In this paper, we have presented a software implementation of paring functions such as *Weil*, *Tate*, *Ate* and *Optimal Ate* in *Jacobean* coordinates, over BN curve at the 128-bits security level. The flexible design is implemented on Xilinx Virtex-5 based on *MicroBlaze* and ZedBoard Zynq-7000platform based on *ARM* Cortex-A9 hardcore processor. According to the flexibility, our software system can be performed for any curve parameters.

References

1. Boneh, D., Lynn, B., Shacham, H.: Short signatures from the Weil pairing. In: Advances in Cryptology ASIACRYPT 2001, vol. 2248, pp. 514–532. Springer (2001)
2. Barreto, P.S.L.M., Naehrig, M.: Pairing-friendly elliptic curves of prime order. In: Selected Areas in Cryptography (SAC 2005). Lecture Notes in Computer Science, vol. 3897, pp. 319–331. Springer (2005)
3. Barker, E., Barker, W., Burr, W., Polk, W., Smid, M.: Recommendation for key management-part 1: General (revised). In: Published as NIST Special Publication 800-57 (2007)
4. Freeman, D., Scott, M., Teske, E.: A taxonomy of pairing-friendly elliptic curves. J. Cryptol. **23**, 224–280 (2010)
5. Miller, V.: The Weil pairing and its efficient calculation. J. Cryptol. **17**, 235–261 (2004). LNCS
6. Cohen, H., Frey, G.: Handbook of Elliptic and Hyperelliptic Curve Cryptography, Discrete Mathematics and Its Applications. Chapman & Hall/CRC (2006)
7. Scott, M.: Computing the tate pairing. In: Topics in Cryptology CT-RSA. LNCS, vol. 3376, pp. 293–304 (2005)
8. Matsuda, S., Kanayama, N., Hess, F., Okamoto, E.: Optimised versions of the ate and twisted ate pairings. In: Cryptography and Coding. LNCS, vol. 4887, pp. 302–312 (2007)

9. Vercauteren, F.: Optimal pairings. IEEE Trans. Inf. Theory **56**(1), 455–461 (2010)
10. Beuchat, J.L., González-Díaz, J.E., Mitsunari, S., Okamoto, E., Henríquez, F.R., Teruya, T.: High-speed software implementation of the optimal ate pairing over Barreto–Naehrig Curves. In: International Conference on Pairing-Based Cryptography, pp. 21–39. Springer, Heidelberg (2010)
11. Scott, M., Benger, N., Charlemagne, M., Dominguez, P.L.J., Kachisa, E.J.: On the final exponentiation for calculating pairings on ordinary elliptic curves. In: Third International Conference in Pairing-Based Cryptography, USA, pp. 78–88 (2009)
12. Arene, C., Lange, T., Naehrig, M., Ritzenthaler, C.: Faster computation of the tate pairing. In: Cryptology ePrint Archive, Report 2009/155 (2009). http://eprint.iacr.org/2009/155.pdf
13. Hankerson, D., Menezes, A., Scott, M.: Software implementation of pairings. In: Joye, M., Neven, G. (eds.) Identity-Based Cryptography, Amsterdam, The Netherlands. IOS Press (2008)
14. Ghosh, S., Mukhopadhyay, D., Roychowdhury, D.: Secure dual-core cryptoprocessor for pairings over Barreto-Naehrig Curves on FPGA Platform. IEEE Trans. Very Large Scale Integr. Syst. **21**(3) (2013)
15. Devegili, A.J., Scott, M., Dahab, R.: Implementing cryptographic pairings over Barreto-Naehrig curves. In: International Conference on Pairing-Based Cryptography. LNCS, vol. 4575, pp. 197–207 (2007)

Implementation of Multi-bin CABAC Decoder in HEVC/H.265 on FPGA

Menasri Wahiba[1,2](✉), Skoudarli Abdellah[1], and Belhadj Aichouche[1]

[1] Faculty of Electronic and Computer Science, Laboratory of Image Processing and Radiation, USTHB, BP 32 El Alia, Bab Ezzouar, Algiers, Algeria
{omenasri,askoudarli,abelhadj}@usthb.dz

[2] CRD/MSS/DFM-Department of Electronic and Automatic, Sidi-abdelkader, Blida, Algeria

Abstract. Context-based adaptive binary arithmetic coding (CABAC) is specified as the single operation mode for entropy coding in the newest standard High Efficiency Video Coding (HEVC). While it provides high coding efficiency, the data dependencies in H.265/HEVC CABAC make it challenging to parallelize and thus, limit its throughput. This paper proposes a multi-bins CABAC decoder architecture adaptive to HEVC syntax elements with small FSM (finite state machine) for the control of SE order. In order to reduce the critical path delay, we exploit different techniques of optimization such as a speculative decoding, logic balancing techniques and our proposed technique RLpsLZpStateIdx LUT. The parallel implementation can process 1,34 bin/cycle when operate at 134,23 MHz and improved high throughput of 179.86 Mbin/s with an optimized path delay compared to the serial process. The architecture is coded using VHDL ISE language, simulated and synthesized using Xilinx tools with virtex4 xc4vsx25-12ff668 board.

Keywords: HEVC · Entropy coding · CABAC · FPGA · HDL

1 Introduction

Video compression is the key enabling technology for multimedia communications. With the increasing demand of high-quality video on handheld devices, we will see widespread use of the state-of-art video compression standard H.265/HEVC.

HEVC (High Efficiency Video Coding) is the most recent video coding standard developed by the Joint Collaborative Team on Video Coding (JCTVC) [1, 2]. HEVC allows video compression with the same perceptive quality as its predecessor H.264/AVC [3] while requiring only half the bitrates [4].

HEVC uses CABAC (Context-based Adaptive Binary Arithmetic Coding), a form of entropy coding, to achieve high coding efficiency. The CABAC is comparatively much more complex than traditional entropy coding. For hardware implementation, parallel processing is a commonly used technique to improve performance [3]. But in H.265/HEVC, the well-known bottleneck owing to highly serialized calculating of context adaptive binary arithmetic coding is hard to be parallelized. Complex steps

O. Demigha et al. (Eds.): CSA 2018, LNNS 50, pp. 113–123, 2019.
https://doi.org/10.1007/978-3-319-98352-3_13

were involved to decode each one bit or bin for CABAC, including context modeling, binary arithmetic decoding and de-binarization.

This work proposes the multi-bins CABAC decoder without forwarding context. The throughput is increased with acceptable resources costs, also a small FSM (Finite State Machine) is used for the control of the SEs decoding order at the (TU) transform unit level. The context selection and bin decoding can be processed in parallel under some conditions. In our works, some of optimized techniques are exploited with our proposed technique of rearrangement of memory for decreasing the critical path delay.

The rest of paper is organized as follow: In Sect. 2, we described the architecture of CABAC algorithm by introducing the primary steps of CABAC decoding for the regular mode process. In Sect. 3, we proposed a multi-bin pipeline CABAC decoder and we specify the optimization techniques exploited and proposed in our work:

- Logic balancing technique [5];
- Speculative decoding technique [5];
- The Memory rearrangement.

We focus on the simulation and synthesize of our architecture in Sect. 4. Finally, in Sect. 5, we summarize the conclusions and future work.

2 CABAC Algorithm Architecture

The entropy coder/decoder plays an important role and can save significant bit rate in several applications. The design and implementation of the CABAC is difficult due to its inherent bit-serial nature and the context models [6] of the current SE (syntax element) are closely related to the results of its neighboring macro blocks, which leads to frequent memory access. The CABAC decoder [7] reads in bit stream and outputs meaningful information.

The H.265/HEVC standard defines this meaningful information as SE. Three types of decoding modes are used in HEVC standard (Fig. 1):

- Bypass mode (BM);
- Regular mode (RM) when the context modeling is needed;
- Terminate mode (TM) is applied at the end of block or slice.

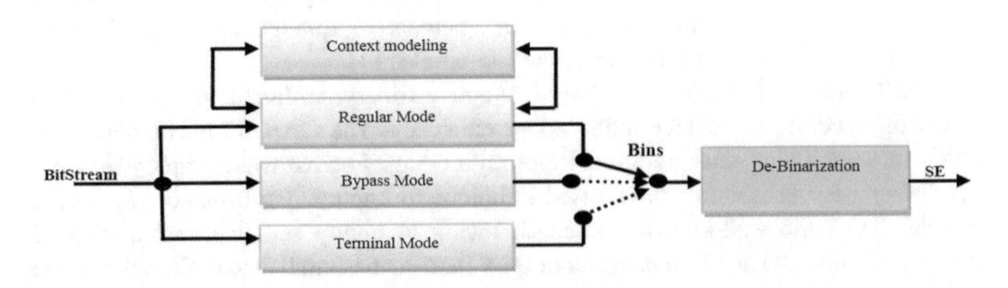

Fig. 1. Diagram block of HEVC CABAC decoder [5].

At the last de-binarization block is applied for recovering the SEs from the decoding bins.

CABAC decoder defines fives variables namely: Offset (of 9 bits) and Range (of 9 bits) are initialized at the beginning with the first bytes of the bit stream, RMPS (Range of the Most Probable Symbol of 9 bits), RLPS (Range of the Less Probable Symbol of 9 bits) and pStateldx (probability state index of 6 bits). The CABAC decoder determines the value of decoded bin according to the values of Offset and RMPS. For example, when RMPS is larger than Offset, the value of output bit will be equivalent to that of MPS (Most Probable Symbol) and when offset is larger than RMPS, the value of output bit will be equivalent to that of LPS as shown in Fig. 2.

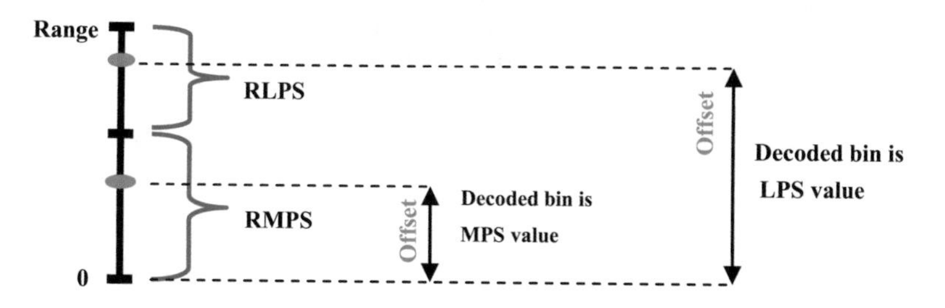

Fig. 2. Sub-interval of Arithmetic decoding engine.

The H.265 CABAC uses five tables (for the regular mode) namely five tables: The context table (is constructed from initial tables), the initial table (contain the intValue (initial values)), the qCodIRangeldx table (contain the LPS Range values), the transIdxLPS table (contain the pStateldx values, used in the LPS decoded bin case) and transIdxMPS (contain the pStateldx values, used in the MPS decoded bin case).

In the RM decoding process, five steps are executed for decoding one bin:

- Context selection (CS);
- Context modeling (CM);
- BAD (binary arithmetic decoding);
- Renormalization;
- De-binarization (DB).

The intValue is used to calculate the pStateldx and MPS values, which are using as inputs for the BAD with OffsetIdx (Offset Index) for decoding one bin, in this step the Range is updated by looking up the range TabLPS table and update the corresponding entry in the context table by looking up the transIdxLPS table and the transIdxMPS table. In the end of regular process renormalization of Range and Offset values are needed by reading from the bit stream. This process is executed for each SEs will be decoded by regular mode process, only modified the inputs parameters. The ctxIdx (context Index) is determined for each SE to be decoded. The Fig. 3 shows as all steps executed for decoding one bin, the inputs parameters are BinIdx (bin index), logtafoSize ($\log_2$ of TU size), ctxIdxOffset (context index offset) and others information's specified for each SEs. The context selection consists to select the intValue of 8 bit by looking up the context table with the address ctxIdx.

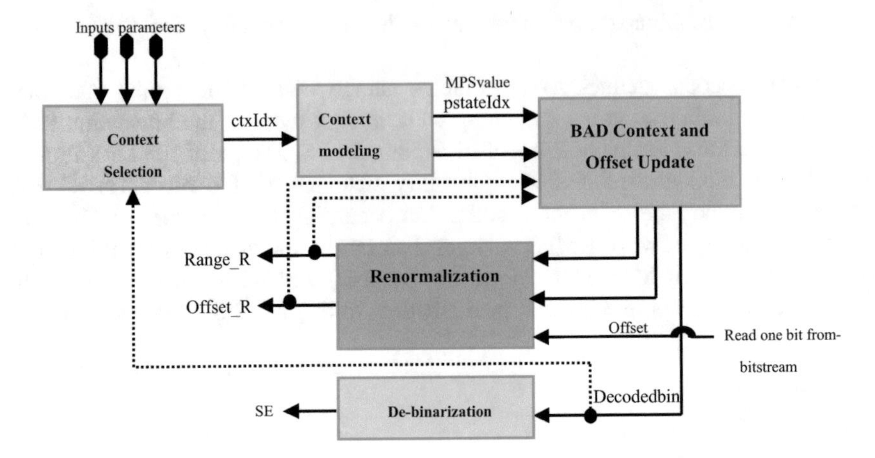

Fig. 3. Five operations executed at the CABAC decoder side.

2.1 SEs at the TU Level

Table 1 shows all SEs used at TU for each sizes and their specific decoding mode Bypass mode (BM) or Regular Mode (RM).

Table 1. Residuals SEs with TUs sizes and decoding mode.

TUs sizes	SEs	Decoding mode
32 × 32, 16 × 16, 8 × 8	transform_skip_flag[][][]	RM
32 × 32, 16 × 16, 8 × 8, 4 × 4	last_sig_coeff_x_prefix	RM
32 × 32, 16 × 16, 8 × 8, 4 × 4	last_sig_coeff_y_prefix	RM
32 × 32, 16 × 16, 8 × 8, 4 × 4	last_sig_coeff_x_suffix	BM
32 × 32, 16 × 16, 8 × 8, 4 × 4	last_sig_coeff_y_suffix	BM
32 × 32, 16 × 16, 8 × 8, 4 × 4	coded_sub_block_flag[][]	RM
4 × 4	sig_coeff_flag[][]	RM
4 × 4	coeff_abs_level_greater1_flag[]	RM
4 × 4	coeff_abs_level_greater2_flag[]	RM
4 × 4	coeff_abs_level_remaining[]	BM
4 × 4	coeff_sign_flag[]	BM

2.2 Context Selection

First, in the CS step we calculate the ctxInc (context increment) variable for deduce the context address (ctxIdx) which is obtained by the following equation:

$$ctxIdx = ctxInc + ctxIdxOffset \tag{1}$$

The ctxIdxOffset value depends on type of SEs, intType and slice type. The Table 2 show the ctxIdxOffset values of X, Y position SEs as example.

Table 2. ctxIdxOffset values for X, Y position [8].

Slice type	I	P	B
cabac_init_flag	0/1	0/1	0/1
initType	0	1/2	2/1
ctxIdxOffset	0	18/36	36/18

2.3 Context Modeling

The CtxIdx Is Used at the Input of Context Modeling:

- From the context Tab select the intValue (initial value of 8 bits) with ctxIdx address;
- Calculate the two variables pStateldx and ValMps, these two values are used at the input of BAD.

These two variables are used for recovering the decoded bins from the BAD stage.

2.4 Binary Arithmetic Decoding and Renormalization

In the arithmetic decoding engine, the Offset value is compared with the RMPS sub-interval. However, according to this comparison the decoded binVal is obtained. The pStateldx and MPSVal variables are using to get the decoded bin and the variables (Range, Offset, pStateldx) are updated according to the bin decision result. At the end if needed the renormalization is applied by shifting the Range and the Offsetlike shows in Fig. 9-6 and Fig. 9-7 of HEVC standard [2].

2.5 The Exploited and the Proposed Technical Optimization

A technique of Logic Balancing [5]. Can be adopted for reduce the path delay. This method consists of split the RLPS look-up into two stages. The first one is reserved for loaded four possible RLPS values selected by pStateldx address and the second stage get the correct result of RLPS according to bits 7 and bit 6 of the Range value.

Speculative Decoding Technique [5]. We can perform these updating process based on both possible bin decisions in parallel, and choose the correct result in the end. As circuit transform demonstrated in [5], the MPS and LPS decoding path are calculated in same time of the bin decision process.

Rearrangement of Memory. We propose to store the RangeLps and LZ values in the same memory and the two transition tables of context in the same memory. This technique we permit to decrease the critical path delay by **3%** approximately, resource slices by **8,1%** and keep the same amount of other resources (show the Fig. 4).

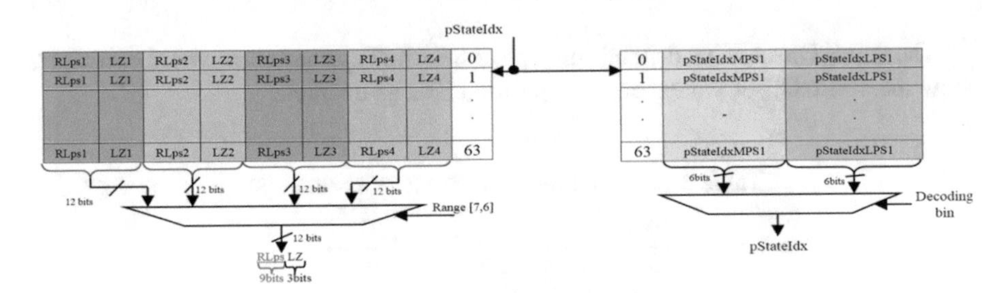

Fig. 4. RLpsLZpStateIdx architecture.

2.6 De-binarization

Some of the decoded bins are needed to be de-binarized. So these bins are mapped to SEs according to their binarization process DB. The de-binarization is the inverse operation of the binarization. At the transform unit level only the X or Y position and the residuals coefficient are needed to be binarized. Table 3 shows the binarization of (X, Y) coordinates with the truncated unary code for prefix part and fixed length for suffix, but the suffix part is not existed for 4 by 4 TU size, the number (1) into bracket exists only when TU size is greater than the largest last position that the code can represent and x means 0 or 1 [9].

Table 3. Binarization of (X, Y) position for 4 × 4 block.

Position value (x or y)	Prefix truncated unary code	Suffix fixed length code
0	1	–
1	01	–
2	001	–
3	000(1)	–
4–5	00001	X
6–7	00000(1)	X
8–11	0000001	XX
12–15	0000000(1)	XX
16–23	000000001	XXX
24–31	000000000	XXX

3 Multi-bin CABAC Decoder

In case of decoding SE composed of multi-bins (last_significance_X/Y, residual coefficients), we can used the parallel process for RM or BM for decoding multi-bins successively like show Fig. 5, so two context selection, context modeler and BADs are applied at the same time for RM process and two BAD bypass for the second one.

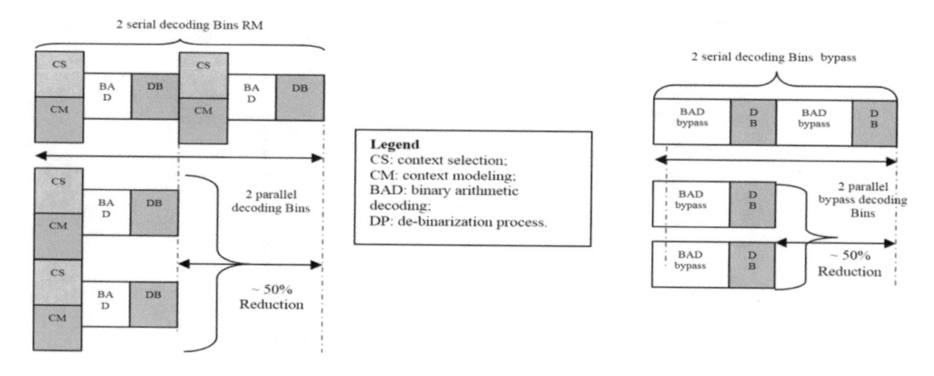

Fig. 5. Parallel process decoding of 2 bins for multi-bins SE.

Finally the de-binarization blocks select the valid SE value of the current series of bins. If a SE value is valid, the next SE must be decoded with initialization of bin index and SE respectively otherwise two bins decoded in the same SE (the last_significance_X/Y can go to 9 bins per SE), as shown in Table 3.

4 Experiment Results

This work is coded using HDL language; we have used structural level implementation on FPGA board type **virtex4 xc4vsx25-12ff668** of XLINX, with maximum frequency of **134.23 MHz**. The RTL tops level and detailed of RTL schematic are shown in the Figs. 6 and 7 respectively.

The Table 4 shows the comparison of materials cost, frequency and path delay between serial and the parallel solutions for processing 2 bins. So we observed that the maximum frequency is increased and the path delay is decreased by 59%. In this work, 2-bin are considered as a baseline for a balance between hardware costs, a reasonable system clock frequency and an optimized critical path delay, although the optimizations to be proposed in this paper can also be applied to other cases (4-bin or more). The biggest size of transform block is 32×32 in HEVC standard and the maximum binarized value of prefix last position SE is represented by 9 bits, so we can used five parallel processes of 2 bins for obtained SE in one clock cycle.

The simulation result of the proposed work is shown in the following Figure when 2 bins are obtained at the output for each clock with an average of 1.34 bin/cycle (Fig. 8).

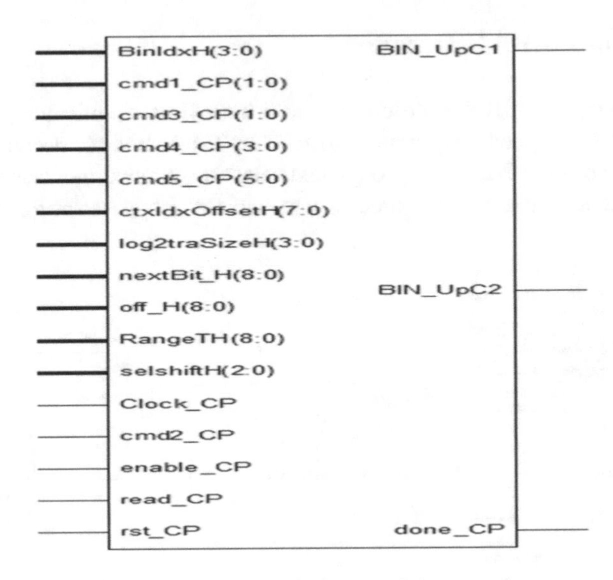

Fig. 6. Top level of 2 bins parallel processing.

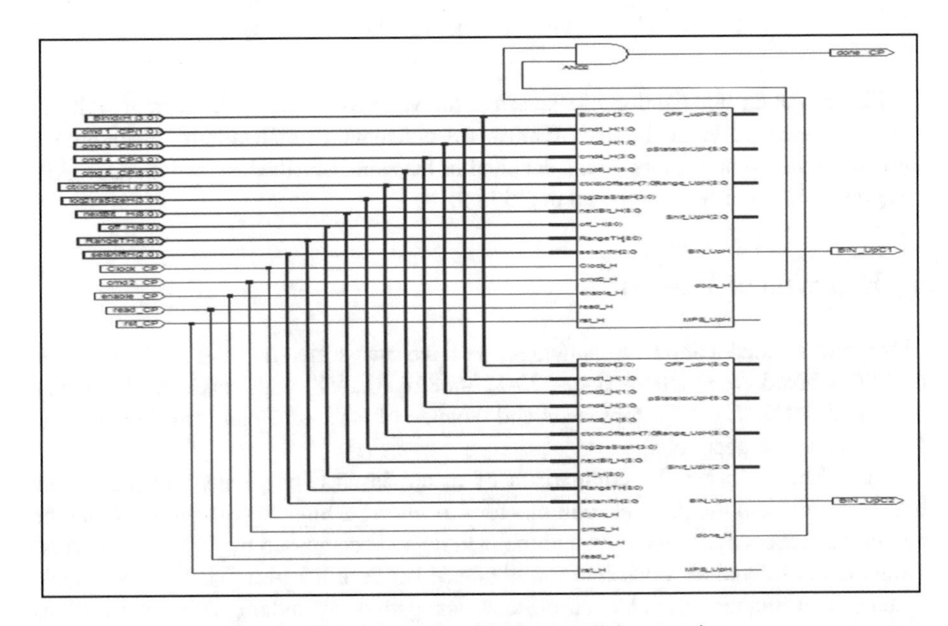

Fig. 7. RTL Schematic of 2 bins parallel processing.

Table 4. Comparison of processes (serial and parallel).

Process type	Bins number	Frequency (Mhz)	Path delay (ns)
Serial	2 bin	132.47	11.638
Parallel	2 bin	134.23	6.892

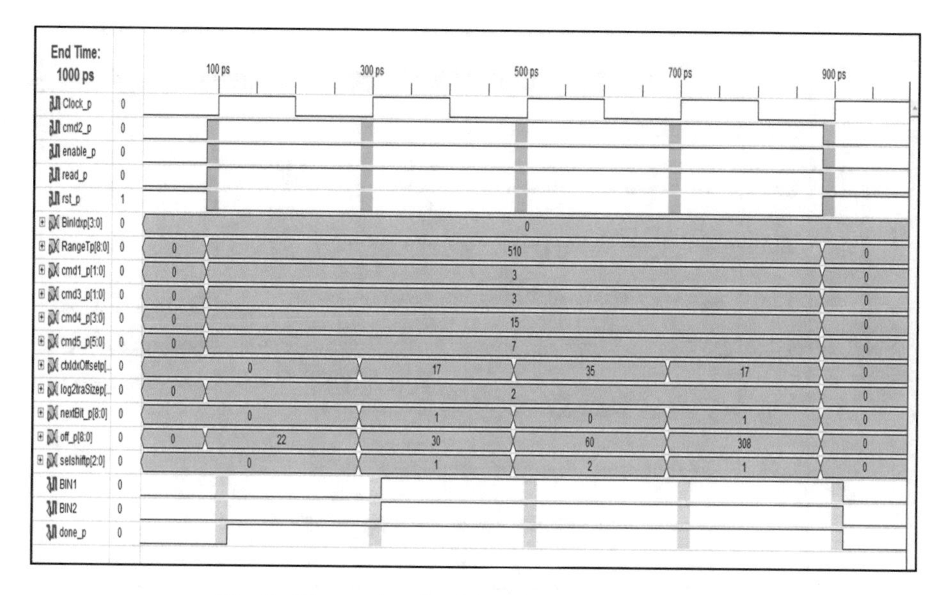

Fig. 8. The test bench of decoding 4 bins in parallel.

In our work, acceptable hardware cost is used and the maximum frequency is increased compared to all works. In [12–14], only 1 and 0.5 bin per cycle is proposed with acceptable values of frequency but an optimized hardware cost in [14], otherwise an average of 1,34 bins/cycle is proposed in our work for multi-bins SE type less than 2 bins/cycle proposed in [11] but without specify the SE type. High throughput is improved in our proposed compared [10]. The context table is implemented with the same method on RAM but our proposed method give an optimized path delay with high frequency clock.

Table 5 shows the results and the comparisons with the most relevant works found in literature (implemented on FPGA board).

Table 5. Comparison with other works.

	W1 [11]	W2 [12]	W3 [13]	W4 [14]	W5 [9]	W6 (Proposed)
Standard	H264	H264	H264	H264	H265	H265
Type of FPGA	Virtex 4	Altera Stratix II	Virtex 4	Virtex-4	Virtex-5	Virtex 4
Average (bin/cycle)	0,5	1	1	1	1.38	1,34
Frequency (MHz)	100	105	105	100	125.1	134.23
Throughput (Mbins/s)	50	105	105	100	172.64	179.86
Context memory architecture	RAM	RAM	RAM	RAM	RAM	RAM
hardware costs	302 slice	389 slice	273 slice	1372 LUT	n.a	365 slice
SE type	n.a	n.a	n.a	n.a	Multi-bin	Multi-bin

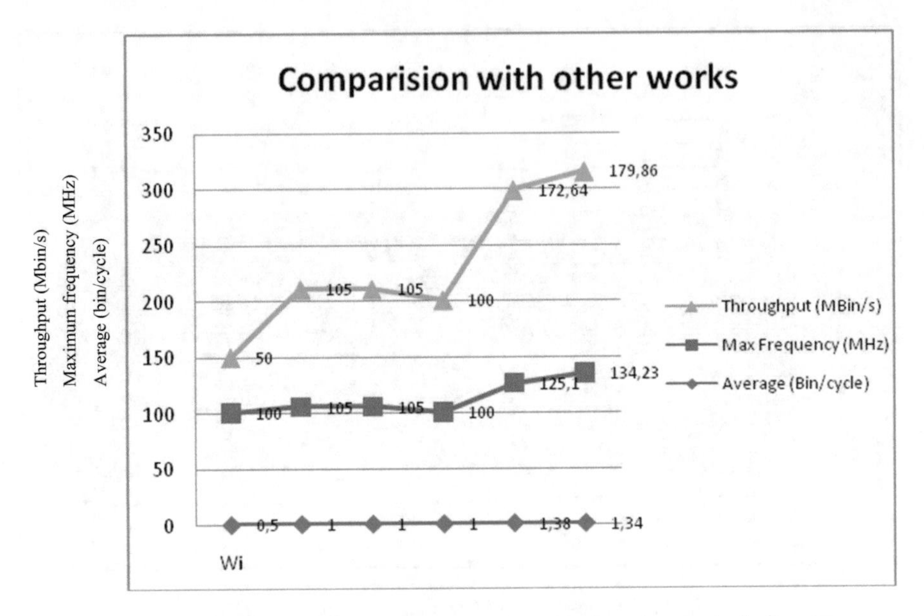

Fig. 9. The diagram of comparison of our work with the other works in literature.

Figure 9 shows the comparison of the maximum frequencies, throughput and average bin/cycle obtained in our implementation W6 (our proposed work) and all other works (W1, W2, W3, W4, W5) in literature.

5 Conclusion

In this work we have verified, simulated and synthesized the parallel decoding process for multi-bins SEs type on FPGA board using (ISE) Xilinx tool **with virtex4 xc4vsx25-12ff668 device**. The path delay is decreased by **59%** approximately compared to the conventional process (serial) and optimized with our proposed technique of rearrangement memory by **3%** and the slices is reduced by **8,1%** approximately. **1.34 bins/cycle** en average is obtained when operate at **134.23 MHz** for multi-bins SEs when process 2 bins in parallel. As future work we are going to implement entire HEVC decoder algorithm on FPGA platform. In our future implementation, the tradeoff between high throughput and coding efficiency will be the challenging task.

References

1. Sullivan, G.J., Ohm, J., Han, W., Wiegand, T.: Overview of the High Efficiency Video Coding (HEVC) standard. Copyright (c) 2012 IEEE, Pre-publication Draft, to Appear Trans. On circuits and system for video technology, December 2012
2. Telecommunication Standardization Sector of ITU High efficiency video coding. H 256 04/2013

3. Wieg, T., et al.: Overview of the H.264/AVC video coding standard. IEEE Trans. Circ. Syst. Video Technol. **13**(7), 560–576 (2003)
4. Sze, V., Budagavi, M.: High throughput CABAC entropy coding in HEVC. IEEE Trans. Circ. Syst. Video Technol. **22**(12), 1778–1791 (2012)
5. Zhang, P.: Fast CABAC decoding architecture. Electron. Lett. **44**(24), 1394–1395 (2008)
6. ISO/IEC 14496-10: Generic Coding of Audio-Visual Objects, Part 10: Advanced Video Coding, March 2006
7. Marpe, D., Schwarz, H., Wiegand, T.: Context-based adaptive binary arithmetic coding in the H.264/AVC video compression standard. IEEE Trans. Circuits Syst. Video Technol. **13**(7), 620–636 (2003)
8. Karwowski, D., Domański, M.: Optimized architectures of CABAC codec for IA-32-, DSP- and FPGA-based platforms. Poznan University of Technology, Chair of Multimedia Telecommunications and microelectronics
9. Bae, B.-H., Kong, J.-H.: A design of pipelined-parallel CABAC decoder adaptive to HEVC syntax elements. Department of Computer Engineering Kwangwoon University. IEEE ISCE 2014 1569954413
10. Choi, J.A., Ho, Y.-S.: Differential pixel value coding for HEVC lossless compression (2012)
11. Nunez-Yanez, J.L., Chouliaras, V.A., Alfonso, D., Rovati, F.S.: Hardware assisted rate distortion optimization with embedded CABAC accelerator for the H.264 advanced video codec. IEEE Trans. Consum. Electron. **52**(2), 590–597 (2006)
12. Eeckhaut, H., Christiaens, M., Stroobandt, D., Nollet, V.: Optimizing the critical loop in the H.264/AVC CABAC decoder. In: 2006 IEEE International Conference on Field Programmable Technology (FPT2006), pp. 113–118 (2006)
13. Petrovsky, A., Stankevich, A., Petrovsky, A.: Pipeline processing in real-time of CABAC decoder based on FPGA. In: 2012 International Conference on Signals and Electronic System 978-1-4673-1711-5/12/$31.00 c 2012 IEEE
14. Osorio, R.R., Bruguera, J.D.: High-speed FPGA architecture for CABAC decoding acceleration in H264/AVC standard. J. Sign. Process. Syst. **72**, 119–132 (2013)

An Incremental Approach for the Extraction of Software Product Lines from Model Variants

Mohammed Boubakir(✉) and Allaoua Chaoui

MISC Laboratory, Department of Computer Science and Its Applications, Faculty of NTIC, University Constantine 2-Abdelhamid Mehri, Constantine, Algeria
boubakirmohamed@yahoo.fr, a_chaoui2001@yahoo.fr

Abstract. In practice, a large amount of Software Product Lines (SPLs) are developed using a bottom-up process. In this case, an SPL is synthesized from similar product variants that are developed for SPL using ad hoc reuse techniques such as copy-paste-modify. In this paper, we present an approach for migrating existing product variants into an SPL. This approach is applied on models and it takes as input a set of models that abstract the product variants. The result of the approach is a software product line represented by the SPL model and the variability model. SPL model is the result of merging input product models. The variability model is a Feature Model (FM) allowing the specification of the variability on the SPL model. We propose to construct the SPL in an incremental way. After an initialization step, the set of input products are integrated in the SPL one after another. To integrate a new product, we first compare the input product model with the SPL model in order to identify the variability, and then we update both the SPL model and the variability model. The approach is implemented and evaluated on a case study.

Keywords: Variability · Feature model · Software product line · SPLE

1 Introduction

Due to the considerable benefits that it provides, Software Product Line Engineering (SPLE) is emerging as an important software development paradigm. Benefits of SPLE are attested both in academia and industry [1, 2]: they improve both productivity and quality, and reduce time-to-market. SPLE aims to promote the systematic reuse of software artefacts [3]. But, its main objective is to produce a family of systems rather than individual systems. The main idea of SPL is to analyze the domain of a family of products to identify Commonalities and Variabilities. These two concepts represent respectively artefacts that are shared by all products of the SPL, and those that are shared by only some products [2]. Feature Models (FMs) [4] are widely used to describe commonality and a variability of an SPL [5–8]. An FM allows representing the information of all product line members in terms of features and relationships between them. Each member of the SPL is identified by a unique and valid combination of features called configuration [8].

O. Demigha et al. (Eds.): CSA 2018, LNNS 50, pp. 124–134, 2019.
https://doi.org/10.1007/978-3-319-98352-3_14

SPLE implementation can be done as a top-down process that consists of firstly modeling the variability, then deriving different products. However, developing an SPL from scratch is a high cost activity. In practice, it is common for companies to develop a set of similar software product variants from existing ones using ad hoc techniques such the clone-and-own approach [2, 9]. These product variants are developed and maintained in a separate way without any prior consideration of the concept of SPL. However, as their number increases, several problems arise. For example, the task of maintaining these product variants becomes increasingly difficult and expensive [9]. In this case, it is worth reengineering similar product variants into an SPL and therefore taking advantage of the benefits of the SPL concept [10, 11]. This way of developing SPLs using a bottom-up process is called extractive approach [12].

A number of approaches have been proposed to support the extraction of SPLs from exiting products. However, each of them focuses on a specific type of artifacts. For example, [13] considers product description, while [10] considers architectural artifacts. Furthermore, most of these approaches make strong assumptions about their input. For example, [14] assumes that variability is implemented using conditional compilation mechanism, while [5] requires that features have previously been identified. Finally, almost all of these approaches assume that all input product variants are available at the beginning of the SPL extraction process. Consequently, they do not propose any solution to evolve an existing SPL by integrating new product variants.

In this paper, we present our approach for the extraction of SPLs from existing product variants. This approach is applied on a set of models that abstract the input product variants. These models can be reverse engineered from source code. The result of the approach is an SPL represented by (1) the SPL model that gives a compact representation of all models of the input product variants, and (2) the variability model that specifies the variability in the SPL. The SPL is constructed in incremental way by adding the set of input models one after another. The approach is implemented and evaluated on three real SPLs.

The remainder of this paper is structured as follows: Sects. 2 and 3 present respectively our approach and the case study used to evaluate it. Section 4 presents work related to our approach. We conclude the paper and give an outlook on our future work in Sect. 5.

2 Our Approach

Figure 1 gives an overview of our approach that aims to construct a software product line from existing product variants. This operation is performed in an incremental way by integrating the set of input product variants one after another. The approach is applied on models that represent the static architecture of the product variants. The result of the approach is a software product line represented by the two kinds of models: (1) SPL model and (2) Variability model. The SPL model gives a compact representation of all models of the product variants. The variability model is an FM that allows specifying the variability in the SPL.

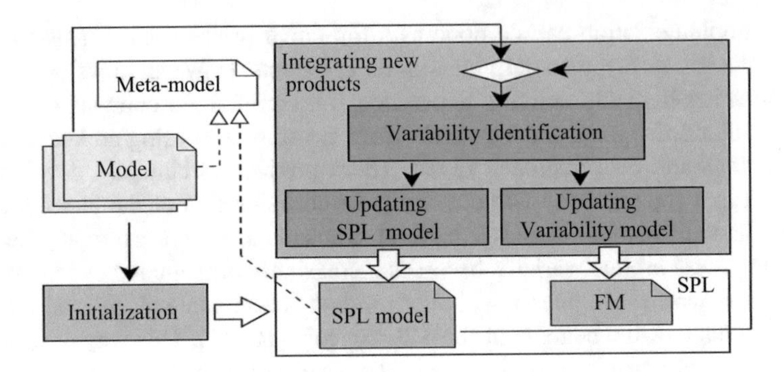

Fig. 1. Overview of our approach

Before presenting in detail each step in our approach, we will firstly present an illustrative example. Then, we will present our representation of the SPL. After that, we will describe how to identify different features of the FM.

2.1 An Illustrative Example

As illustrative example, we consider, a set of products variants related to a simple family of text editor tools called NotePad-SPL obtained from [15]. The variability in this SPL concerns the functionalities *copy/cut/paste*, *undo/redo* and *find* that are optional. Figure 2 illustrates three UML class diagrams that represent three product variants in this family. *NotePad1* contains a set of classes: *About*, *Actions*, *Center*, *ExampleFileFilter*, *Fonts*, *NotePad* and *Print*, regrouped in the *editor* package. Each of these classes contains a set of attributes and operations. The *NotePad* class is mandatory because it is present in all products, whereas the two classes *RedoAction* and *UndoAction* are optional because they are only present in *NotePad2*. Similarly, the *copy* operation of the *Actions* class is optional because it is not present in *NotePad2*.

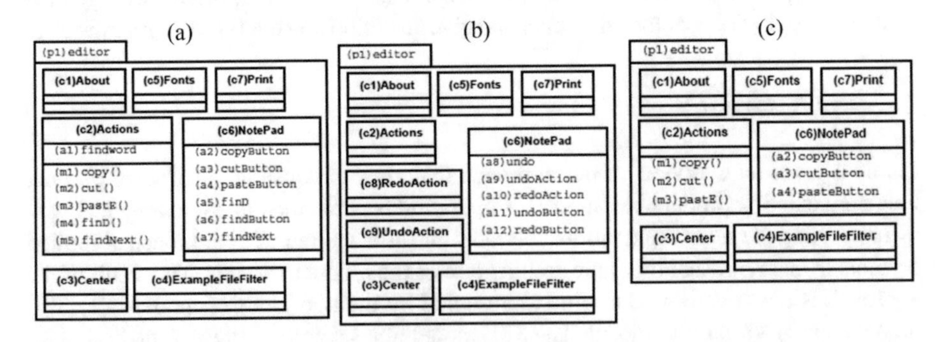

Fig. 2. Example of product models (The text between parentheses before the name of an element corresponds to its identifier. Furthermore, for simplicity, we don't represent mandatory attributes and operations of the *NotePad-SPL*.): (a) *NotePad1*, (b) *NotePad2*, and (c) *NotePad3*

2.2 SPL Representation in Our Approach

As illustrated above, our approach produces an SPL represented by the SPL model and the variability model (a feature model). The SPL model is a product model allowing a compact representation of all models of the product variants. It is obtained by merging the input product models. Each element of the SPL model can be either mandatory or optional. The FM contains a mandatory feature that we call the *base* feature and a set of optional features. The mandatory feature contains all mandatory elements of the SPL, while each optional feature contains a set of optional elements. Each feature is represented as a set of elements.

Figure 3 illustrates the SPL model and the variability model of the NotePad-SPL reconstructed using our approach from the three models of Fig. 2. The SPL model is the result of merging these three models. The FM contains the *base* feature and three optional features denoted *f1*, *f2* and *f3*.

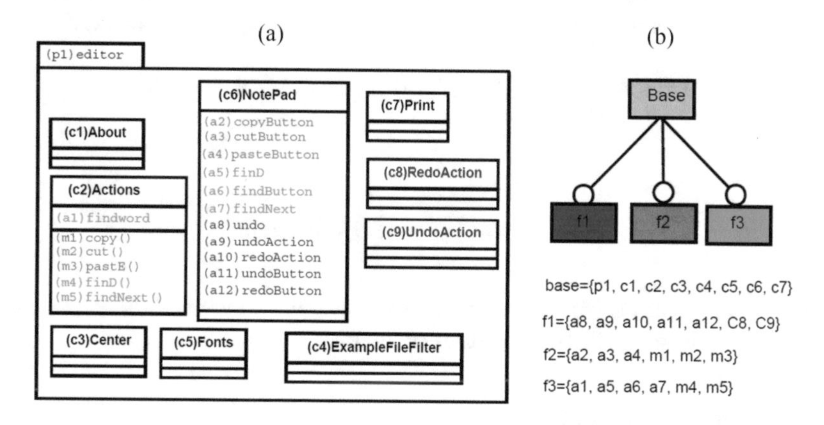

Fig. 3. Example of our SPL representation: (a) the SPL model, and (b) the FM of NotePad-SPL

2.3 Feature Identification

We describe here how to identify different features of the FM, i.e., how to decide whether a set of elements can be considered as a feature. We assume that each feature has exactly the same implementation in all products that support it. This means that the elements that implement a feature must always appear together in different product variants. More formally: $\forall(e_1, e_2) \in F \times F, \forall m \in M_{spl}(e_1 \in m \Rightarrow e_2 \in m)$[1], where F is a feature, e_1 and e_2 are two elements, M_{spl} is the set of all product models of the SPL and m is a product model. This formula can be expressed as follows:

$$\exists M_F \in M_{spl}\left(M_F \neq \emptyset \wedge F \in_{all} M_F \wedge F \notin_{all} M_F^c\right) \quad (1)$$

[1] In our work, we consider features and product models as sets of elements, so all usual operations on sets (such as union, intersection, and difference, etc.) are also applicable on them.

Where:

- M_F^c is the complement of M_F (the set of product models that are not in M_F).
- Let A be a set of elements, and B a set of product models, the following notations: $A \in_{all} B$, and $A \notin_{all} B$, mean respectively $\forall e \in A, \forall m \in B(e \in m)$, and $\forall e \in A, \forall m \in B(e \notin m)$.
- M_F contains models of all products supporting the feature F.

We define a function denoted by ψ on a set of elements denoted by E, and a set of product models denoted by M such that $\psi(E, M) = M_0$, where $M_0 \subseteq M$, $E \in_{all} M_0$, and $E \notin_{all} M_0^c$. Set E satisfies the formula 1 if and only if $\psi(E, M_{spl}) \neq \emptyset$. This formula is satisfied by each feature denoted by F and by all of its non empty subsets denoted by F_0. Moreover, we have $\psi(F, M_{spl}) = \psi(F_0, M_{spl})$. This means that if a set of elements satisfies the formula 1, then it is either a feature or a subset of a feature. Therefore, if we have a set of n product models denoted by M and a set of elements denoted by E. E is considered to be a feature if and only if it satisfies the following constraint:

$$\psi(E, M) \neq \emptyset \wedge \forall E_0 \in E_{all}((\psi(E_0, M) = \psi(E, M) \Rightarrow E_0 \subseteq E) \quad (2)$$

Where:

$$E \subseteq E_{all} \text{ and } E_{all} = \{\bigcup_{i=1}^{n} mi | m_i \in M\}$$

In the present paper, we refer to the above constraint as the *feature constraint* and we rely on it to identify different features of the FM.

2.4 The Different Steps of the Approach

Initialization. The initialization of the SPL consists of initializing both the SPL model and the variability model. This operation is achieved by integrating a first product into the SPL. The model of this product becomes the SPL model. The feature model contains only the *base* feature that contains all elements of the SPL model.

Integrating New Products. After the initialization step, new product variants are iteratively integrated into the SPL. In each iteration, the approach takes as input, a model variant denoted by M_p, the SPL model denoted by M_{spl}, and the FM. As a result, the SPL model and the FM are updated such that the new product can be represented. We present below the three steps of the process that allows integrating a new product into an SPL. These steps are: variability identification, updating the SPL model, and updating the variability model.

Variability Identification. In this step, we compare the model of the new product (M_p) and that of the SPL (M_{spl}). The comparison result is represented by three sets of elements denoted, respectively, by $E_{p \cap spl}$, E_{p-spl}, and E_{spl-p}, and contain, respectively, the elements that belong both to M_p and M_{spl}, the elements that belong only to M_p, and the elements that belong only to M_{spl}. All the elements of a given type in the first model are compared with the elements of the same type in the other model. Two elements are said to be similar if the algorithm used for comparison decides that the first element in the first model is the same as the second one in the other model. Similar elements are included in $E_{p\cap spl}$, where each pair of similar elements is represented by a single element. An element is said to be unmatched if it is not similar to any other element. The unmatched elements of M_p and those of M_{spl} are respectively included in E_{p-spl}, and E_{spl-p}. The current implementation of our approach uses a simple comparison algorithm. This later considers that:

- Two packages are similar if they have the same name.
- Two classes are similar if they have the same name and their packages are similar.
- Two attributes are similar if they have the same name, their classes and their types are similar.
- Two operations are similar if their classes are similar and they have the same signature (they have the same name, their return types are similar, they have the same number of parameters, and their parameter types are similar).

Updating SPL Model. In order to permit the representation of the new product model, we update the SPL model by applying a simple merge algorithm. This later adds to the SPL model, the elements introduced by the new product (the elements of E_{p-spl}).

Updating the Variability Model. Integrating a new product may modify the variability in the SPL. Therefore, we update the variability model to ensure that it represents correctly the variability. To this end, we apply the Algorithm 1 which has been inspired from [16]. This algorithm takes as input the *base* feature, a set of optional features denoted by F and the two sets E_{p-spl}, and E_{spl-p}. Then, modifies both the *base* feature and F so that: (1) the *base* feature contains only all mandatory elements of the SPL model, and (2) each elements of F satisfies the *feature constraint* seen above. The algorithm involves three main steps: In the first step (lines 5–8), a new optional feature is created from the elements of E_{p-spl}. In the second step (lines 9–14), new optional feature is created from the elements of the *base* feature that are not present in the new product. In the third step (lines 15–30), the set F is updated so that all its elements satisfy the *feature constraint*. The algorithm uses two functions: (1) *createNF()* that allows to create a new feature from a set of elements, and (2) *get()* that returns a feature from F. It returns the empty set if F is empty or all its features were treated.

3 Case Study

In this section, we assess the feasibility of our approach. To this end, we evaluate it on the three following SPLs: *NotePad*, *GameOfLife* and *GPL*. These SPLs are written in Java and are obtained from the *FeatureIDE* framework [15]. The evaluation includes the following three steps. First, we generate a set of product variants from the SPL used. While each generated product variant is a Java application, we obtain in the second step models from source code using reverse engineering technique. Finally, we apply our approach on these models to obtain an extracted SPL. Similarly to [17], we divided the evaluation into two parts. In the first part, we evaluate the ability of our approach to extract an SPL according to the original one. In the second part, we assess whether our approach ensure a minimum level of quality.

Algorithm 1. Updating Variability Model

1: **input:** F, *base*, $E_{spl\text{-}p}$, $E_{p\text{-}spl}$
2: **output:** F, *base*
3: **var:** f, f': *feature*
4: **begin**
5: **if** $E_{p\text{-}spl} \neq \phi$ **then**
6: $f \leftarrow createNF(E_{p\text{-}spl})$
7: $F \leftarrow F \cup \{f\}$
8: **end if**
9: **if** $E_{spl\text{-}p} \cap base \neq \phi$ **then**
10: $f \leftarrow createNF(E_{spl\text{-}p} \cap base)$
11: $F \leftarrow F \cup \{f\}$
12: $base \leftarrow base \backslash f$
13: $E_{spl\text{-}p} \leftarrow E_{spl\text{-}p} \backslash f$
14: **end if**
15: $f \leftarrow get(F)$
16: **while** $f \neq \phi$ **and** $E_{spl\text{-}p} \neq \phi$ **do**
17: **if** $E_{spl\text{-}p} \cap f \neq \phi$ **then**
18: **if** $f \subseteq E_{spl\text{-}p}$ **then**
19: $E_{spl\text{-}p} \leftarrow E_{spl\text{-}p} \backslash f$
20: **else**
21: $f' \leftarrow createNF(f \cap E_{spl\text{-}p})$
22: $F \leftarrow F \cup \{f'\}$
23: $F \leftarrow F \backslash f$
24: $f \leftarrow f \backslash f'$
25: $F \leftarrow F \cup \{f\}$
26: $E_{spl\text{-}p} \leftarrow E_{spl\text{-}p} \backslash f'$
27: **end if**
28: **end if**
29: $f \leftarrow get(F)$
30: **end while**
31: **end**

3.1 First Part

We check here whether the FM of the extracted SPL and that of the original one are sufficiently similar. For this, we use the NotePad-SPL seen above. This SPL is a relatively small system containing 9 classes. Its FM contains one mandatory feature and three optional features. By combining these three optional features, we obtain 8 different product variants. We apply our approach on these product variants and manually compare the features of the extracted SPL with those of the original one. We find that our approach produces the same FM as the original one.

3.2 Second Part

We check here whether the following two conditions are satisfied: (1) the extracted SPL allows generating the product variants that have been used to extract it, and (2) the mandatory feature of the extracted SPL includes all mandatory features of the original one. To perform this part of the evaluation, we use the two following SPLs: *GameOfLife* and *GPL*. Table 1 gives more details about these two SPL, columns 1, 2, and 3 give respectively, the number of classes, the number of features, and the number of all product variants in each SPL. The evaluation is performed using different numbers of input product variants. The numbers of the input product variants used for each SPL are given in column 4. We apply our approach 50 times for each SPL and for each number of input product variants, except the last number that represents all members of the original SPL. In each time, the input product variants are randomly selected among all the members of the original SPL.

Table 1. SPLs used in the second part of the evaluation

	Classes	Features	All product variants	Product variants used
GameOfLife	19	23	65	1, 2, 5, 10, 20, 30, 40, 50, 65
GPL	14	38	156	1, 2, 5, 10, 20, 30, 40, 50, 60, 70, 80, 90, 100, 156

To verify whether the first condition is satisfied, we automatically verify whether the extracted SPL can generate the product models that have been used to extract it. The evaluation shows that for the two SPLs used and for each number of products and for each run of our approach, the first condition is satisfied. To verify whether the second condition is satisfied, we automatically compare the mandatory feature of the extracted SPL with all mandatory features of the original one. Similarly to the first condition, the evaluation shows that the second condition is satisfied for the two SPLs used and for each number of products and for each run of our approach.

4 Related Work

The approach proposed in [10] considers component architecture information to ease the extraction of an SPL from a collection of product variants. Unlike our approach which is fully automatic, this approach contains a semi-automatic step. Furthermore, it assumes that all input product variants are available at the beginning of the SPL extraction process. Koschke et al. [9] propose also to consider architectural-level. Their approach requires pre-existing module architecture. The approach proposed by Rubin et al. [6] aims to extract an SPL from existing model variants. The FM produced by this approach is less expressive than that produced by ours. It contains only a set of alternative features where each feature contains the elements of a model. Ryssel et al. [18] propose an approach for extracting variability from a collection of similar model variants. This approach works only on function-Block models. Martinez et al. [19] propose an approach for automating the migration of existing similar MOF[2]-based model variants into an SPL. Contrary to our approach that proposes to extract the SPL in an incremental way, this approach assumes that all input product variants are available at the beginning of the SPL extraction process.

Other approaches do not address the feature identification problem as our approach does, but assume that the features have been already identified. For example, Xue et al. [11] present an approach for feature location in a collection of product variants by using software differencing and Formal Concept Analysis (FCA). For each product variant, this approach requires as input the set of features associated to the product variant. Ryssel et al. [20] propose to use FCA to extract a FM from an incidence matrix describing common and variable artifacts of the input product variants.

Contrary to our approach that aims to extract SPLs from models, other approaches concentrate on the extraction of SPLs from source code. For example, Fenske et al. [21] propose a semi-automated approach for extracting an SPL from the code of a collection of product variants, basing on the clone detection technique. Zhang et al. [14] propose a framework solution to extract variability from source code. They only consider variability implemented using conditional compilation mechanism. An approach for refactoring SPL from Java code source is presented in [22].

Some other approaches like for instance [23, 24] focus on model comparison and model merging which are two key tasks in the process of extracting an SPL from existing product variants.

5 Conclusion and Future Work

In this paper, we proposed an approach for the extraction of software product lines from existing product variants. Giving a set of models abstracting the input product variants, our approach produces automatically an SPL. This later is constructed in an incremental way by integrating the set of input product variants one by one. The fact that our approach performs the extraction of SPLs in an incremental way provides the advantage of evolving an existing SPL by integrating new products into it. Furthermore, the user can intervene after each iteration, in order to review and, if necessary, adjust the result. We implemented and successfully applied our approach on three real systems. As future work, we plan to apply our approach to more case studies, and extend it in order to support any kind of EMF-based model.

[2] Meta-Object Facility (MOF) Core Specification, http://www.omg.org/spec/MOF/2.0/.

References

1. Czarnecki, K., Eisenecker, U.: Generative Programming: Methods, Tools, and Applications. Addison-Wesley, Boston (2000)
2. Pohl, K., Böckle, G., van der Linden, F.: Software Product Line Engineering: Foundations, Principles and Techniques. Springer, New York (2005)
3. Acher, M., Cleve, A., Collet, P., Merle, P., Duchien, L., Lahire, P.: Extraction and evolution of architectural variability models in plugin-based systems. In: SoSyM, pp. 1–28 (2013)
4. Kang, K., Cohen, S., Hess, J., Nowak, W., Peterson, S.: Feature oriented domain analysis (FODA) feasibility study, Technical report. CMU/SEI-90-TR-021 (1990)
5. She, S., Lotufo, R., Berger, T., Wasowski, A., Czarnecki, K.: Reverse engineering feature models. In: Proceedings of ICSE 2011, pp. 461–470. ACM (2011)
6. Rubin, J., Chechik, M.: Combining related products into product lines. In: Proceedings of FASE 2012, pp. 285–300. ACM (2012)
7. Acher, M., Baudry, B., Heymans, P., Cleve, A., Hainaut, J.L.: Support for reverse engineering and maintaining feature models. In: Proceedings of VaMoS 2013. ACM (2013)
8. Batory, D.: Feature models, grammars, and propositional formulas. In: Proceedings of SPLC 2005, pp. 7–20 (2005)
9. Koschke, R., Frenzel, P., Breu, A.P.J., Angstmann, K.: Extending the reflexion method for consolidating software variants into product lines. Softw. Qual. J. **17**(4), 331–366 (2009)
10. Klatt, B., Küster, M.: Respecting component architecture to migrate product copies to a software product line. In: Proceedings of WCOP 2012, pp. 7–12. ACM (2012)
11. Xue, Y., Xing, Z., Jarzabek, S.: Feature location in a collection of product variants. In: Proceedings of WCRE, pp. 145–154 (2012)
12. Assunção, W.K.G., Vergilio, S.R.: Feature location for software product line migration: a mapping study. In: Proceedings of SPLC 2014, pp. 52–59. ACM (2014)
13. Davril, J.-M., Delfosse, E., Hariri, N., Acher, M., Cleland-Huang, J., Heymans, P.: Feature model extraction from large collections of informal product descriptions. In: Proceedings of ESEC/FSE, pp. 290–300. ACM, Saint Petersburg, Russia (2013)
14. Zhang, B., Becker, M.: Recovar: a solution framework towards reverse engineering variability. In: PLEASE, pp. 45–48 (2013)
15. The FeatureIDE framework. https://featureide.github.io/
16. Ziadi, T., Frias, L., da Silva, M.A.A., Ziane, M.: Feature identification from the source code of product variant. In: Proceedings of CSMR, pp. 417–422 (2012)
17. Ziadi, T., Henard, C., Papadakis, M., Ziane, M., Traon, Y.L.: Towards a language-independent approach for reverse-engineering of software product lines. In: Proceedings of SAC 2014, pp. 1064–1071. ACM (2014)
18. Ryssel, U., Ploennigs, J., Kabitzsch, K.: Automatic variation point identification in function-block-based models. In: Proceedings of GPCE 2010, pp. 23–32. ACM (2010)
19. Martinez, J. Ziadiy, T., Bissyandé, T.F., Klein, J., Traon, Y.L.: Automating the extraction of model-based software product lines from model variants. In: ASE (2015)
20. Ryssel, U., Ploennigs, J., Kabitzsch, K.: Extraction of feature models from formal contexts. In: Proceedings of SPLC 2011, vol. 2. ACM, Munich, Germany (2011)
21. Fenske, W., Meinicke, J., Schulze, S.: Variant-preserving refactorings for migrating cloned products to a product line. In: SANER (2017)

22. Kim, J., Batory, D., Dig, D.: Refactoring java software product lines. In: Proceedings of SPLC, vol. A, pp. 59–68. ACM (2017)
23. Boubakir, M., Chaoui, A.: A pairwise approach for model merging. In: MISC 2016, pp. 327–340. Springer, Constantine, Algeria (2016)
24. Martinez, J., Ziadi, T., Klein, J., Traon, Y.: Identifying and visualizing commonality and variability in model variants. In: Proceedings of ECMFA, vol. 8569, pp. 117–131. ACM (2014)

Domain-Level Topic Detection Approach for Improving Sentiment Analysis in Arabic Content

Bilel Kaddouri and M'hamed Mataoui(✉)

Information Systems Engineering Laboratory, Ecole Militaire Polytechnique, BP 17, Bordj El Bahri, Algiers, Algeria
kaddouribilel@gmail.com, mataoui.mhamed@gmail.com

Abstract. Social networks are considered today as the most popular interactive media where people can communicate, share information and express opinions without any limitation. The interest of the scientific community towards social contents has increased due to their importance in various fields such as marketing, sociology and politics. Several research areas related to social networks have emerged namely, community detection, sentiment analysis and topic detection. In this paper, we propose a domain-level topic detection approach for improving sentiment analysis in Arabic social content. The proposed approach is based on a supervised learning technique on Arabic collected data. Training dataset is mainly composed of Arabic press articles, while the test dataset is represented by posts and comments extracted from Arabic Facebook pages. Experimental evaluation showed that the proposed approach achieves good performances with precision values between 75.36% and 97.89%.

Keywords: Topic detection · Social networks · Sentiment analysis
Arabic social content

1 Introduction

Since their existence, humans have naturally organized themselves into communities in order to communicate, collaborate and share points of view in various fields. This social character has always been following the development of civilizations. With the advent of Web 2.0, these social groups have given rise to the concept of digital social networks. Digital social networks are seen as open spaces where people can communicate, share information and express opinions without any constraint.

The interest of the scientific community towards these social contents has increased because of their importance in various fields such as marketing, sociology and politics. The handling of these contents has been performed mainly by using natural language processing (NLP) techniques. Several areas related to social networks emerged, namely: community detection, sentiment analysis (SA) and topic detection.

O. Demigha et al. (Eds.): CSA 2018, LNNS 50, pp. 135–146, 2019.
https://doi.org/10.1007/978-3-319-98352-3_15

Topic detection, also called classification or categorization of texts, is often considered as an unavoidable passage in various applications related to the social media monitoring. Thus, in the case of morphologically rich languages such as Arabic, the fact of being able to predict, with a certain degree of confidence, the topic of a social content, can eliminate ambiguities relating to the meaning of the words which will considerably improve the performance of SA systems.

This paper deals with the problem of topic detection in Arabic social content by proposing a domain-level approach for improving sentiment analysis systems. The proposed approach will be integrated into the EMP's SA system [1]. This integration aims mainly to take into account the topic during the semantic orientation computation phase of Arabic social content. Indeed, the aforementioned SA system does not consider the topic of a social content, which prevents it from dealing with some issues such as the expression of opinions belonging to different domains (economy, sociology, sport, religion, culture, etc.) given the use of a common lexicon (general SA lexicon). For example, a term like "رخيص" has a negative polarity in the current system, whereas in economy it should have a positive polarity.

To address this issue, we propose a topic detection approach based on a supervised learning process. The proposed approach builds a vector representation for each domain of knowledge. These representations are then compared to social content representation using similarity computations to determine the degree of social content to belong to each domain.

The rest of the paper is organized as follows. In Sect. 2, we present topic detection field related work. In Sect. 3, we describe our domain-level topic detection approach for Arabic content. In Sect. 4, we present the experimental evaluation with special emphasis on used resources, datasets, the experimental protocol and obtained results. Finally, In Sect. 5, we conclude the paper.

2 Related Work

Text mining is one of the most sought-after fields of research because of the growing number of electronic documents available on the Web from various sources. These documents are of different forms, i.e. structured, unstructured or semi-structured. The main purpose of text mining is to allow extraction of information from textual content. Topic detection represents a very important phase in the text mining process and has attracted the attention of many researchers in the field [2].

According to Zhu et al. [3], the problem of domain-level topic detection from social content can be compared to a classic text categorization process where the goal is to construct a prediction system that serves to classify a set of documents according to a set of predefined categories. Ahmed et al. [4] distinguish two types of categorization: assigning a document to a single category; or assigning a document to several categories at the same time depending on a certain degree of membership.

Zhu et al. [3] developed a classification system for articles to be published on social networks. They extracted the key-phrases of an article to produce a summary. Then, a vector representation of these summaries was used to construct a neural network capable of predicting the category of a given article.

Bairagi and Tapaswi [5] proposed a new learning method based on probabilities of terms' categories and a set of rules for each category using fuzzy logic. The terms of the test documents have been combined with the prediction model rules to compute the membership degree of a document to a category.

Faqeeh et al. [6] apply the supervised methods using various classification algorithms to categorize small text content from Facebook social network. Authors were interested to Facebook posts written in two different languages, namely: Arabic and English. They collected 4000 posts divided equally into the two languages. Two categories have been defined: weather and food. Two thirds of the data were used for learning and the rest were used for the test. Many classification algorithms under WEKA toolkit have been tested, among which: SVM, KNN, decision trees and naive Bayesian. The results obtained in terms of accuracy vary between 70% to 82% for English and from 75% to 89% for Arabic.

Unsupervised methods in the context of social networks are less used in the literature. These methods are generally based on semantic relationship between terms established via tools such as WordNet. This type of approach is very difficult to implement when the text does not follow language rules which is the case of social networks. However, Ezzat et al. [7] proposed a multi-label categorization system of Arabic social texts, based on unsupervised methods, called TopicAnalyzer. They used a dataset composed of 51917 posts from Arabic social networks. TopicAnalyzer approach split categorization process into three main steps: (1) the construction of the ontology, (2) the computation of terms' weights and (3) the categorization. The evaluation of the system showed good performance with an f-measure of 84.59%.

3 Domain-Level Topic Detection Approach for Arabic Social Content

The general problem addressed by our work is to develop a domain-level topic detection system for Arabic social content that aims to improve EMP's SA system [1] performance. To deal with this problem, we adopt supervised methods to develop a system based on an evolutionary prediction model. Our approach, described in Fig. 1, is divided into two phases: training and test. The training phase aims to construct the prediction model of the form "*terms-categories*" graph. The test phase consists of classifying one or more documents according to one or more categories and comparing the results with the manual assessments. This classification will always be followed by a prediction model update to ensure the evolution and adaptation of the developed system.

3.1 Training Phase

The training phase is carried out using a corpus of documents already divided into a number of predefined categories. Thus, the membership of a document d_j in a category c_i is already known. A vector representation "*Bag of words (BOW)*" of each category is produced based on its documents through the following steps: pre-processing, document representation and dimensionality reduction. The pre-processing of documents in a text classification process aims to transform textual content into a representation consisting of standardized vectors of "*terms-frequencies*". We propose to add a new pre-processing sub-step for the context of Arabic content which consists mainly to remove opinion words represented generally by adjectives. This idea has been applied by Gutierrez-Batista et al. [8] in their proposed topic detection approach for English social texts.

Document representation denotes the basic unit for the category representation. All documents d_{ij} of a category c_i are processed in order to compute DF (document frequency) and the TF (term frequency) for each term t_{ik}. Thus, the data structure representing a category (set of documents) is composed mainly of terms. For each term we will retain its DF and TF for each of these documents.

The last step of the this phase is the construction of the "*terms-categories*" graph composed of two kinds of nodes: categories and terms. Each node representing a term is connected at least to one category. The "category" nodes must contain the information relating to the number of documents classified respectively in each category. At the end of the training phase, each node NC_i will contain a number of documents n belonging to the category c_i. The degree of importance of the term t_{ik} within a category c_i is measured by a weight computed by summing up the weights TF * IDF in the set of documents d_{ij} belonging to the category c_i, in which it appears, see formula (2).

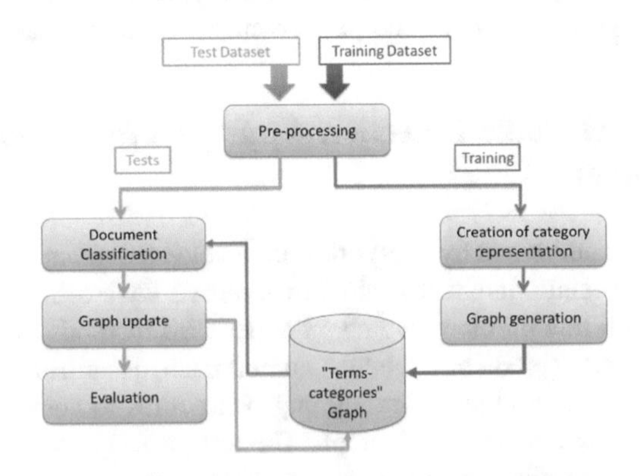

Fig. 1. The architecture of the proposed domain-level topic detection approach.

3.2 Test Phase

It is important to mention that our system carries out "*multiclass categorization*". During the process of text categorization, the text passes through all the following steps: pre-processing, document representation, similarity computation, text labeling and graph update.

In order to be able to detect the domain-level topic of a text, we compute the degree of similarity between the text and the set of categories represented in the "*terms-categories*" graph. The degree of similarity between a document and a category is given by the sum of the weights of the terms in common between them.

$$\text{degree of similarity}\,(\mathrm{d}, c_i) = \sum_{\mathrm{j=1}\,|\,\mathrm{t}\in\mathrm{d}}^{n} \mathrm{w}\,(t_{ik}) \tag{1}$$

In order to observe the impact of the dimensionality reduction on the behavior of the proposed approach, we apply the $\chi 2$ "Chi-square" method to compute the weights of terms within the categories. We define a reduction parameter *S1* (representation threshold). Then, the term t_{ik} will be taken into consideration only in the *S1* first categories. The weight of the term t_{ik} is given by the following formula:

$$\mathrm{W}\,(t_{ik}) = \sum_{\mathrm{j=1}\,|\,\mathrm{t}\in\mathrm{d}}^{n} \mathrm{TF}(t_{ik}) * \mathrm{IDF}\,(t_{ik}) \tag{2}$$

Once the weight of a term is computed, two situations occur: if the term t_{ik} is connected to several "category" nodes, then its weight will be reduced by a factor *F* to give it less importance. Otherwise, it will be amplified by the same factor.

The "document-category" degree of similarity computation algorithm is described by the algorithm 1.

Algorithm 1. "document-category" degree of similarity computation algorithm

INPUT: *GR* : The graph "terms-categories", *T* : list of tokens and their *TF* in the document to be classified, *S1:* representation threshold, *C* : list of categories, *F*: amplification/reduction factor.
OUTPUT: *Vector_DS* : similarity degrees between the document and each category.
INIT: *Threshold* ← *S1*; *Vector_DS* ← 0.

```
    Begin
1.      For (any term t_k of T)
2.          If (t_k belongs to GR)
3.              Vector_indice = Compute_chi_square (t_k, Threshold)
4.              For (any category c_j belonging to C)
5.                  For (any v_i element belonging to Vector_indice)
6.                      If (c_j = v_i)
7.                          Weight (t_k, c_j) = GR (TF (t_k, c_j))) * log (Nc_j / GR (DF (t_k, c_j)))))
8.                          If (t_k belongs to a single category)
9.                              Weight (t_k , c_j) = Weight (t_k , c_j) * F
10.                         Else
11.                             Weight (t_k , c_j) = Weight (t_k , c_j) / F
12.                         EndIf
                            // Add the weight of the term to the sum of the weights for category c_j.
13.                         Vector_DS (c_j) = Vector_DS (c_j) + Weight (t (c_j))
14                      EndIf
15.                 EndFor
16.             EndFor
17.         EndIf
18.     EndFor
    End
```

Indications:
· *GR* (TF (t_k , c_j)): represents the *TF* of the term t_k with respect to category c_j in *GR*.
· *GR* (DF (t_k , c_j)): represents the DF of the term t_k with respect to category c_j in *GR*.
. Nc_j : number of documents already classified in category c_j.
. Weight (t_k , c_j): weight of the term t_k in category c_j.
. *Vector_DS* (c_j): degree of similarity of the document with category c_j.
. *Vector_indices*: vector containing the indices of the *S1* categories with which the term *ti* has the greatest degree of importance.
. *Compute_chi_square*: function that returns a vector of the indices of *S1* categories

The text labeling step consists of assigning a ranked list of categories to the test document. The degree of similarity of a test document with each category is computed by using information from our prediction model (multi-class categorization).

During the graph update step, we will proceed to the update of the "terms-categories" graph and consequently ensure the evolutionary character of our knowledge about category representation, as follows: ***(i)*** First, we increase by 1, the values contained in the "category" nodes. A threshold (denoted *S2*) will be experimentally defined in order to select only the "category" nodes to be updated. To do this, only "category" nodes whose degree of similarity ratio with the document exceeds S2, will be updated. The degree of similarity ratio of a document within a category is defined by formula 3. ***(ii)*** Then, we divide the terms belonging to the test document into two categories: the terms already existing in the graph and those that do not appear (new terms). Regarding an existing term, if it is linked to the list of categories resulting from the labeling step, then we add only its frequency to each of the links and increase the DF value of 1. Otherwise, we add links between the term node and the related category nodes by putting the DF = 1 and the TF values in each of these links. Regarding new terms, links are created between the term node and the category nodes resulting from the labeling by taking into account the *S2* parameter. The value of DF is set to 1 and TF to the frequency of the term in the test document. In Algorithm 2, we present the process of performing the update of the "terms-categories" graph for the two cases.

$$\text{The degree of similarity ratio } (\mathrm{d}, c_i) = \frac{\text{degree of similarity } (\mathrm{d}, c_i)}{\sum_{c_j \in \mathrm{C}} \text{degree of similarity } (\mathrm{d}, c_j)} \quad (3)$$

Algorithm 2. "terms-categories" graph updating algorithm

INPUTS: *GR*: terms-categories graph, *T*: list of tokens and their *TF* in the document to be classified, *S2*: number of categories to be updated threshold, *C*: ranked list of categories.
OUTPUTS: up-to-date graph.
INIT: Threshold ← S2.
Begin

```
1.     For ( any term tk of T )
2.        If ( tk belongs to GR )
3.           For ( any category cj belonging to C whose Percentage(d, cj) >= Threshold)
4.              If ( tk is connected to cj in GR )
5.                 GR (TF (tk , cj)) ← GR (TF (tk , cj)) + TF (tk)
6.                 GR (DF (tk , cj)) ← GR (DF (tk , cj)) + 1
7.              Else
8.                 Create a connection between tk and cj within GR
9.                 GR (TF (tk , cj)) ← TF (tk)
10.                GR (DF (tk , cj)) ← 1
11.             End If.
12.          End For.
13.       Else
14           Add the term tk to GR.
15.          For ( any category cj belonging to C whose Percentage(d, cj) >= Threshold )
16.             Create a relationship between tk and cj within GR
17.             GR (TF (tk , cj)) ← TF (tk)
18.             GR (DF (tk , cj)) ← 1
19.          End For.
20.       End If.
21.    End For.
```

End.

4 Experimental Evaluation

4.1 Resources and Developed System

We use several tools for the development of our Domain-level topic detection system, namely: Facebook4j API to retrieve the textual content available on Facebook (Posts and their comments) to build our test dataset; Oracle database to store all collected data; and *Lucene Arabic Stemmer* API with its two variants: *khoja stemmer* and *light stemmer*. The architecture of our system is presented in Fig. 2.

4.2 Datasets Description and Evaluation Protocol

To evaluate the performance of our system, we perform the tests using two datasets: ***(i)*** training phase dataset composed of 2000 textual documents from Algerian newspaper articles written in Arabic which are: "*El Chourouk*", "*El Khabar*", "*Echaab*" and "*El Fadjr*" organized according to 4 categories: Sport, Economy, Religion and Politics (500 textual documents within each category); ***(ii)*** and test phase dataset (social networks content) consists of posts and comments from various Facebook pages whose subscribers use mainly the Arabic language.

We carried out the tests according to the protocol described in Fig. 3. The first stage of this evaluation protocol was conducted on newspaper articles whose authors practice a well-structured, Modern Standard Arabic (MSA). This stage has two main objectives: system setting; and validation of the training dataset.

System setting aims to determine the optimal values of the various parameters of the system. The list of parameters to be estimated includes: Amplification or reduction factor, *S1* and *S2*. We perform a series of tests by varying one parameter while setting the values of the others. This allowed us to identify the values that give the best performance scores. These values will be used during the second test phase. To validate the training data, we use the K-Fold cross-validation method.

The second stage of the evaluation protocol consists of using data from social networks to measure the effectiveness of our system.

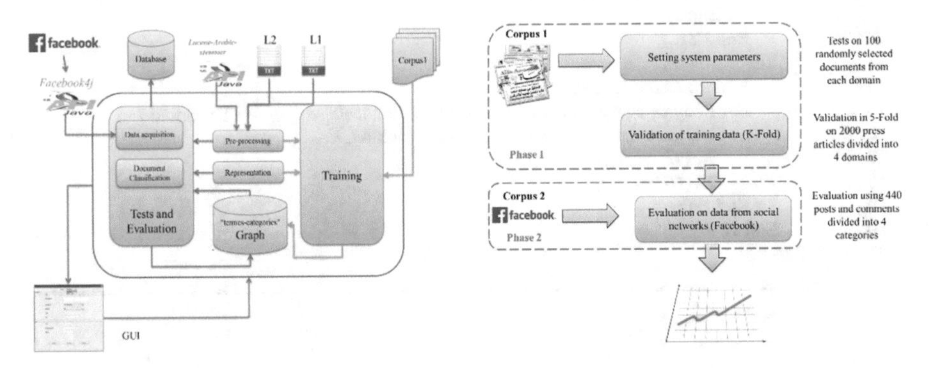

Fig. 2. The architecture of the system.

Fig. 3. Evaluation protocol of our approach.

4.3 Results and Discussion

Our results have been divided into two main phases: the first phase consists of estimating the parameters of the classification system; and the second phase consists of applying the best configuration obtained during the first phase to content from social networks.

System Parameters Setting. The first obtained results represent the system parameters setting stage, namely: Amplification or reduction factors, *S1* (The number of categories represented by a term) and *S2* (Threshold for the number of categories to be updated).

Amplification or Reduction Factor (F). We vary the values of F by fixing the other two parameters as follows: S1 = 1; S2 = 50% (the minimum percentage);

The obtained results are presented in Fig. 4. We note that the best values for precision, recall and F-measure are obtained for F equals to 10. Therefore, this value will be used during the rest of the experiments.

The Number of Categories Represented by a Term (S1). To determine the optimal value of S1, we conduct tests for the different possible values of *S1 = {1, 2, 3, 4, 5}*, while fixing the other parameters as follows: F = 10 and S2 = 50%. The evaluation results obtained for this configuration are depicted in Fig. 5. We note that the best results are obtained for S1 equals to 4 categories. This implies that each term will be considered as a descriptor for up to 4 categories during the rest of the experiments.

Threshold for the Number of Categories to be Updated (S2). The results obtained for the variation of *S2* are described in Fig. 6. They clearly show that updating the categories is a very important operation and requires a high degree of confidence for the classification of the documents. The best value for this parameter is obtained for S2 equals to 70%.

To summarize the above-presented results, we adopt the following configuration: *F* = 10, S1 = 4 and *S2* = 70%.

Cross-Validation Results. The results obtained in this experiment are presented in Fig. 7 (precision values between 84.89% and 98.26%, with an average precision of 91.78%). They show the validity of our choice of training corpus and the technique used to create category representations.

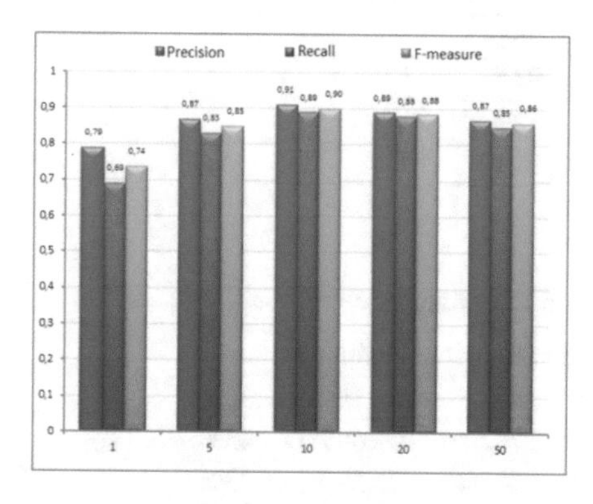

Fig. 4. Results obtained by varying *F* parameter (*S1 = 1 & S2 = 50%*).

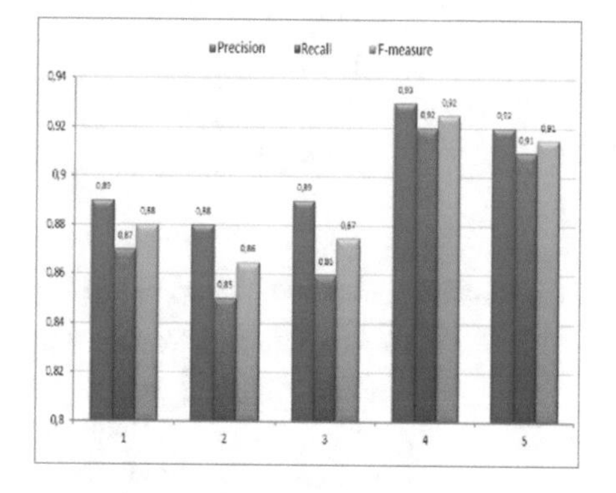

Fig. 5. Results obtained by varying *S1* parameter (*F = 10 & S2 = 50%*).

Impact of Opinion-Words Removal. In this experiment, we study the impact of eliminating opinion-words on the quality of classification results. The first configuration includes opinion-words and the second eliminates them. The results obtained are detailed in Table 1.

These results show that the configuration that eliminates the opinion-words outperforms the configuration that considers them. Thus, an improvement of 2% in the precision is realized. This precision can be considered marginal because the documents used in the training phase represent press articles and generally contain few opinion expressions contrary to content from social networks.

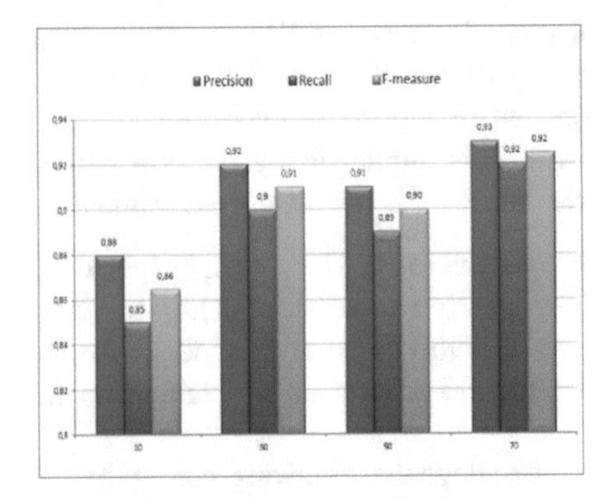

Fig. 6. Results obtained by varying *S2* parameter (*F = 10 & S1 = 4*).

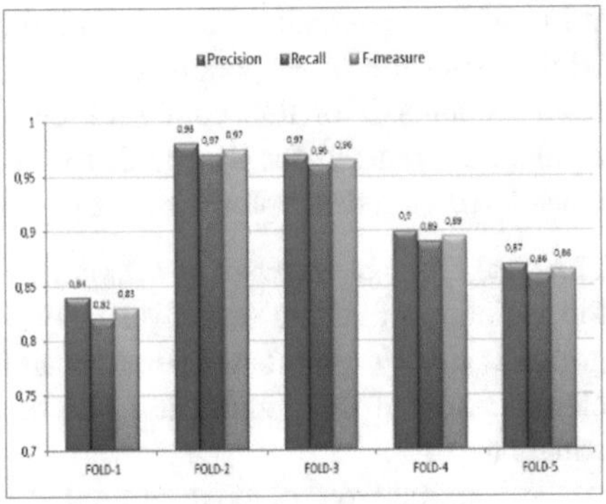

Fig. 7. Precision, recall and F-measure values obtained for the different folds.

Table 1. Results related to the impact of opinion-words removal.

	With opinion-words	Without opinion-words
Graph size	9847	9623
Precision	92%	**94%**
Recall	90%	92%
F-measure	91%	93%

Table 2. Results obtained by the application of our approach on social content.

Measure/Category	Sport	Economy	Religion	Politics	Global
Precision	86,11%	97,89%	85,85%	75,36%	86,30%
Recall	83,78%	83,78%	76,57%	93,69%	84,45%
F-measure	84,93%	90,29%	80,94%	83,53%	85,36%

Evaluation on Social Content. The parameters used are set according to the best configuration obtained during phase 1. In this case: $F = 10$, $S1 = 4$ and $S2 = 70\%$.

The results presented in Table 2 show that our proposed domain-level topic detection approach achieves very good precision rates (from 75.36% to 97.89%, giving an average precision of 86.30%). We note that this rate is lower than that obtained in the first test phase because of the nature of social content characterized by a less rich vocabulary and a relatively small text size.

5 Conclusion

In this paper, we proposed a domain-level topic detection approach in Arabic social content to improve the sentiment analysis task. The proposed approach is based on a supervised learning technique applied to Arabic social content. The training phase based on a dataset composed mainly of Arabic press articles, generates a "*terms-categories*" graph representation. The test phase is performed using a dataset consisting of posts and comments extracted from Arabic Facebook pages. Two algorithms have been proposed, namely, "document-category" similarity computation and the "terms-categories" updating. Three parameters, namely: F, S1 and S2 have been defined for which the best configuration has been set. Additionally, we tested our proposal to measure the impact of opinion-words removal and noticed that the configuration based on the elimination of the opinion-words outperforms that considering them.

Experimental evaluation by the application of our approach on social content from Facebook Arabic pages showed good performances with precision values from 75.36% to 97.89%.

References

1. Mataoui, M., Zelmati, O., Boumechache, M.: A proposed lexicon-based sentiment analysis approach for the vernacular algerian arabic. Res. Comput. Sci. **110**, 55–70 (2016)
2. Sebastiani, F.: Text categorization text mining and its applications. In: Text Mining and its Applications, A, pp. 109–129 (2005)
3. Zhu, X., Huang, J., Zhou, Z., Han, Y.: Chinese article classification oriented to social network based on convolutional neural networks. In: IEEE International Conference on Data Science in Cyberspace (DSC), pp. 33–36 (2016)
4. Ahmed, M.S., Khan, L., Rajeswari, M.: Using correlation based subspace clustering for multi-label text data classification. In: 2010 22nd IEEE International Conference on Tools with Artificial Intelligence (ICTAI), pp. 296–303 (2010)
5. Bairagi, V., Tapaswi, N.: Social network comment classification using fuzzy based classifier technique. In: Symposium on Colossal Data Analysis and Networking (CDAN), pp. 1–7 (2016)

6. Faqeeh, M., Abdulla, N., Al-Ayyoub, M., Jararweh, Y., Quwaider, M.: Cross-lingual short-text document classification for facebook comments. In: 2014 International Conference on Future Internet of Things and Cloud (FiCloud), pp. 573–578 (2014)
7. Ezzat, H., Ezzat, S., El-Beltagy, S., Ghanem, M.: Topicanalyzer: a system for unsupervised multi-label arabic topic categorization. In: 2012 International Conference on Innovations in Information Technology (IIT), pp. 220–225 (2012)
8. Gutierrez-Batista, K., Campaña, J.R., Martinez-Folgoso, S., Vila, M.A., Martin-Bautista, M.J.: About the effects of sentiments on topic detection in social networks. Int. J. Des. Nat. Ecodyn. **11**, 387–395 (2016)

Towards the Paradigm of Information Warehousing: Application to Twitter

Hadjer Moulai(✉) and Habiba Drias

LRIA, USTHB, BP 32, El Alia Bab Ezzouar, Algiers, Algeria
{hamoulai,hdrias}@usthb.dz

Abstract. Over the last decade, social media have dominated our lives. The exploding number of data produced by these platforms triggered a wave of research works that mainly focus on the storage and analysis of this data. In this paper, we propose an original information warehouse architecture for the storage and analysis of social media information. A multidimensional model is defined and the information is extracted, transformed and loaded in the warehouse using ETL (Extract, Transform, Load). The described framework is implemented for Twitter and a data mining analysis is performed on the collected tweets using a clustering algorithm to uncover most discussed topics. The preliminary results are satisfactory and the proposed paradigm can be applied for various information sources such as newspapers and scientific papers.

Keywords: Information warehouse · Social media
Multidimensional model · Data mining · Twitter

1 Introduction

An impressive amount of data is produced daily by web applications, mainly social media. Because of their popularity, millions of users generate a huge amount of data at a very high speed which contributes to the big data phenomenon. The latter has attracted the attention of many research communities who exploited this data for analysis and modeling purposes. Moreover, since it is difficult to operate on such a big amount of data, data warehousing technology is often exploited to store and analyze this large-scale data.

However, the scope of the generated data has not been addressed. In fact, among this huge amount of data we find: data, information and knowledge. These three entities are, in our sense, important to distinguish, especially when using technologies such as warehousing.

This is what motivated our approach on introducing and considering an information warehouse paradigm instead of the existing data warehouse technology for the storage and analysis of information in order to preserve its value.

The rest of this paper is organized as follows. The next section reviews the related work. Section 3 introduces the paradigm of information warehousing.

O. Demigha et al. (Eds.): CSA 2018, LNNS 50, pp. 147–157, 2019.
https://doi.org/10.1007/978-3-319-98352-3_16

In Sect. 4, a general social media information warehouse framework and a multidimensional schema are presented and described. Section 5 illustrates the implementation of the proposed architecture for Twitter. Section 6 explains our tweets' content clustering technique. In Sect. 7 the results are presented and discussed. Section 8 concludes this work and discuss future research.

2 Related Work

The warehousing technology in general has been widely exploited into the analysis and storage of big amounts of data in different applications.

The author of [1], proposed a universal information warehouse system and described it as a method and tool for collecting, storing and organizing information using a new classification system. The main purpose of the warehouse was the storage of information in databases consisting of a set of database tables.

In [2], the authors proposed a framework and a method for developing an information warehouse. The goal was to provide a standalone component that includes both the data warehouse and other tools for integrating technical data and transforming data into relevant business views.

In [3], the authors proposed an analytical warehouse information platform for the detection of clinical phenotypes in electronic health record (EHR) data to improve the quality of investigations.

The authors of [4], proposed an information warehouse for the screening of patients for clinical trials. A functional prototype is described that uses data in the information warehouse to select the patients who meet the eligibility criteria for clinical trials. The warehouse includes clinical information such as lab results, diagnosis codes, pathology reports and clinical notes.

These efforts for proposing a storage and analysis model for information prove its importance and relevance in organizations in general. However, we observe that information and data are often treated equally as the same entity and are usually part of the same infrastructure.

In our opinion, it is very important to treat information separately from data or even knowledge in order to propose an effective tool. These three entities have different properties and importance in the analysis process and decision making.

In this paper we focus on introducing the concept of information warehousing in general. The latter is applied for the case of social media where a multidimensional model is proposed. Such a paradigm can be applied for other information sources such as news and scientific articles.

3 Introducing the Paradigm of Information Warehousing

In [5], the author observed a difference between data, information and knowledge and provided a hierarchy between these entities as shown in Fig. 1a.

We are constantly exposed to signals that allow us to perceive and interpret the outer world. Signals are different perceptions of the real world, through sights, sounds and other sensory phenomena. Depending on beliefs and the

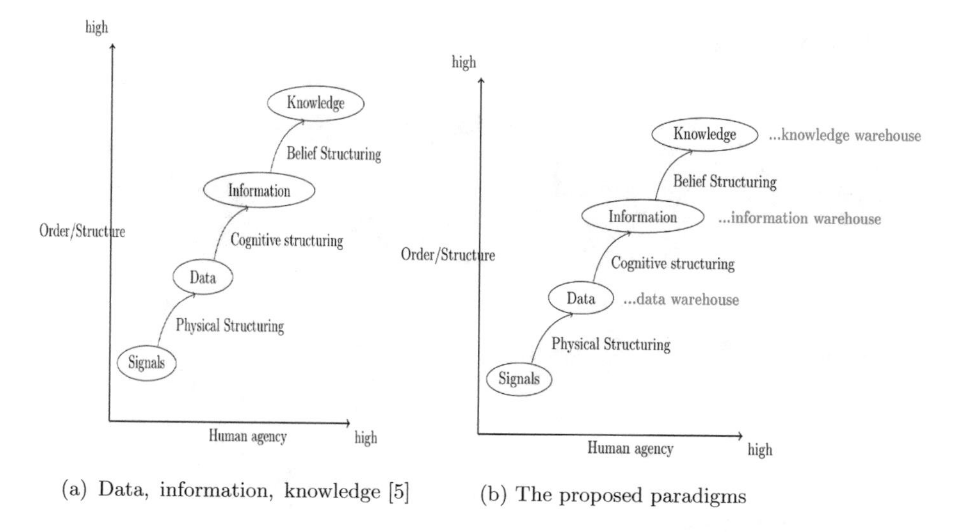

(a) Data, information, knowledge [5] (b) The proposed paradigms

Fig. 1. Data, information, knowledge and warehousing paradigms

material environment (noise, lightning), signals are selected and organized into packets of data. For example marks on a paper are recognized as words.

Then, data become information when they constitute a content from several data, giving it a meaning and significance. Examples are text, images and videos.

Finally, information is transformed into knowledge when it becomes a justified true belief. Three conditions are thus necessary for an information to become knowledge: truth, justification and belief.

The scaling of these concepts is very relevant in our sense and treating them similarly is not effective, especially when integrating the warehousing technology. Therefore we proposed as illustrated in Fig. 1b, the paradigms of information warehousing and knowledge warehousing, corresponding to the notion of information and knowledge, respectively.

In the present study, we introduce the paradigm of information warehousing, which is a level up than data warehousing. While the latter stores data, the former holds information which is data with meaning and significance. This kind of information includes what the users communicate and exchange in social media platforms. An example of information in social media is a tweet.

4 Information Warehouse

An information warehouse is an infrastructure that would gather all types of information (text, image and video) in one repository to support decision making and knowledge discovery. This information can be found in various sources such as newspapers, scientific papers and social media, etc.

In this paper we propose a generic information warehouse framework for social media (Facebook, Twitter, Instagram, Snapchat, etc.). This warehouse would gather all these platforms' information whether it is text, image or video.

The proposed framework in Fig. 2 is composed of the following layers:

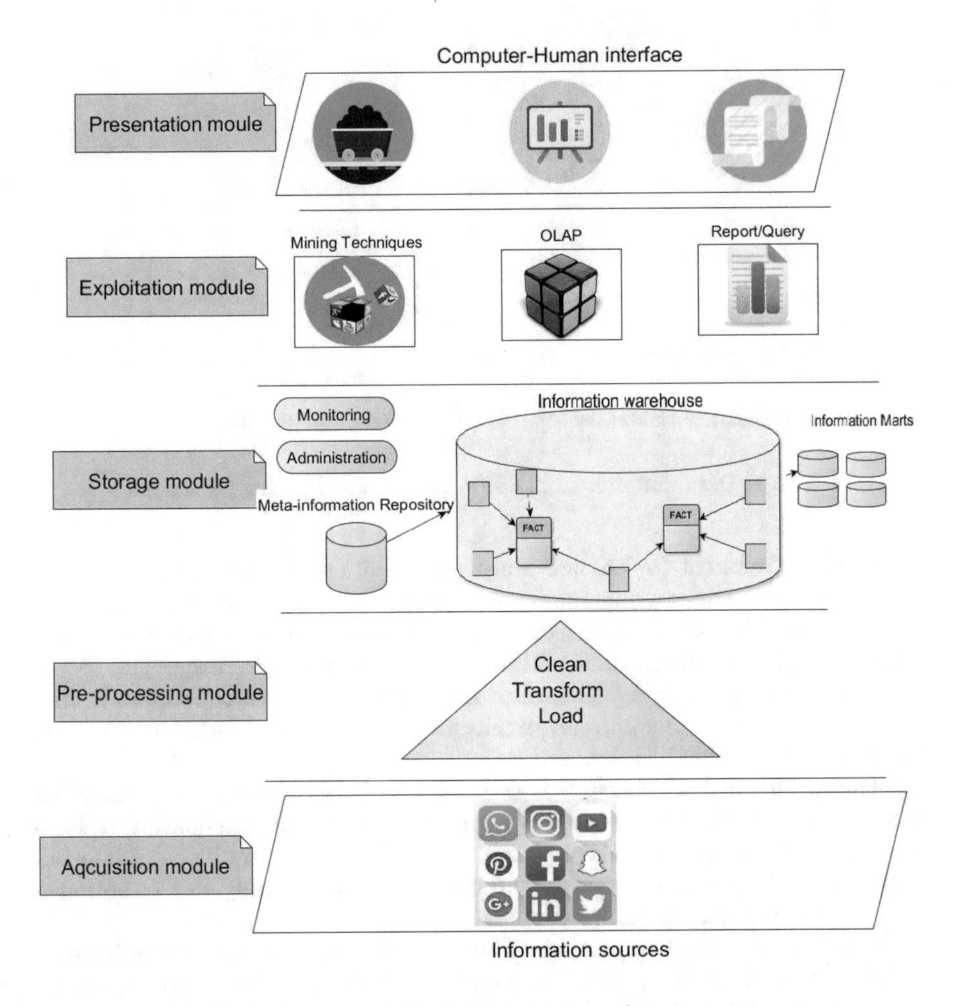

Fig. 2. Architecture of the information warehouse prototype

Information Source Layer: This is the bottom layer of the architecture. It gathers all social media sources such as Twitter, Facebook and Instagram.

ETL (Extract, Transform, Load) Layer: The ETL tool is one of the key layers in the warehouse. As social media data can be unstructured, the ETL is responsible for capturing and cleaning the raw information stream, and transforming it into a format congruent with the target database before feeding it automatically into the multidimensional model.

Information Warehouse Layer: This is the core layer of the architecture where the information is stored in data dimensions and information fact tables.

It also includes information marts and a meta-information repository, which is a summary of information in the warehouse to help for its exploration and exploitation. It contains information on the state, evolution, location, etc., of information in the warehouse as well as information about the warehouse itself.

Information Analysis Layer: The analysis layer includes all of the analysis tools such as OLAP (OnLine Analytical Processing) tools, query tools, reporting tools, and data mining techniques.

Presentation Layer: This layer consists of a front-end tool that is responsible for presenting the obtained results from the analysis layer in the most appropriate way. It could be a computer-human interface (CHI).

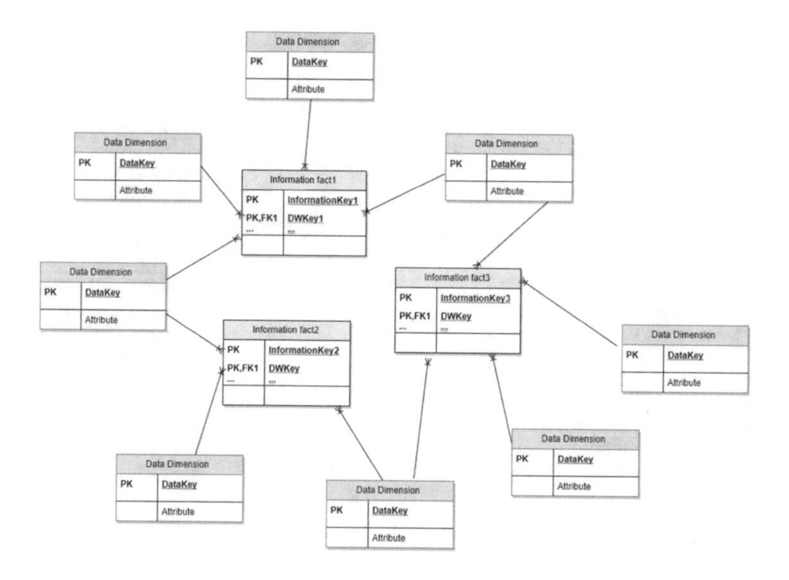

Fig. 3. Multidimensional schema of an information warehouse

In the multidimensional schema, information represents the fact while dimensions represent the data by which the information is described and constituted.

Another important characteristic of an information warehouse's schema is the fact that data dimensions can be shared between different information fact tables. Indeed, these dimensions can be used alongside other dimensions in order to represent other information facts. This characteristic is illustrated in Fig. 3.

5 Implementation of an Information Warehouse for Tweets

Twitter is one of the most popular social media platforms. It allows the users to share information mainly through messages. Its main advantage is that almost every user's tweets are public and exploitable through Twitter APIs.

The information source is represented by Twitter APIs stream where tweets are extracted using the package *twitteR* in R language. They are then cleaned by converting their text to lower case and removing undesirable content such as stop words, special characters, urls, punctuation, hashtags, extra whitespace and numbers.

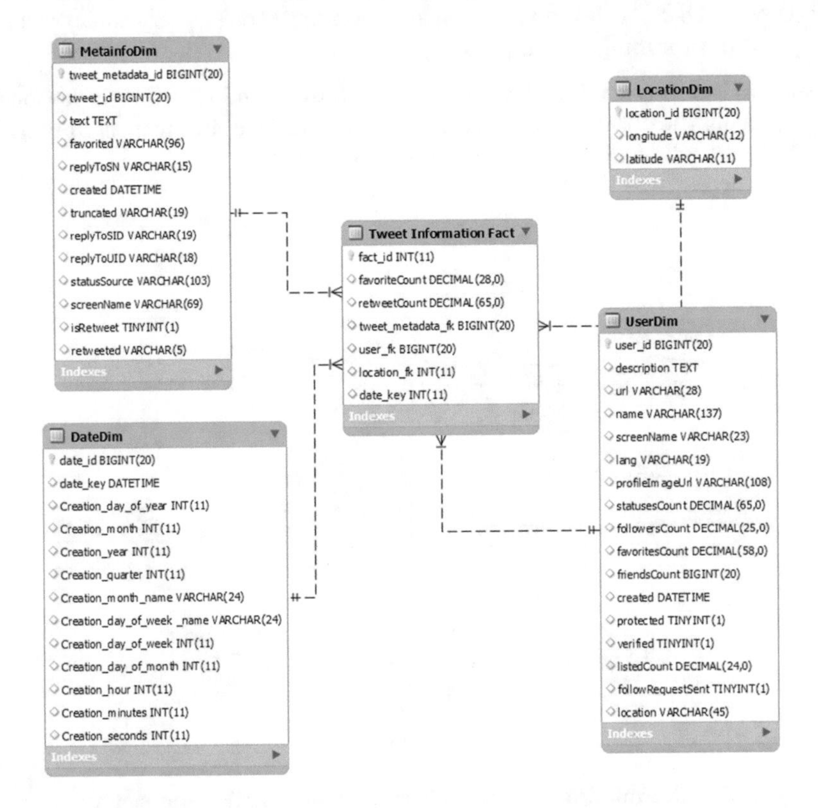

Fig. 4. Information warehouse schema for Twitter

To build a multidimensional schema for Twitter, we need to identify Twitter's data model fields that can be mapped into facts and dimensions. Numeric attributes such as the counters of a user's profile (number of followers, etc.) and the tweet record (number of retweets, etc.) can be treated as measures in the OLAP cube schema. The rest of the attributes are descriptive characteristics about a user and his tweet, and can thus be considered as dimensions or dimensions' hierarchies [6]. We therefore propose a star schema that evolves around a tweet fact table as illustrated in the Fig. 4.

Tweet Fact: A tweet fact table contains identifiers of the tweets' fact stored in the warehouse and foreign keys to dimension tables. It also includes the following measures: RetweetCount (the number of times the tweet was retweeted) and FavoriteCount (the number of times the tweet was favoriated).

Meta-information Dimension: This dimension includes the tweet text itself and all of the characteristics and attributes by which it is described such as its creation date and the name of its author.

User Dimension: This dimension includes information about the tweet's author such as his name, description, language, etc., and information about their account such as its creation date, if it is verified or protected, etc. Note that a single user can be the author of different tweets in the tweet fact table.

Location Dimension: This dimension gives information about the geographic coordinates of the place from which the tweet was posted.

Date Dimension: This dimension is responsible for attributing a date key to each record in the information warehouse. It has attributes organized in hierarchies, from the finest level (second) to the most general one (year). It allows the summarization of information according to date.

6 Tweets' Content Clustering

In order to analyze the flow of information collected in the warehouse, we've chosen a data mining technique which is clustering. Clustering is an unsupervised classification method that gathers tuples in clusters depending on their degree of similarity. The selected algorithm for this study is BSO-CLARA [7], which is based on the Bee Swarm Optimization method and k-medoids partitioning. It is proven to be effective for large-scale data.

6.1 Tweets Pre-processing

Tweets are often written in an informal style and consist of simple phrases, sentence fragments, ungrammatical text and sparse words. It can include any kind of alphanumeric characters as well as emojis, URLs, abbreviations, Internet slang and misspelled words.

Therefore, it is important to pre-process their content before starting any analysis for effectiveness. At the end of this process the number of tweets will decrease, as redundant and empty tweets will be removed.

The first step of pre-processing is to remove non-relevant content. In our case this includes: english stop words, emojis, URLs and users' notifications.

The second step is stemming, which consists of transforming all words to their base format. For example words like: *start*, *starting*, *started* are all considered as the same category of word *start*.

6.2 Tweets Representation

Tweets are represented in the vector space model using TFIDF, such as each tweet is represented by a vector of weights of size n; with n being the total number of terms in tweets corpus.

$$TFIDF(t, d, D) = TF(t, d) \times IDF(t, D) \tag{1}$$

where t is a term, d is a tweet and D is the total set of tweets.

$$IDF(t, D) = \frac{log \mid D \mid}{d_t} \quad (2)$$

where d_t is the number of tweets in which the term t occurs.

6.3 Similarity Measure

Cosine similarity measure is used to calculate the similarity degree between a tweet and a cluster's medoid in order to determine the cluster to which the tweet should be affected. A high cosine value indicates big similarity and hence a tweet is affected to the cluster that maximizes this measure.

$$cos(x, y) = \frac{x \cdot y}{||x|| \cdot ||y||} = \frac{\sum_{i=1}^{n} x_i y_i}{\sqrt{\sum_{i=1}^{n} (x_i)^2} \sqrt{\sum_{i=1}^{n} (y_i)^2}} \quad (3)$$

6.4 Fitness Function

The fitness function is given by the sum of the cosine similarity measure between every tweet and its corresponding cluster. The computed value is to maximize.

7 Experimentations and Results

For the implementation of our Twitter information warehouse we have used the following tools: Pentaho Data Integration (ETL), MySql (storage), R language and environment (tweets extraction and pre-processing), Java programming language (tweets' clustering).

Description of the Information Set. We have collected a total of 131487 tweets about the recent catastrophe that mainly affected California: the wildfires. These tweets were loaded in the information warehouse using the multidimensional schema presented in Sect. 5.

Intensive tests have been carried out in order to fix the empirical parameters of BSO-CLARA:

- Number of bees: 10
- Maximum number of iterations: 10
- Maximum number of chances when stagnation: 3
- Number of flips: 2
- Number of clusters: 12

Table 1 shows the obtained clusters such that each cluster is represented by its size (number of tweets) and its medoid which references a specific topic. The latter is extracted according to the medoid's terms.

Table 1. Clustering results

Cluster	Medoid	Size	Topic
1	"California wildfires: Crews search for traces of victims as fires rage on"	15558	Victims' search
2	"Marijuana Crop Is Burning in the California Wildfires"	26867	Marijuana farms burning
3	"Wildfires now up to 100 miles wide as death toll reaches 40"	4441	Wildfires progress and deaths' toll
4	"Nine dead as wildfires ravage northern Portugal, Spain"	1864	Spain and Portugal wildfires
5	"A lot of natural disasters are happening all over the world, Let them speak to us by reminding us that life is short"	3228	Natural disasters
6	"Three hurricanes Continuing deaths in Puerto Rico, Massive wildfires"	5474	Hurricanes and wildfires
7	"This is a good story to share this Odin's day"	4108	Random
8	"Northern California Wine Country Wildfires: Death Toll Rises to 35"	17505	Rise in deaths
9	"NFL decides to keep game in Oakland, to monitor air quality from wildfires"	3235	NFL games
10	"At Least 23 Dead, Hundreds Missing as Winds Fan California Wildfires"	10966	Deaths and missing people
11	"Not one word from Trump on the Ca wildfires. Not one! Is it because Ca votes democrat"	4068	Trump not reacting to wildfires
12	"Portugal wildfires kill at least 27; 4 dead in Spain"	29004	Deaths in Spain and Portugal

A graphic illustration of the clustering results is given in Fig. 5.

We observe that the tweets collection is partitioned into 12 different clusters. Among the uncovered topics we have: reports on deaths and missing people in the California wildfires and its progress, Spain and Portugal wildfires, Marijuana farms burning in California, Hurricanes and wildfires, NFL games. Another interesting and relevant topic is president Trump not reacting to the California wildfires at that time.

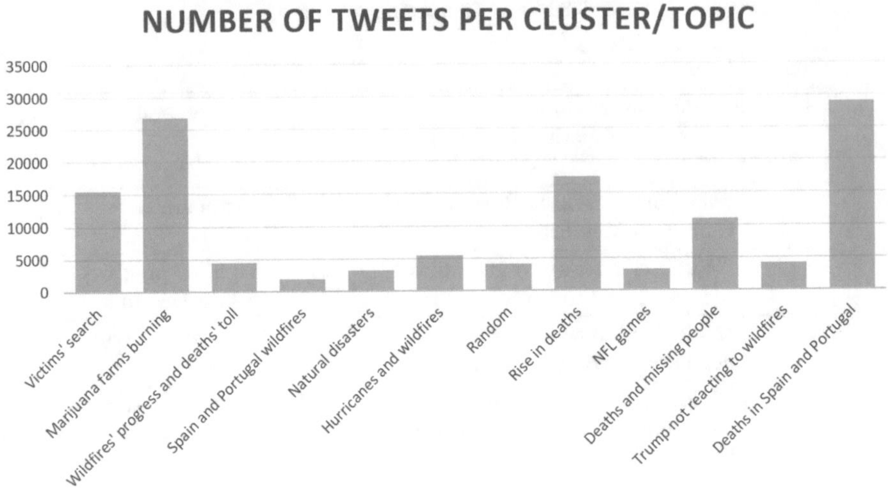

NUMBER OF TWEETS PER CLUSTER/TOPIC

35000
30000
25000
20000
15000
10000
5000
0

Victims' search
Marijuana farms burning
Wildfires' progress and deaths' toll
Spain and Portugal wildfires
Natural disasters
Hurricanes and wildfires
Random
Rise in deaths
NFL games
Deaths and missing people
Trump not reacting to wildfires
Deaths in Spain and Portugal

Fig. 5. Number of tweets per cluster/topic.

The results on this information set prove the effectiveness of the warehouse for information analysis.

8 Conclusion and Future Work

In this work, we presented a novel warehousing paradigm, namely information warehousing, for the storage and analysis of massive volumes of information. The paradigm is illustrated for social media where a generic information warehouse architecture is described and then implemented for the case of Twitter, supported by a multidimensional model. Finally, we performed a large-scale clustering analysis on the warehouse's tweets to uncover the discussed topics.

Our perspectives aim at improving the architecture components and implement it for other information sources such as newspapers. Also, performing further analysis on different sets of information (other tweets, newspapers, scientific papers, etc.) would confirm the performance and effectiveness of this paradigm.

References

1. Khan, J.: Universal information warehouse system and method, 11 May 2004
2. Holten, R.: Framework and method for information warehouse development processes, pp. 135–163. Physica-Verlag HD, Heidelberg (2000)
3. Post, A.R., Kurc, T., Cholleti, S., Gao, J., Lin, X., Bornstein, W., Cantrell, D., Levine, D., Hohmann, S., Saltz, J.H.: The analytic information warehouse (AIW): a platform for analytics using electronic health record data. J. Biomed. Inform. **46**(3), 410–424 (2013)

4. Kamal, J., Pasuparthi, K., Rogers, P., Buskirk, J., Mekhjian, H.: Using an information warehouse for clinical trials: a prototype. AMIA Ann. Symp. Proc. **2005**, 1004–1004 (2005)
5. Choo, C.W.: The Knowing Organization: How Organisations Use Information to Construct Meaning, Create Knowledge, and Make Decisions. Oxford University Press, New York (2006)
6. Rehman, N.U., Mansmann, S., Weiler, A., Scholl, M.H.: Building a data warehouse for twitter stream exploration. In: 2012 IEEE/ACM International Conference on Advances in Social Networks Analysis and Mining, pp. 1341–1348 (2012)
7. Aboubi, Y., Drias, H., Kamel, N.: BSO-CLARA: bees swarm optimization for clustering large applications. In: Prasath, R., Vuppala, A., Kathirvalavakumar, T. (eds.) Mining Intelligence and Knowledge Exploration. Lecture Notes in Computer Science, vol. 9468. Springer, Cham (2015)

An FSM Approach for Hypergraph Extraction Based on Business Process Modeling

Bouafia Khawla(✉) and Bálint Molnár(✉)

Information Systems Department, EötvösLoránd University, ELTE, PázmányPétersétány 1/C, Budapest 1117, Hungary
{bouafia, molnarba}@inf.elte.hu

Abstract. The high way of presentation of entities or activities and relations which enterprises needed for their higher success and the powerful descriptive and representation method because of their complexity needed more and more in our days. In this paper, an approach for Business Process Modelling (BPM) using Hypergraphs representation based on Finite State Machines (FSM) will be presented.

The main goal motives this work is the using of hypergraphs as a flexible mathematical structure describing Information System (IS) from various viewpoints to express the best graphical way for Business Process (BP).

The model proposed use FSM as an automated formal model by defining concepts and components of the hypergraph elements based on different FSM patterns (simple, complex) to get representations can be analyzed by either using more traditional tools as logic and inference rules or by a set of tools belonging to data science later.

Keywords: Hypergraph · FSM · BP · IS · IS modelling · Information Patterns

1 Introduction

Nowadays, companies and enterprises need to make available their business and data, in a format that designers can use for models to represent BP. Modelling and analyzing in BP is becoming a challenging research topic because modelling tools, do not offer enough formal specification of time constraints at the earlier phases of design.

Hypergraph model has been widely used such as solving the problems of community detection [1], classification [2], social networks [3] and so on, FSM also is largely used in various fields such as: networks [4] BPM [5], But the best of our knowledge, none has considered to model BP mapped in FSM to hypergraphs representation.

This paper organized as follows. First, we start with an introduction. In Sect. 2 we will survey application areas and literature review, the mathematical background of hypergraphs, formal FSM background and their important patterns. Section 3 gives our proposal. Finally, conclusions and future work will be shown in the last section.

O. Demigha et al. (Eds.): CSA 2018, LNNS 50, pp. 158–168, 2019.
https://doi.org/10.1007/978-3-319-98352-3_17

2 Literature Review

An IS supports BPM and is usually tightly coupled to another IS, a standard way to model BP is the application of BPM methods [6, 7]. Enterprises aspiring for high success need to present entities, activities and relations with a high granularity.

The paper [8] aims to review main current techniques in translation from graph-oriented workflow languages to block-oriented ones in general, and from Business Process Model Notation (BPMN) to Process Execution Language (BPEL), that helps to clarify major existing achievements. An automated approach described in [9]. for synthesizing a hierarchical state machine [10] from UML activity diagram that specifies a BP referencing a stateful object to discover an object life cycle. Another approach [11] deriving more precise process models by leveraging a process to petri net compiler, to produce a most precise verifiable Petri net process model. A proposition of an algorithm extended Petri networks to the linear hybrid automata presented in [12].

A functional syntactic, vertical and exogenous approach described to transform a BPEL programs to an abstract specifications explained in [13]. Hypergraphs have been widely and deeply studied with different names and its instances which can be found in [14]. To solve problems which arise in areas such as propositional logic.

2.1 Hypergraph

As the literature proves it, hypergraphs as representational structures are apt to system modeling, analysis and specification, especially to information and software systems Moreover, formal tools consistency checking is available by test algorithms. Hypergraph is a structure describe complex relationships can be explored among models during analysis and design of IS, it is a generalized graph theory plays a very important role in discrete mathematics [15]. We start with its basic definitions (Fig. 1):

Definition 1. Hypergraph H is pair (V, E) of a finite set $V = \{v_1, v_2, \ldots, v_n\}$ and a set E. The elements of V are called vertices (nodes) and the elements of E are called edges [16]. Formally, let $G(V, E, w)$ denote a hypergraph, where: V denotes a finite set of nodes v, E denotes the set of hyperedges e, w is a weight function defined as $w: E \rightarrow R$. Each hyperedge $e \epsilon E$ is a subset of V and is assigned a positive weight $w(e)$ [17].

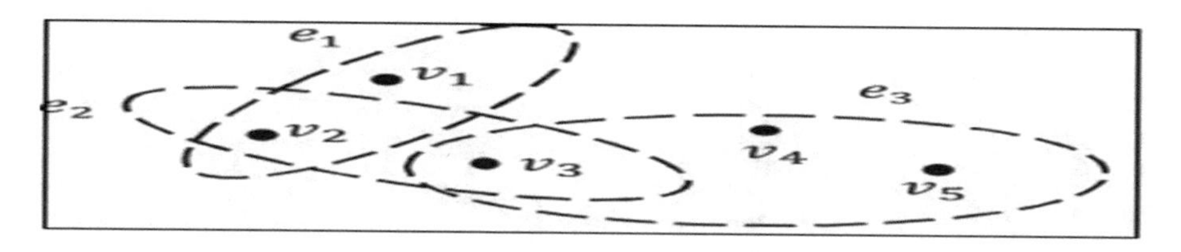

Fig. 1. Hypergraph structure

Definition 2. A generalized or extended hypergraph is that some of the hyperedges are denoted – in certain cases – as vertices, thereby a generalized hyperedge e may consist of both vertices and hyperedges as well. The hyperedges that are contained within the hyperedge e should be different from hyperedges *e* [16].

Definition 3. Directed hyperedge (hyperarc) is an ordered pair $E = (X, Y)$ of (possiblyempty) disjoint subsets of vertices; X the tail (E) of E while Y is the head $H(E)$ [25].

A hypergraph is a generalization of an ordinary graph, let $G(V, E, w)$ denote a hypergraph, where V denotes a finite set of nodes v, E denotes the set of hyperedges e, w is a weight function defined as: $w : E \Rightarrow R$. Each hyperedge $e \in E$ is a subset of V and is assigned a positive weight $w(e)$ [17].

Definition 4. A forward hyperarcs (F-arc) is a hyperarc $E = (T(E), H(E))$ with $|T(E)| = 1$, A backward hyperarcs B-arc, is a hyperarc $E = (T(E), H(E))$ with $|H(E)| = 1$. A Forward-backward graph (or BF-hypergraph) is a hypergraph whose hyperarcs are either B-arcs or F-arcs [18].

Definition 5. A directed path P_{sd} from a source s to a destination d in $H = (V, E)$, is a sequence of nodes and hyperarc $p_{sd} = v_1 = s, E_{i_1}, v_1, E_{i_2} \ldots, E_{i_q}, E_{q+1} = d$ where: s $\epsilon T(E_{i_1})$, $t \epsilon H(E_{i_q}).P_{sd}$ is said to be a cycle, when $t = s$ [19].

Definition 6. Let $H = (V, E)$ be a hypergraph, s and t be two distinguished nodes, the source and the sink respectively. A cut $Q_t = (V_s, V_t)$ is a partition of V into two subsets V_s and V_t such that $s \epsilon V_S$ and $t \epsilon V_t$ given the cut Q_{st} [20].

2.2 FSM Representation

A FSM is a well-known model in formal system specification, it is also a standard model used in the mathematical foundation of computer science. This concept is defined and discussed in innumerable books and papers. Such as the Gomez paper [21]. This section gives basic concepts of FSM that are used in the rest of the paper.

Definition 7. A FSM is composed of a set of states, set of actions and set of transitions labeled between states. There are initial state final ones [22]. It is a quintuple structure $\langle \Sigma, S, s_0, \delta, F \rangle$ where: Σ is the input alphabet (finite non-empty set of symbols), S: finite non-empty set of states. s_0 an initial state, included in S, δ: the state transition function: $\delta : S \times \Sigma \rightarrow S$ and F is final states set (possibly empty) [13].

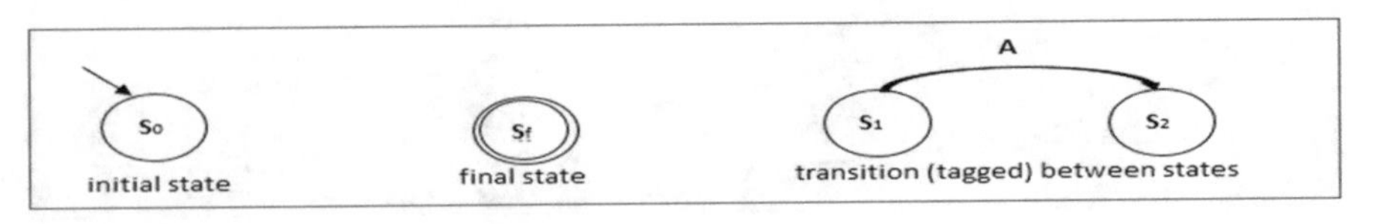

Fig. 2. Initial state, final state, transition (tagged) between states

The semantic, syntactic and structural description of FSM presented in [23] which prove that it easy to understand by users, popular and suitable tools for describing reactive behavior, dynamic systems and performance control services (Fig. 2).

2.3 Basic FSM Patterns

Workflows are a well-established concept to formalize BP and support their execution by a workflow management system. Workflows are "the automation of a BP, in whole or part, during which documents, information or tasks are passed from one participant to another for action, according to a set of procedural rules" [24]. In this section we will present various FSM patterns of which define the basic modeling patterns of BP:

Sequence: is defined as being an ordered series of activities, with one activity starting after a previous activity has completed. for example: in case of 3 activities *A*, *B* and *C*: activity *B* is available after activity A is finished his execution. In the same way, activity *C* is available once activity *B* is complete (Fig. 3).

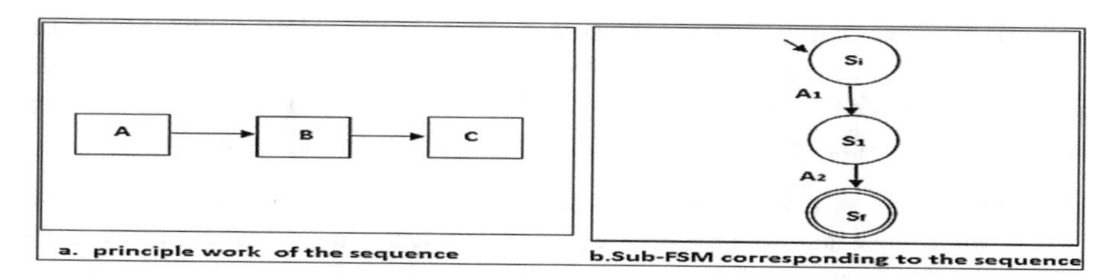

Fig. 3. Pattern of Sequence

Parallel Split: is defined that two or more activities will start at the same time and it is a point in the process where a simple control link is split into multiple control links running in parallel. Generally, parallelism, when finishing the execution of activity *A*, the control link goes to activities *B* and *C* at the same time (Fig. 4).

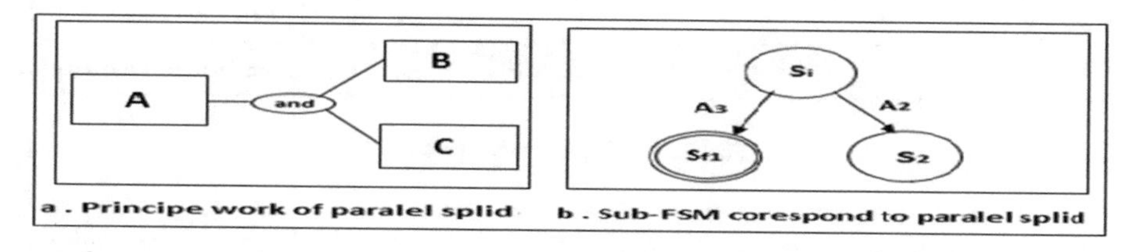

Fig. 4. Parallel split pattern

Exclusive Choice: defined as check point allows to make a choice of the execution path, the process path is chosen using a decision or condition taken at runtime based on information or management data. unlike Parallelism, only one activity in the control layer is invoked. Generally, after activity *A* is completed, there is a selection for the control flow to decide on the continuity of the process: either activity *A* or activity *B* (Fig. 5).

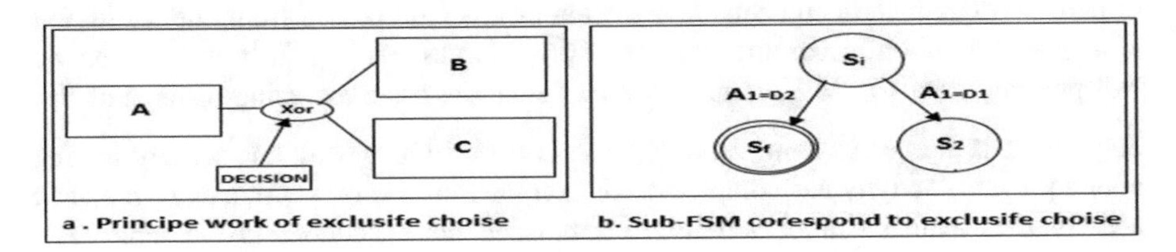

Fig. 5. Exclusive choice pattern

Multiple Choice: is differs from exclusive choice pattern in that the multiple-choice pattern allows from one to all the alternative paths may be chosen at performance time. Technically, a Multiple-Choice pattern may allow zero paths chosen, but this could be considered an invalid situation where the process flow stops unexpectedly.

Multiple Merge: is a location in the process where multiple paths merging, but without any control and is a point in the process in which one or more branches of the control thread join without the need for synchronization. Generally, At the time of the completion of *A* or *B* or both, the process continues with activity *C* (Fig. 6).

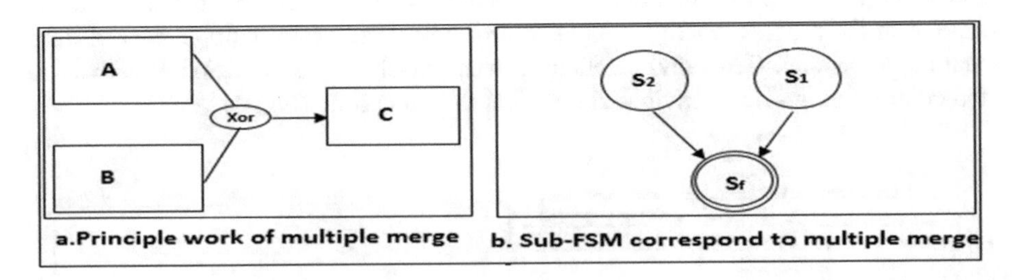

Fig. 6. Multiple merge pattern

Cycle: is a mechanism repeat the same instructions multiple times with conditions. Generally, we execute from top to bottom and, once at the end of the loop, it starts again at the first instruction, then it starts again and repeat the same way (Fig. 7).

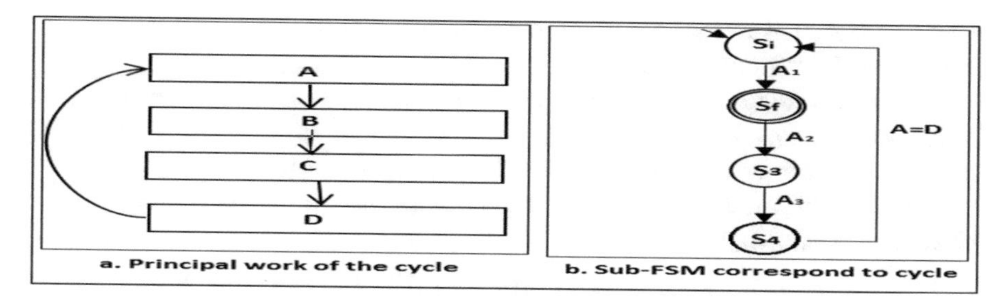

Fig. 7. Cycle pattern

3 Proposed Method

FSM and Hypergraph are completely different representation which are aimed at diverse things. But we will try to suggest our own approach of how the problem of mapping between them could be solved. In this work we will describe a hypergraph to model all types of element and patterns presented by FSM.

Firstly, we will present FSM elements with hypergraph elements, the hypergraph is used to model diverse types of objects or elements (states transitions). We suppose that each kind of vertices corresponds to a state, and each kind of hyperedge corresponds to relation between them, hyperedge can be labelled depend on transitions in FSM.

Secondly various FSM patterns will have represented by hyperedge can be divided into sub-sets of present a special pattern. After we will explain the different concepts with details and describe theoretically how and What are the rules of mapping.

Part One: as we have outlined previously, the hypergraph should to create a flexible structure express diverse types of sub-FSM. This hypergraph can have represented as nodes embedded into hyperedges and hyperarcs along with their interrelationship to describe the roles that are played by nodes and hyperedges. The direction of the hyperarc shows the direction of transition where input and output are vertices (head and tail of hyperarc) which present states. The definition given above explain:

Definition 8. The FSM's elements are represented by hyperedge h_i where $h_i \in E_{FSM}$ the set of hyperedges. The elements of h_i are:

- The finite set of states S are represented by n vertices $V_s = \{v_1, v_2, \ldots v_n\}, V_s \subseteq S$
- The finite set of transitions T can be described by hyperarcs belonging to A_T where $\vec{h}_i \in A_T, A_T \subseteq T = \{\vec{h}_1, \vec{h}_2, \ldots, \vec{h}_m\}$. The variables, place-holders or attributes can be denoted as a label of a transition that is described by variables that belong to *an attribute type, $A_{ttr} = \{T_1, T_2, \ldots T_j\}$ the set of types $h_i \in V_{i_a} \cup V_{inter_{ba}} \cup V_{f_c}$, where $a, b, c, d \in \mathbb{N}$.

NB: vertices can be labelled with integer by $n \in \mathbb{N}$ or denominated with arbitrary strings that can be represented by a binary string $s \in \{0, 1\}*$ (variable of domain).

After we mapped places and transitions onto concepts and structures of hypergraph, we will present the various their sub-sets:

The set of vertices $V \subseteq S$ can be divided into three sub-sets of vertices depend on FSM states: $V_i, V_{inter}\, and\, V_f$ where: V_i describes the initial state, V_{inter} defines the intermediate states V_f denotes the final(s) state(s) of the FSM.

The set of arcs (directed edges) A is partitioned into two subsets: A_{cond} and A_{sim} where: $h_i \in A_{sim} \cup A_{cond}$:

- $\boldsymbol{A_{sim}}$: represents a simple transition (labelled without condition), $h_i \in A_{sim}$ can be used between vertices where: $A_{sim} \subseteq V_{tail} \times V_{head}$ where:

$$V_{tail} = \begin{cases} \{v_k | v_k \in V_i\} XOR \\ \{v_k | v_k \in V_{inter}\} \end{cases}; V_{head} = \begin{cases} \{v_k | v_k \in V_{inter}\} XOR \\ \{v_k | v_k \in V_f\} \end{cases} \tag{1}$$

- $\boldsymbol{A_{cond}}$: represents a transition labeled with condition used between vertices where: $A_{cond} \subseteq V_{tail} \times V_{head}$ where:

$$V_{tail} = \begin{cases} \{v_k | v_k \in V_i\} XOR \\ \{v_k | v_k \in V_{inter}\} XOR \\ \{v_k | v_k \in V_f\} \end{cases}; \quad V_{head} = \begin{cases} \{v_k | v_k \in V_i\} XOR \\ \{v_k | v_k \in V_{inter}\} XOR \\ \{v_k | v_k \in V_f\} \end{cases} \tag{2}$$

NB: Eqs. (1), (2) not true if both vertices $(V_{tail} \times V_{head})$ be finals, i.e.:

$$V_f \times V_f \cap A_{cond} = \emptyset; V_f \times V_f \cap A_{sim} = \emptyset \tag{3}$$

Definition 9. Labelled Hypergraph is a generalized hypergraph that can be extended by some functions and operations: $H = (V, E, label)\ where\ |V| < \infty, E \subseteq V \times V$, $|E| < \infty$. i.e. V, E finite, $\Sigma = \{\sigma | \sigma \in \{0, 1\}*\}$ set of labels, as binary strings.

- $\boldsymbol{label : V \rightarrow \Sigma}$: is a function that assign a label by function $label(v)$ to vertex v so that $L_{node} \subseteq \Sigma; L_{edge} \subseteq \Sigma$.
- $\boldsymbol{label_{node} : V \rightarrow L_{node}}$: where L a set of labels, it is a vertex labeling function.
- $\boldsymbol{label_{edge} : E \rightarrow L_{node}}$: where L set of labels, it is an edge labeling function.
- $\boldsymbol{source_E : E \rightarrow V}$: functions return the source vertex of an edge.
- $\boldsymbol{target_E : E \rightarrow V}$: function return the source and target vertices of an edge E.
- $\boldsymbol{att : E \rightarrow V}$: Attribute assignment function.
- $\boldsymbol{source_{Attr} : attr \rightarrow L_{node}}$: the vertex that owns the attribute is returned.
- $\boldsymbol{target_{Attr} : Attr \rightarrow D}$: data values of attributes are yielded, D the set of data.

The set of data values D can be grasped (efficiency of the representation is left out of the investigation) again as vertices within the hypergraph and it can be interpreted as variables. Over D as the values of sets of variables, sets of operations (OP) can be defined that can be used to describe constraints and rules within formulas.

Definition 10. Limb is either the head or the tail of hyperarc and designated by $\vec{e}_i^{\mp}$.

The set of condition represented by A_{cond} partitioned into two sets of its edges and vertices by the values of attributes as they belong to either T_{bool} or T_{vdom} respectively:

$$Attr \supseteq (Attr_{Bool} \cup Attr_{vdom}); \{Attr_{Bool} \cap Attr_{vdom}\} = \emptyset \tag{4}$$

- T_{bool}: can take tree logical value as variables (Boolean): true false or null where:

$$h_{Bool} \epsilon (A_{cond} \subseteq V_{tail} \times V_{cond}) \text{ and } attr(h_{Bool}) \in T_{Bool} = \{true, false, true\} \tag{5}$$

- T_{vdom}: represents the type that a variable v can take up as value, i.e. the variable attribute can be valuated where by $\forall$ *vvariable* $Dom(v) \subseteq T_{vdom}$, T_{vdom} can be defined as string type $T_{vdom} = \{s \epsilon \{0.1\}*\}$ of binary strings; or $X = \{i \leq n \epsilon \mathbb{N}\}$ a finite alphabet and $T_{vdom} = \{s \epsilon \{x_i\}*\}$, finite series of alphabet $\forall s, \exists n \epsilon \mathbb{N}, s \leq n$.

NB: The relation between the two transitions are:

$$A_{sim} \cup A_{cond} = A \text{ and } A_{sim} \cap A_{cond} = \emptyset \tag{6}$$

Part two: in this step we will mapped FSM patterns presented before in hyperedges:

Definition 11. Each FSM's pattern can be represented by hyperedge $h_i \epsilon E_i$, where E_i the set of hyperedges, E can be divided into sub-sets to present patterns:

- E_{seq}: present sequence pattern which can use all type of vertices (the tail of the first hyperarc). It can start from any $v_k \epsilon V_i$ if the sequence starts from the first vertex in the ordering relations, or $v_k \epsilon V_{inter}$ if not, but cannot start from *any* $v_k \in V_f$. It can be finished (the head of the last hyperarc) with any vertex $v_{lm} \in V_f$ *where* $|\{v_{l1}, \ldots, V_{lr}\}| \leq n, n \in \mathbb{N}$ (typically $n < 3$). The directed hyperarc used to express relationship between vertices V_i, V_{inter}, V_f, use only simple hyperarcs $h_{seq_1}, h_{seq_2}, \ldots, h_{seq_p} \epsilon A_{sim} \subseteq A_{seq}$ between vertices because is a simple structure. The sequence vertices order expressed with pair (tail and head of hyperarc), where the head of the first will be the tail of the next, till the sequence reaches the last head.

$$\vec{h}_i^- \cap \vec{h}_i^+ = v_{ij} \text{ where } v_{ij} \in (h_i, i), v_{ij} \epsilon (h_i, j) \tag{7}$$

The function *label* can designate the order of vertices belonging to an instance of FSM through numbering and labelling the hyperarcs that connect them, and being the element in a $h_{seq} \epsilon E_{seq}$. If the sequence is complex one, it can be represented with another hyperedge $h_m \in E_{seq}$ if that vertices is not V_f or no cycle from *Vf*.

- E_{ps}: describe parallel split pattern, It can represent a parallel vertices included in the same hyperedge: $h_{par} \epsilon E_{ps}$ various parallel vertices $\{v_1 v_2, \ldots v_n\} \subseteq E_{ps}$ where $V_j \neq V_i$ if $i \neq j \epsilon \mathbb{N}$ i.e. only V_{inter} *and* V_f with same tail from the main vertex of different hyperarcs $h_{ps} \in E_{ps}$, without connection between them; where:

$$\forall k_i, k_l, i \neq l, h_{ki} \cap h_{kl} = \emptyset \; and \; h_{ps} \cap h_{ki} = v_{pc}, \forall k_i \epsilon \mathbb{N}, k_1, k_2, \ldots, v_{start_{ps}} \epsilon \vec{h}_{ps}^{\mp} \quad (8)$$

This structure uses only simple hyperarcs connected with another type of hyperedges exists if that vertices are not V_f or no cycle from V_f.

- E_{ec}: represent exclusive choice pattern, a hyperedge $h_{ec} \epsilon E_{ec}$ contain all vertices $\{V_1, V_2, \ldots, V_n\} \subseteq E_{ec}$ i.e. it can contain only two V_{inter} and V_f there are no arcs or hyperedges embedded into any arbitrary $h_{ec_i} \epsilon E_{ec}$ it have the same tail from the main vertex of different hyperarcs with no connection because it is a simple structure where

$$\forall k_i, k_l, i \neq l, h_{ki} \cap h_{kl} = \emptyset \, and \, h_{ec} \cap h_{ki} = v_{ec_{ps}}, \forall k_i \epsilon \mathbb{N}, k_2, \ldots, v_{ec_{ps}} \epsilon \vec{h}_{ec}^{-} \quad (9)$$

To connect the main vertex of the h_{ec} to other vertices we can designate the conditions $A_{cond} \subseteq A_{ec}$ with values that will be assigned to element of conditions: T_{Bool} *exceptnull* (value) because it cannot take up, and attribute values from T_{vdom} are allowed to describe the available options, to denominate the choice thereby can be connected to another structures represented

- $h_{mc} \epsilon E_{mc}$: represent the pattern of multiple choice so that all vertices hyperedge contained in h_{mc} represent different states of exclusive choice of FSM, it can contain only two types of vertices *VinterandVf* have the same tail from the main vertex of $h_{mc} \in E_{mc}$ without connection because it is a simple structure where:

$$\forall k_i, k_l, i \neq l, h_{ki} \cap h_{kl} = \emptyset \, and \, h_{ec} \cap h_{ki} = v_{mc}, \forall k_i \epsilon \mathbb{N} k_1, k_2, \ldots, v_{mc} \epsilon \vec{h}_{mc}^{-} \quad (10)$$

To connect the main vertex and h_{mc} use only $h_{mc_i} \epsilon A_{cond} \in$ to express conditions $A_{cond} \subseteq A_{ec}$, attributes can be used: T_{bool} and *includsnull* (difference between E_{mc} and E_{ec}), moreover attribute values to describe available options, thereby $h_{mc} \in E_{mc}$ can be connected to another structures represented by hyperedge that is allowed in the case if those vertices belong *neitherto* $v \in V_f$ nor a cycle from $v_{cycle} \in V_f$.

- E_{mm}: present the multiple merge, this patterns is more complicated than to others because of multiple states merging vertices are included in the same hyperedge, it can contain all type vertices exists $\{V_1; V_2 \ldots . V_n\} \subseteq E_{ps}$ which should have the same head from different vertices merged *hmm* $\in$ *Emm* to the main vertex which can be only *VinterorVf* and there is no connection merged vertices; where:

$$\forall k_j, k_l, i \neq l, h_{jk} \cap h_{lk} = \emptyset \; and \; h_{i_{mm}} \cap h_{jk} = \emptyset, \exists k_i \in N \, k_1, \ldots k_2 \quad (11)$$

This structure uses only the simple hyperarc to $A_{sim} \subseteq A_{mm}$ connect E_{mm} and the main vertex (the main tail). That main vertex can be connected too with another type of hyperedges exists if it is not *Vf* or no cycle from *it*.

- $\boldsymbol{E_c}$: present cycle pattern which is a simple structure presented only with only a hyperarc $\boldsymbol{h_{ec}} \in \boldsymbol{E_c}$ from hyperedge which presented the head $\boldsymbol{E}$ or vertex $\boldsymbol{V}$ to the tail which can be the same as the head if it is presented with a self-transition in FSM will be described the other case that the tail can be another hyperedge $\boldsymbol{E}$ or vertex **V**.

4 Conclusion

In this paper, we proposed a describing hypergraph as representation for a FSM, the suggested method takes advantages of the basic properties of generalized hypergraphs. there are some distinguished features:

- The structure of the hypergraph can be interpreted as an ontology, thereby it opens the way for semantic methods and reasoning. And considers the time constraints and can be used to analyze the system, and to check the conformance, compliance, and consistency of the set of models.
- It would be a typical specification for a coded solution where we assume that during the coding we find an implementation and an example and will benefit from results of experiments to show its merits.

The ontology as an extra layer on the architecture describing hypergraph provides logical tool set for formal verification, namely the exploitation of description logics.

This approach can also be considered as a formal background to analyse and design another complexes BP models or complexes IS and shows the hypergraph-based approach offers the chance to apply further mathematical tools for assistance in design.

Acknowledgement. The project has been supported by the European Union, co-financed by the European Social Fund (EFOP-3.6.3-VEKOP-16-2017-00002).

References

1. Garcia, J.O., et al.: Applications of community detection techniques to brain graphs: algorithmic considerations and implications for neural connection. Proc. IEEE **99**, 846–867 (2018)
2. Wang, S., et al.: Spectral locality constrained elastic net hypergraph for hyperspectral image clustering. Int. J. Remote Sensing **38**, 7374–7388 (2017)
3. Amato, F., et al.: Centrality in heterogeneous social networks for lurkers detection: approach based on hypergraphs. Concurrency Comput. **30**(3), e4188 (2018)
4. Pola, E.D.S., Benedetto, M.D.D., Pezzut, D.: Design of decentralized critical observers for networks of FSM: a formal method approach. Automatica **86**, 174–182 (2017)
5. Borger, E., Fleischmann, A.: Abstract State Machine Nets: Closing the Gap Between BP Models and Their Implementation, 1st edn. ACM, Kiel (2015)

6. Molnár, B., Béleczki, A., Benczúr, A.: An ISM based on graph theoretic background. J. Inf. Telecommun., 1–23 (2017)
7. Kő, A., Ternai, K.: A development method for ontology-based BP. In: E Challenges e-2011 Conference, Florence, Italy (2011). http://scholar.google.com/scholar?hl=en&q=Kő,+A.,+&+Ternai,+K.+(2011).+A+development+method+for+ontology+based+business+processes.+eChallenges+e-2011+conference,+Florence,+Italy
8. Nguyen, B.T., Huu, D., Thanh, T.: Translation from BPMN to BPEL, Current Techniques and Limitations. ACM, USA (2014)
9. Etouati, Y., Yeddes, M., Alouane, N.H., Alla, H.: From extended time petri network to linear hybrid automata for systems analysis. In: IEEE International Francophone Conference of Automatic CIFA 2015, pp. 01–04 (2015)
10. Eshuis, R., Van Gorp, P.: Synthesizing object life cycles from BP model. Softw. Syst. Model. **15**(1), 281–302 (2016)
11. Heinze, T.S., Amme, W., Moser, S.: Static analysis and process model transformation for an advanced BP to Petri net mapping. Softw. Pract. Exp. **48**, 161–195 (2017)
12. Bouarioua, M.: A graph-based transformation approach for generating Petri nets models that can be parsed from UML diagrams. Ph.D. thesis (2013)
13. Bouafia, K., Molnár, B., Khebizi, A.: Functional approach for transformation to abstract specifications for Web services from BPEL programs characteristics of the approach. In: Proceedings of WSPS-4, Pecs, Hungary (2017)
14. Raman, M.G., Somu, N., Kirthivasan, K., Sriram, V.S.: A hypergraph and arithmetic residue-based probabilistic neural network for classification in intrusion detection systems. Neural Netw. **92**, 89–97 (2017)
15. Cui, K., Yang, W., Gou, H.: Experimental research and finite element analysis on the dynamic characteristics of concrete steel bridges with multi-cracks. J. Vibroeng. **19**(6), 4198–4209 (2017)
16. Bretto, A.: Hypergraph Theory: An Introduction, pp. 111–116. Springer, Heidelberg (2013)
17. Li, D., et al.: Link prediction in social networks based on hypergraph. In: Proceedings of the 22nd International Conference on WWW, pp. 41–42. ACM, USA (2013)
18. Tran, T., Scaparra, M.P., O'Hanley, J.R.: A hypergraph multi-exchange heuristic for the single-source capacitated facility location problem. Eur. J. Oper. Res. **263**(1), 173–187 (2017). https://doi.org/10.1016/j.ejor.2017.04.032
19. Ausiello, G., Italiano, G.F., Laura, L., Nanni, U., Sarracco, F.: Structure theorems for optimum hyper paths in directed hypergraphs. In: Proceedings of the 2nd International Symposium on Combinatorial Optimization, ISCO 2012, vol. 7422, pp. 1–14 Springer (2012)
20. Chekuri, C., Xu, C.: Computing minimum cuts in hypergraphs. In: Proceedings of the Twenty Eighth Annual ACM-SIAM Symposium on Discrete Algorithms (SODA 2017), pp. 1085–1100. Society for Industrial and Applied Mathematics, USA (2017)
21. Gomez, M.: Embedded systems programming feature, vol. 13, no. 13 (2013)
22. Walkinshaw, N., Taylor, R., Derrick, J.: Inferring extended FSM models from software executions. Empir. Softw. Eng. **21**, 811 (2016)
23. Bouafia, K, Molnár, B.: Adaptive case management and dynamic bp modeling a proposal for document-centric and formal approach. In: 12th AIS 2017 (2017)
24. Minor, M., Bergmann, R., Görg, S.: Case-based adaptation of workflows. Inf. Syst. **40**, 142 – 152 (2014)
25. Molnár, B., Benczúr, A.: Facet of modeling web IS from a document-centric view. Int. J. Web Portals (IJWP) **5**(4), 57–70 (2013)

An Automatic Muscle Activation Detection Using Discrete Wavelet and Integrated Profile: A Comparative Study

Ahlem Benazzouz(✉) and Zine-Eddine Hadj Slimane(✉)

Biomedical Engineering Research Laboratory, Faculty of Technology, University Aboubekr Belkaid, Tlemcen, Algeria
ahlembenazzouz@gmail.com, hadjslim@yahoo.fr

Abstract. The surface electromyogram (SEMG) is an electrophysiology signal that can be used in many fields such as biomechanical engineering, medicine and sport. It's applied to analyze and study the human movement. The SEMG recordings can be contaminated by spurious background spikes, quiescent baseline. These artifacts and noises produce false muscle activation detection. The muscle activation detection during movement depends on several parameters such as the beginning and the end of an activity, the nerve conduction velocity, the on-off interval, etc. In this paper, we conduct a study to detect the activation interval from the biceps brachial muscle using discrete wavelet transform (DWT) for SEMG signal denoising based on thresholding method. We compare our method with the integrated profile method. The results show that our method can effectively reduce the detection error.

Keywords: Biceps brachial muscle · Denoising · Discrete wavelet transform Integrated profile · Muscle activation detection · Surface electromyography

1 Introduction

The motion of the human body is the perfect integration of the brain, nervous system and muscles. From these muscles the EMG signals are obtained by measuring the electrical activity. Careful registration and study of electrical signals in muscle (electromyograms) can thus be a valuable aid in discovering and diagnosing abnormalities not only in the muscles, but also in the motor system as a whole [1].

Additionally, the Surface electromyography (SEMG) is receiving significant attention as a non-invasive technique; it is found in certain applications particularly in clinical and sports medicine such as muscle activities. While recording, the onset and cessation of a movement can be identified from the raw signal. As the muscle turns on, the SEMG becomes denser with larger amplitude. In healthy case, the muscle in regular conditions turns off if it is not needed, but if it still turns on, then it indicates a muscular hypertonicity, active muscle spasm or joint instability (stress, bad muscle coordination).

In the literature, many recent works have been proposed to detect the muscle activation. Jie Liu et al. used the integrated profile for muscle activation using EMG signals

O. Demigha et al. (Eds.): CSA 2018, LNNS 50, pp. 169–178, 2019.
https://doi.org/10.1007/978-3-319-98352-3_18

with spurious background spikes. The results show that the integrated profile method with presence of spike artifacts achieved a comparable performance with the sample entropy analysis but in shorter running time [2].

The SEMG signal can be influenced by various types of noise or artifact sources such as: Interfering power, baseline offset and shifts, spurious involuntary spikes produced by physiological and extrinsic/accidental origins, etc. These noisy effects decrease the performance of computerized analysis e.g.: the detection of contraction onset/offset and the activation interval. For the correct diagnosis, the denoising of SEMG signals is very necessary.

In previous studies, Hussain et al., de-noised SEMG signals using Discrete Wavelet Transform (DWT) with thresholding method. The results showed that, SEMG becomes less Gaussian with the increasing of MVC [3]. Also N. M. Sobahi used DWT denoising with manual soft minimax thresholding method whereas un-scaled white noise structure was selected. The results showed that the db6 at level 4 is chosen and found to be effective for noise removal [4].

Furthermore, Hussain et al. applied WT denoising method with measuring the RMSE (Root Mean Square Error) and the mean power frequency. The results showed that wave function db2 presents less RMSE compared to the other WFs (DB6, DB8 and dmey) and the effectiveness was observed more clearly while analyzing the power spectrum properties of the SEMG signals [5].

The aim of our work is to develop an algorithm of muscle activation detection based on discrete wavelet (DWT) denoising with thresholding method and the envelop detection. The determination of the wavelet transform and the choice of thresholding parameters are considered. The results are obtained by computation of the activation interval, the onset/offset time and the detection error, and are compared with integrated profile method.

2 Materials and Methods

2.1 Data

A group of 20 individuals (old and young subjects) was used to evaluate the quality of the research studies carried out in this work. All recordings were obtained from volunteers with subjects of different ages and genders: 6 elderly (72–85 years old) with 2 diabetic (1 male and 1 female), 4 healthy elderly (1 male and 3 females), 8 healthy adults male subjects (34–46 years old, 2 males and 6 females), 3 healthy young (19–22 years old, 2 males and 1 pregnant woman) and 3 children (8–12 years old).

The SEMG signals were detected from the brachial biceps muscle. These signals were first amplified by an instrumentation amplifier (AD620) and filtered by analog filters: A passive high pass filter with cut-off frequency of 0.04 Hz was applied to reduce the baseline wander noise, and a Sallen–Key active low-pass filter was applied to cut-off frequency of 600 Hz. A DC output voltage was added to the output signal to make it positive. The analog signals were sampled and converted into digital values by Arduino Uno card at frequency of 2000 Hz. Finally, a software interface with Matlab environment was developed. All data was saved in formats MAT-file and TXT- file [6].

2.2 Algorithm Implementation

The overall block diagram of the algorithm implementation is shown in Fig. 1. This proposed algorithm includes four sequential steps: (1) denoising signal using discrete wavelet transform, (2) rectification of denoised signal, (3) envelope detection and smoothing process, and (4) conversation to binary signal (output: 0/1).

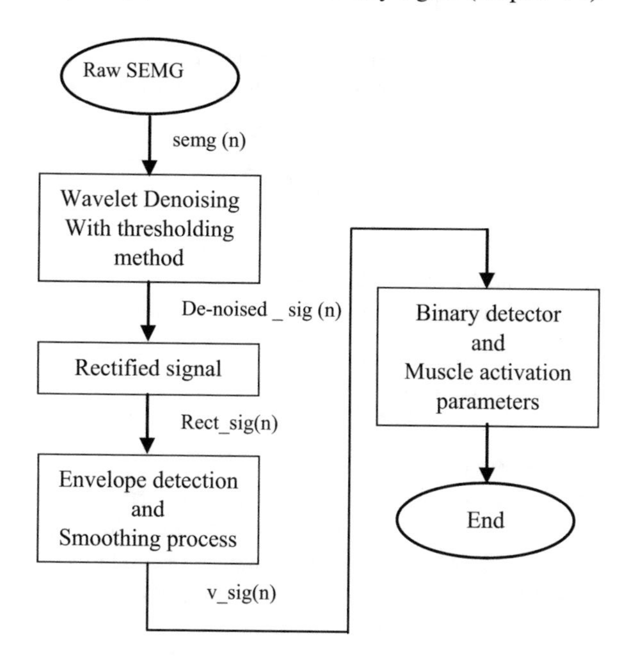

Fig. 1. Block diagram of proposed detecting muscle activation algorithm.

2.3 Wavelet Transform Method

The denoising EMG signal step is very important to reduce the influence of the baseline wander which typically appears from wire movement artifacts in the muscle tissue and other noise contamination interference in 50 Hz; thus making the contractile actions of the muscle easier to observe and interpret. Figure 2 illustrates the block diagram of wavelet denoising method which contains three steps: wavelet decomposition, thresholding method and wavelet reconstruction.

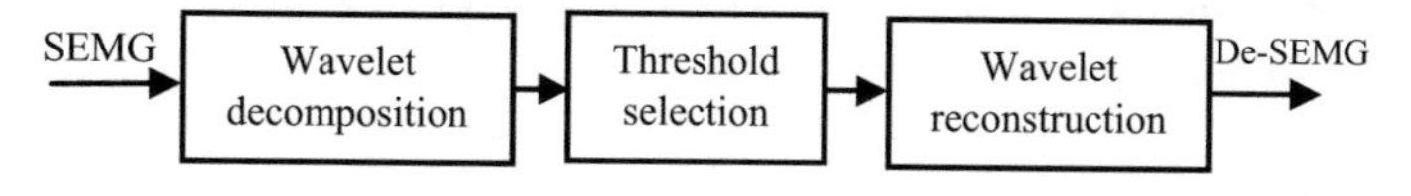

Fig. 2. Block diagram of denoising signal method based on wavelet

Wavelet Transform

The wavelet transform divides the signals through the dilated and translated wavelet. A mother wavelet Ψ (t) is a function of zero average according to the following formula [7]:

$$\int_{-\infty}^{+\infty} \Psi(\mathrm{t})\mathrm{dt} = 0 \tag{1}$$

It is also normalized (||Ψ|| = 1) and centered at t = 0.

Furthermore, wavelet functions are local in both frequency/scale through a dilatation and in time by translation of mother wavelet as follows:

$$\Psi_{a,b}(t) = \frac{1}{\sqrt{|a|}}\Psi\left(\frac{t-b}{a}\right) \tag{2}$$

Where a, b are the scaled and position parameters.

The wavelet transform of the function f (t) is given by the following equation:

$$w_f = \int_{-\infty}^{+\infty} f(t)\Psi_{a,b}(t)dt \tag{3}$$

Discrete Wavelet Transform

DWT is discrete in time and scale; it is used usually in practice to transform discrete (digital) signals to discrete coefficients in the wavelet domain. The discrete wavelet transform (DWT) requires less space and uses space-saving coding based on orthogonal or bi-orthogonal function to produce less redundant analysis. A typical DWT decomposition equation can be formulated as follows [8]:

$$DW(j,k) = \frac{1}{a}\sum_{n=0}^{N-1} s(n)\Psi\left(\frac{l-b}{a}\right) \tag{4}$$

Where a = 2j, b = 2jk, s(n) is the original signal, and DWT(j, k) is a sampling of Ψ(a, b) at discrete points j and k.

Denoising Algorithm

Decomposition: In the wavelet decomposition, a wavelet function and a level N are chosen. The first step is the selection of the mother wavelet such as: Meyer, Mallat and Daubechies wavelets. The wavelets are generally chosen based on the similarity of their shapes to those of the MUAP. The second step is the decomposition of the signals at level N, which is implemented by a filter bank that decomposes the signal into successive coarser approximations and details (see Fig. 3).

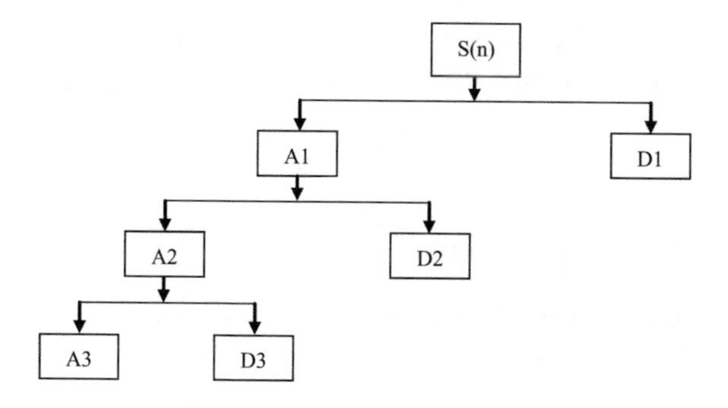

Fig. 3. The wavelet decomposition tree with level N = 3.

Threshold selection: Select a threshold in each level and apply the soft/hard thresholding to the detail coefficients. The hard thresholding is the simplest method but the soft thresholding is smoother.

Hard thresholding: setting to zero the elements whose absolute values are lower than the threshold δ.

$$S_{hard}(t) = \begin{cases} s(t), & s(t) > \delta \\ 0, & s(t) \leq \delta \end{cases} \tag{5}$$

Soft thresholding: setting to zero the elements whose absolute values are lower than the threshold δ and shrinking the nonzero coefficients towards 0.

$$S_{soft}(t) = \begin{cases} Sign(s(t))(|s(t) - \delta|), & s(t) > \delta \\ 0, & s(t) \leq \delta \end{cases} \tag{6}$$

The choice of threshold δ: There are four threshold selection rules

Rigrsure: adaptive threshold selection using principle of Stein's Unbiased Risk Estimate.
Heursure: heuristic variant of the first option.
Sqtwolog: threshold is sqrt (2*log (length(s))).
Minimaxi: minimax thresholding.

Reconstruction: Applying the inverse discrete transform using the original approximation coefficients of level N and the modified detail coefficients of level 1 to N.

Rectified SEMG

We use full-wave rectification, which means that the absolute values of each data point are kept and are converted to positive. The rectified signal is given by Eq. (7):

$$\text{Rect\_SEMG} = \text{abs(SEMG)} \tag{7}$$

Envelope Detection

We use the Hilbert transform. Mathematically, the envelope e(t) of a signal s(t) is defined as the magnitude of the analytic signal. An analytic signal is a complex signal, where the real part is the original signal and the imaginary part is the Hilbert transform of the original signal as shown by the following Eq. (8):

$$e(t) = \sqrt{s(t)^2 + \hat{s}(t)^2} \tag{8}$$

Where $\hat{s}(t)$ is the Hilbert transform of $s(t)$.

Smoothing Process: to smooth the SEMG envelope signal, we apply the moving average filter with a span of 1.5%.

Binary Detector: The SEMG signals are converted to binary signal which means that the output signal is 0/1: 0 for non-active muscle and 1 for an active muscle.

2.4 The Integrated Profile Method

A zero-phase high-pass Butterworth filter with cut-off frequency of 30 Hz is used to reduce the influence of the baseline wander which typically appears from wire movement artifacts in the muscle tissue and other noises contaminated in surface EMG signal. Then, the integrated profile of the EMG signal is applied to determine the onset/offset. The discrete integration of all rectified samples of signal (x) is calculated as follows:

$$IP(t) = \sum\nolimits_{i=1}^{t} |x(i)| \tag{9}$$

With t = 1, 2… Length(x).

The linear line L(t) is a linear function that has the same maximum value of IP(N). It can be defined as:

$$L(t) = IP(N) * t/N \tag{10}$$

Where N is the Length of signal (x).

The difference between L(t) and IP(t) is calculated to determine the onset and offset, where D(t) reaches its minimum and maximum values, respectively.

$$D(t) = IP((t) - L(t) \tag{11}$$

2.5 Computation of the Parameters and Performance Evaluation

The activation interval and onset/offset time are very important indicators especially in clinical application. The onset time is the time of starting the muscular activation, and the activation interval is the interval between onset and off set time. These parameters are calculated visually by two trained observers (mean of two values) and automatically by the implemented algorithm.

The performance of these two methods is evaluated by the latency, defined as the absolute difference between the automatic onset/offset time t_{auto} and the visual onset/offset time t_v.

$$\tau = |t_{auto} - t_v| \quad (12)$$

3 Results and Discussion

The SEMG signals are noisy and contaminated by different noises. In this study, we apply the denoising method based on DWT. As we know, the relation between the length of signal (L) and the level of decomposition is $L = 2^N k$, where k is a natural number. Since L = 2000, we choose N = 4. From different tests, the effective wavelet function is db6 with level N = 4. Figure 4 illustrates the decomposition of signal recording with 4 details (D1, D2, D3, and D4) and one approximation A4.

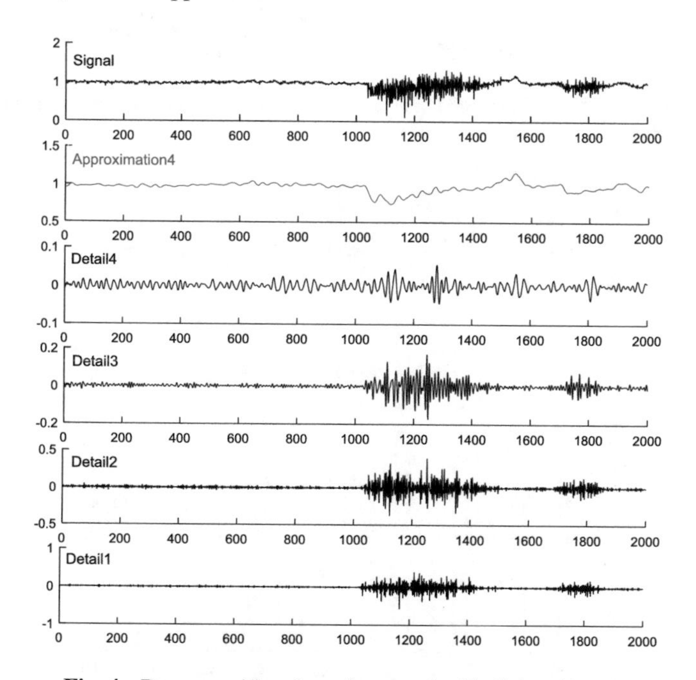

Fig. 4. Decomposition the noisy signal with db6 and level4

The dominant energy of EMG is concentrated in range 20–500 Hz [9]. Using Welch's power spectrum density with Hamming window of 256 points and default overlap of 50%, we demonstrate that the approximation A4 is the baseline wander and the dominant energy is concentrated in the range of (0–1000 Hz) as shown in Fig. 5.

After testing different thresholds, the optimal threshold "minimaxi" is selected. Figure 6 demonstrates an example of comparison between DWT and IP: it is noticed that the muscle activation interval in DWT method is closer to visual detection than IP method.

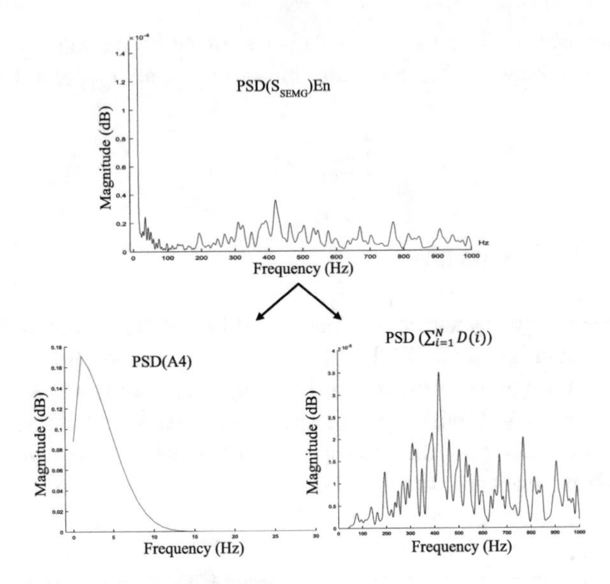

Fig. 5. Power spectral density of the noisy SEMG signal, the reconstructed A4 and details with Hamming window.

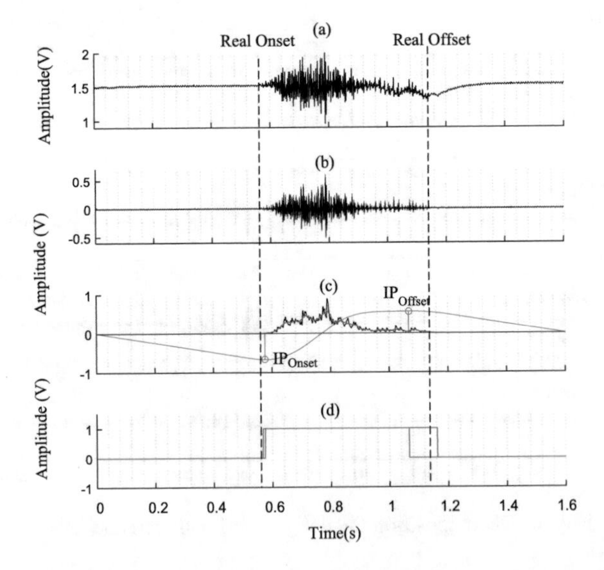

Fig. 6. An example of a comparison between two detection onset/offset methods. (a) The real surface EMG signal from Biceps Brachial with real onset/offset detection (vertical dashed lines). (b) Denoised signal (red signal without baseline wander using Butterworth filter, black after DWT thresholding). (c) The rectified (black curve) and smooth (blue curve) signal, the integrated profile (IP) (red curve). (d) Binary signal of muscle activation using DWT (blue solid line) and IP (red solid line).

To evaluate the performance of discrete wavelet transform method, we have compared it with the Integrate Profile method; we compute the latency between automatic and manual detection. The differences between automatic and the mean of manual detections are presented in Table 1 as mean ± standard deviation.

Table 1. Comparison of onset and offset detection performance of the DWT and integrated profile methods

	Visual detection (t_v)	DWT method	IP method
	|expert1-expert2| (*mean ± SD*)	|tauto-mean(t_v)| (*mean ± SD*)	|tauto-mean(t_v)| (*mean ± SD*)
Onset time(s)	0.0035 ± 0.0018	0.0055 ± 0.0047	0.0099 ± 0.0044
Offset time(s)	0.0037 ± 0.0043	0.0075 ± 0.0063	0.078 ± 0.0089

Using DWT with an optimal threshold improves and eases the onset/offset detection, compared to IP method. It is less sensitive, during the detection of short muscle activation intervals, and limited in reducing the error even though the signals are denoised.

4 Conclusion

The DWT method is an automatic method for detection of activation muscle and denoising SEMG signals. It can be used for the analysis of the surface EMG signal recorded during movement such as in sport and exercise science.

The DWT method is computational efficient compared to Integrate Profile method, because it depend on an optimal threshold selection which can reduce the error.

In this study, the db6 with soft threshold minimaxi was selected.

In the future, the proposed study will be modified and will be improved by integrating other developed methods.

Acknowledgment. The authors would like to thank all the participants and all our colleagues for their volunteering in this experience.

References

1. Gokgoz, E., Subasi, A.: Comparison of decision tree algorithms for EMG signal classification using DWT. International Burch University, Faculty of Engineering and Information Technologies, 71000 Sarajevo, Bosnia and Herzegovina, Elsevier Ltd. (2015)
2. Liu, J., Liu, Q.: Use of the integrated profile for voluntary muscle activity detection using EMG signals with spurious background spikes: a study with incomplete spinal cord injury. Biomed. Signal Process. Control **24**, 19–24 (2016)
3. Hussain, M.S., Reaz, M.B.I., Ibrahimy, M.I., Ismail, A.F, Mohd-Yasin, R.: Wavelet based noise removal from EMG signals. Informacije MIDEM **375**(2), 94–97 (2007)
4. Sobahi, N.M.: Denoising of EMG signals based on wavelet transform. Asian Trans. Eng. **1**(5), 17–23 (2011). (ATE ISSN: 2221-4267)

5. Hussain, M.S., Mamun, Md.: Wavelet denoising and surface electromyography analysis. Res. J. Appl. Sci. Eng. Technol. **4**(15), 2372–2374 (2012)
6. Benazzouz, A., Eddine, H.Z.: Real-time acquisition of surface EMG signal system. In: 3rd International Conference on Embedded Systems in Telecommunications and Instrumentation (ICESTI 2016) (2016)
7. Misiti, M., Misiti, Y., Oppenheim, G., Poggi, J.-M.: Wavelets and their Applications (2006)
8. Michael, W.: Digital signal processing using MATLAB and Wavelets (2007). ISBN 0-9778582-0-0
9. Chu, J.-U., Moon, I., Mun, M.-S.: A real-time EMG pattern recognition system based on linear-nonlinear feature projection for a multi-function myoelectric hand. IEEE Trans. Biomed. Eng. **53**(11), 2232–2239 (2006)

Image Processing and Computer Vision

Virtual and Augmented Reality, from Design to Evaluation

Rachid Gherbi(✉)

Université de Paris-Sud XI, Orsay, France
rachid.gherbi@u-psud.fr

Abstract. The talk will highlight the general idea of Virtual and Augmented Reality (V&AR), from design specifications to evaluation methods. After a short introduction into the history of V&AR, we will have a look at various types and concepts of V&AR known today. We then discuss some experimental cases to highlight the need of specification of all V&AR parts, such as technical and hardware constraints but also those concerning tasks modeling and users needs. We will present some criteria we consider when evaluating a V&AR experiences success. Finally, some Algerian V&AR applications and projects will also be presented in order to show national developments in this area.

O. Demigha et al. (Eds.): CSA 2018, LNNS 50, p. 181, 2019.
https://doi.org/10.1007/978-3-319-98352-3_19

Motion-Based Analysis of Dynamic Textures – A Survey

Ikram Bida(✉) and Saliha Aouat

Laboratory for Research in Artificial Intelligence (LRIA),
Department of Computer Science,
University of Science and Technology Houari Boumediene (USTHB),
Bab-Ezzouar, Algiers, Algeria
{ibida,saouat}@usthb.dz

Abstract. Textures such as grass, trees, mountains, buildings and others occupy large spaces of our visual environment. Numerous researches have been devoted to automatically analyze and characterize textures, where static textures found in single images were the first to be studied. Subsequently, this notion was extended to temporal dimension, known as dynamic texture representing variable properties in time such as flames, swaying trees, moving clouds, crowds in public places and even shadows, etc. Lately, temporal texture research is gaining a lot of attention, due to its importance as an effective component in the interpretation of video content.

This paper presents a research survey that focuses on a very captivating subject: Dynamic texture analysis, characterization and recognition. Its motivation is to give an overview of the most up-to-date analysis approaches that have been proposed to characterize then recognize temporal textures in different fields.

Keywords: Dynamic texture · Motion-based analysis
Motion estimation · Signature extraction
Dynamic texture recognition · Video content interpretation

1 Introduction

Static texture study is an ancient field of image processing and computer vision. Despite the variety and efficiency of its applications, it is still difficult to precisely characterize or to give a universal definition. Hence, expanding it temporally, i.e. studying its progress over time, only makes the analysis task more complicated and accompanied with new challenges.

The authors employed diverse terminologies to refer to dynamic textures based on the definitions they suggested, Nelson [1] first used Temporal Textures labeling, then Peh [2] called them Spatiotemporal Textures. In 2001, Saison [3] introduced the most common nomenclature: Dynamic Textures. After that, Time-varying Textures nomination was proposed by Bar-Joseph [4], a little later wang [5] used the term Moving Textures.

O. Demigha et al. (Eds.): CSA 2018, LNNS 50, pp. 182–192, 2019.
https://doi.org/10.1007/978-3-319-98352-3_20

Finding an accurate definition for dynamic textures appears to be a real issue that carries significant difficulties. While browsing the literature, we came across several of them.

Back in the nineties, Nelson and Polana were the first to approach dynamic texture subject in their paper [1], they defined temporal textures as a class of image motions that are present in natural environment exhibiting spatial and temporal stationarity.

There exist some other definitions found in the literature that are either global or seem to be missing an important dynamic texture characteristic.

Duboi and colleagues [6] proposed a definition which is distinctly the most complete and profound one. They see dynamic textures as a textured structure that can be rigid or deformable. Characterized by a repetitive spatial and temporal phenomena, this structure has a deterministic or anarchic motion induced by a force generated internally, received externally or produced by camera motion. In real scenes, dynamic textures may superimpose, where their origins can differ from natural, artificial or synthesized.

Based on our knowledge, up to now, only two recent surveys addressed dynamic texture analysis. The first released in 2005 [7] by Chetverikov and al, the other published by Rahman and al [8] in 2008. The two surveys gave a general overview on some of the existing techniques at the time. This work is an up to date survey that studies and briefs all the previous and current motion-based approaches in the same theme.

What remains of this article is organized as follows. Section 2 enumerates the vast application domains of dynamic textures. We devote Sect. 3 to review the existing datasets. Section 4 is devoted to study motion-based approaches, we explain the general principle of resolution, the extracted features and the obtained results for each. Furthermore we present the limits. In Sect. 5, we end the survey with a conclusion and some perspectives.

2 Dynamic Texture Application Domains

The analysis methods of dynamic texture are used in a vast field of applications, such as indexing and video retrieval [9–12], where a signature is extracted, it describes the content of videos based on dynamic texture characteristics, these signatures are then used to retrieve videos by forming requests.

Dynamic texture is also used in the domain of video surveillance [13–17] in applications like road accident detection, risk behavior in road traffic and crowds. It is also beneficial in the detection of emergencies beginnings, forest fire tracking, or explosion in factories and vehicles. Likewise in prevention of natural disasters, detecting abnormal activities in parking lots, pools, or amidst seas.

Many works concentrate on segmenting dynamic textures [18–21], its purpose is to help comprehend the scene by isolating each temporal texture aside, additionally it is employed to construct video summaries, i.e. depict a video at different times.

The background of scenes can be disturbing while processing videos such as segmentation and motion analysis, subtracting the dynamic textures [22–26] present in the background improves the effectiveness of treatments and decrease the noise.

A considerable amount of papers have been devoted to dynamic texture synthesis [27–30], synthesizing realistic dynamic texture is used in animated movies, video games and video coloring. Dynamic textures also benefited from 3D rendering that recovers affine geometry [31].

3 Dynamic Texture Datasets

Three purely dynamic texture datasets were found in the literature as illustrated in Fig. 1, available with their corresponding links as follow:
the MIT dataset cited as the Szummer dataset[1] (Fig. 1a), UCLA dataset[2] (Fig. 1b) and DynTex dataset[3] (Fig. 1c).

Fig. 1. Dynamic Texture dataset samples: a-MIT, b-UCLA, c-DynTex

4 Dynamic Texture Analysis, Characterization and Recognition Approaches

Many scientists worked on the analysis of temporal texture, their fundamental purpose is to obtain features in the form of representative signatures, as compact and discriminative as it can be, these chosen elements will maintain both spatial and temporal qualities.

Five families of approaches were distinguished to which belong the existing methods for dynamic texture analysis: Motion-based approaches, Mathematical model-based approaches, approaches calculating geometric properties, Multi-scale (spatiotemporal) transforms approaches, and Deep Learning approaches that are very recent. In this survey, we are only interested in Motion-based approaches; the remaining four families will be addressed in other surveys.

[1] http://alumni.media.mit.edu/~szummer/icip-96.
[2] http://www.bernardghanem.com/datasets.
[3] http://projects.cwi.nl/dyntex/index.html.

4.1 Motion-Based Approaches

Motion-based analysis approaches are very common and widely used. Generally, they follow three phases as shown in Fig. 2. Firstly, the motion is estimated from video images, it is calculated between each two frames. Thereafter, a set of features is extracted from the estimated movement fields to compose a descriptor that represents the video recognition signature. Sometimes, in order to better preserve spatial characteristic, extra features are collected from single images.

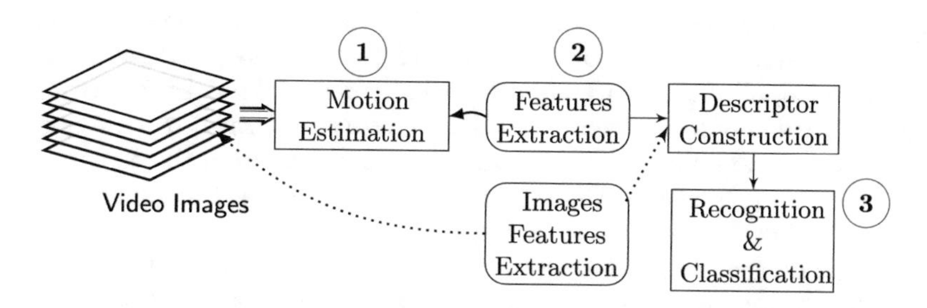

Fig. 2. General process of Motion-based approaches

SFTR. In their papers [1,32] Nelson and Polana proposed the first motion-based method: **S**pacial **F**eature based **T**exture **R**ecognition, they considered to relate temporal texture analysis with motion data present in images sequence, rather than scanning single images containing static textures, as movement areas hold both spatial and temporal information. To estimate motion fields, they used the normal flow, which is an approximated motion measure, due to its simplicity of calculation compared to the optical flow and its preservation of both temporal and spatial criteria.

The following ensemble of features are then extracted from the flow fields: peakness, positive and negative divergence, positive and negative curl, non-uniformity and spatial homogeneity. These characteristics are generally invariant to translation, rotation, temporal and spatial scaling.

They validated their method on 7 different temporal texture videos, applying a nearest centroid classification algorithm. Different combinations of the already cited features were tested, mostly the method was successful.

SCNT. **S**patio-temporal **C**lique **N**eighbourhood **T**echnique was suggested by Bouthemy and Fablet [12,33] to improve Nelson's and Polana's previous works, they sought to take better account of temporal features and spatial characteristics through temporal neighborhood to recognize dynamic textures. SCNT approach uses the normal flow to estimate the motion contained in video images. Then, it constructs both spatial and temporal co-occurrence matrices from the resulting motion vectors. The temporal motion co-occurrence value is pointed out as the probability of two motion frames separated by a distance in time for a certain spatial window.

In the first published paper [12] designed for video indexing, five classical features were pulled out from the co-occurrence matrices, that are: average, variance, dirac, angular second moment and contrast. In trials, Four videos were tested where the method was able to characterize and distinguish them.

SCNT intended for video indexing and retrieval [33] utilized another five slightly different features collected from the co-occurrence matrices: entropy, inverse difference moment, acceleration, kurtosis and difference kurtosis.

MST. Bouthemy and Fablet introduced other works to analyze dynamic textures, some applied to video indexing and retrieval [34], others to motion recognition [35,36]. In their framework, globally named **M**odel-based **S**tatistical **T**echnique, video sequence reconstruction is firstly performed by deleting noisy frames, that are assumed to be having motion resulted from the camera movement. Next, they estimated the motion by computing local measurement from the spatiotemporal intensities derivations in images, which they believe are steadier than the normal flow. After that, temporal co-occurrence matrices were calculated for the local-motion distribution. Finally, purely temporal statistics were pulled out from the co-occurrences using Temporal Gibbs random field Models (TGM) or Temporal Multiscale Gibbs Models (TMGM).

In the recognition process, MST used the maximum likelihood similarity measure in learning and testing phases on different databases containing mostly real and unsegmented scenes.

SSTF. Peh and Cheong developed an approach titled: **S**ynergizing **S**patial and **T**emporal **F**eatures [2] for temporal texture analysis combining spatial and temporal features. Initially, the authors estimated the normal flow, next the direction and magnitude measurements were mapped into gray level intensities resulting in direction plots and magnitude plots (intensity images). Separately, these two series of plot frames were superimposed to trace motion history. They were indicated to as extended magnitude plots (EMP) and extended direction plots (EDP). Afterwards, the extended plots were studied by classical static texture methods: gray level co-occurrence matrices, fourier spectrum and difference statistics. A set of standard features were then extracted such as: inertia, shade, correlation computed from the co-occurrence matrices. Centered energy at 45° and centered energy at 135° both from fourier spectrum, furthermore the mean static. For single images, these descriptors are very quick to compute, they assure an invariance to rotation and translation.

SSTF used a normalized version of features, it was tested on ten classes of 47 video sequences, they employed an unsupervised classification algorithm: K-means clustering, the technique was capable of giving a 87.44% average success on the ten classes.

NFTR. Recognizing dynamic texture utilizing **N**ormal **F**low and **T**exture **R**egularity was another method proposed by Pétri and Chetverikov [46,47] to characterize dynamic textures. First, they pulled from the optical flow the normal flow estimated fields, thereafter a set of criteria was extracted from it. Three of these descriptors are similar to Nelson's and Polana's: divergence, preakness, rotational or curl. Homogeneity of orientation was an additional optical flow feature. Moreover, they introduced novel spatial features pulled straightly from original image sequence (static texture) called regularity features that are: the mean of the MRA (maximum regularity areas) and variance of the MRA. All these descriptors are robust against translation and rotation.

The experimentations were conducted on 80 real videos of the MIT Dataset, adopting a normalized version of the extracted features. To classify the sequences, a leave-one-out technique was performed. The result was 93,8% rate of recognition. In 2011, Pétri [48] utilized three of these features to track dynamic textures, that are: divergence, curl and orientation.

SMHR. Lu and al [37] presented an original approach: **S**patio-temporal **M**ultiresolution **H**istograms **R**ecognition in 2005, it uses the complete optical flow, calculated utilizing the 3D structure tensor-based method. Next, from the spotted motion field vectors, velocity and acceleration are computed building spatiotemporal multiresolution histograms for each. By the end, velocity and acceleration histograms are collected on different scales to characterize the video sequence. SMHR is considered to be invariant to rotation. Tests were proceeded on 7 real videos, where they employed a similarity measure to compare between histograms in the purpose of differentiating dynamic texture.

OTSR. Rahman and al published another series of works based on motion to characterize temporal textures [38–41] by their approach **O**ptimal **T**ime-Space **R**atio **t**echnique. As a first step, OSTR exploits Block Motion technique (full search algorithm) to estimate the motion for the reason of handling a whole block is more effective and faster than handling pixels, which is the case in complete optical or normal flow estimation thus motion frames will be treated as a sequence of K parallel grids of size a $\times$ b. Next, it uses only three of co-occurrence matrices proposed in STCN approach; furthermore, they related spatial and temporal co-occurrence matrices with a causal relationship resulting in each clique a set of co-occurrence matrices and the corresponding Gibbs potential matrices. To compromise both information a weighted ratio formula was proposed.

OSTR was compared to SFTR, SSTF and STCN on 9 sequences from the Szummer dataset and 26 videos of the DynTex dataset. The authors used a k-NN classifier. Experimentations showed that OSTR gave a 97,62% rate of recognition which exceeded only STCN, it is considered as a better approach despite SFTR and SSTF better accuracies for the reason that it is very rapid and can be integrated into real time applications.

In 2009, Rahman and Murshed [41] studied the characterization of multiple dynamic textures present in a scene using co-occurrence matrices without any segmentation.

OFTP. In their approach OFTP **O**ptical **F**low features and **T**emporal **P**eriodicity proposed in 2007 [42–44], Fazekas and Chetverikov studied dynamic textures recognition for classification purposes, their basic idea revolves around local image distortion named: scaling, biaxial shearing, rotation, directed scaling and directed shearing, the set of these images express the spatial derivatives of flows. Firstly, OFTP estimates motion using optical flow, to this end, they tested different implementations, then decomposed the motion field derivatives (velocity fields) to three measures referred to as: divergence, rotation and shearing. Next, a collection of 8 static features were deduced using two indicators on these measures: indicators to characterize direction and others for magnitude (length) of vectors. Another two temporal features expressing periodicity were computed which are: average periodicity and relative absolute variation.

Experimentations were performed using 26 videos of the Dyntex Dataset utilizing a simple leave-one-out test, OFTP gave a 95.2% of recognition rate as a best value from all the different tested architectures.

CCMOF. Andrearczyk and Whelan [45] have introduced in 2015 a novel approach named **C**ombined **C**o-occurrence **M**atrix of **O**ptical **F**low for classification goals. Globally, CCMOF follows five steps, firstly it computes the optical flow between each two frames, therefore magnitude and orientation of motion fields are calculated, these measures were quantified into grey intensities of 8 levels for each. Afterwards, spatial and temporal co-occurrences matrices were constructed both of 8×8 size for each motion frame. Once all done, resulting matrices were summed up and normalized. As a final step, a total set of 19 temporal and spatial slightly modified Haralick's features were extracted for each summed normalized co-occurrence matrix.

CCMOF was validated on a composed dataset that contains 200 sequences taken from the following datasets: Dyntex, Dyntex++ and UCLA. While building the SVM classifier, 100 videos were used for training and the remaining for tests, the approach gave a 68% accuracy rate, but combined them with the LBP-TOP technique, it demonstrated a better rate of recognition that arrived until 84.4%.

4.2 Critics and Limitations

Regardless of the good results that each of the cited methodologies have shown, some still exhibit serious limitations. For instance, the extracted features in SFTR are extremely spatial, consequently when the temporal data is dominant, it will be so poorly characterized or even neglected. Also, SCNT isn't capable to transfer meaningfully spatial motion distribution information by temporal features. The MST approach fails to maintain a balance between passing spatial features through temporal ones. A limitation of SSTF is that some aspects of temporal textures can be lost during the superimposing process, i.e. while constructing EMP and EDP. The method NFTR does not succeed to integrate temporal information and SMHR consumes a lot of time for the reason that it uses complete optical flow. The CCMOF number of features seems to be very huge and not well compacted.

5 Conclusion

In this paper, we have addressed a very compelling topic: Dynamic textures analysis and characterization. Our work provides readers with a summary about all the motion-based methods that exist in literature. In general, we believe that these methods can be improved in several ways for instance: by reducing the size of descriptors, also by enhancing the motion estimation algorithms in term of speed, moreover by extracting discriminant features able to transpose both spatial and temporal information at once, employing meta-heuristics in the phase of feature selection would be a very interesting theory to validate.

References

1. Nelson, R.C., Polana, R.: Qualitative recognition of motion using temporal texture. CVGIP: Image Underst. **56**(1), 78–89 (1992)
2. Peh, C.H., Cheong, L.F.: Exploring video content in extended spatio-temporal textures. In: In 1st European workshop on Content-Based Multimedia Indexing, Toulouse, France, pp. 147–153 (1999)
3. Saisan, P., Doretto, G., Wu, Y.N., Soatto, S.: Dynamic texture recognition. In: Proceedings of the 2001 IEEE Computer Society Conference on Computer Vision and Pattern Recognition CVPR 2001, vol. 2, pp. 58–63 (2001)
4. Bar-Joseph, Z., El-Yaniv, R., Lischinski, D., Werman, M.: Texture mixing and texture movie synthesis using statistical learning. IEEE Trans. Vis. Comput. Graph. **7**(2), 120–135 (2001)
5. Wang, Y., Zhu, S.C.: Modeling textured motion : particle, wave and sketch. In: Proceedings Ninth IEEE International Conference on Computer Vision, vol. 1, pp. 213–220, October 2003
6. Dubois, S., Peteri, R., Menard, M.: Decomposition of dynamic textures using morphological component analysis. IEEE Trans. Circuits Syst. Video Technol. **22**(2), 188–201 (2012)
7. Chetverikov, D., Peteri, R.: A brief survey of dynamic texture description and recognition. In: Proceedings of International Conference on Computer Recognition Systems, pp. 17–26. Springer, Heidelberg (2005)
8. Rahman, A., Murshed, M.: Temporal Texture Characterization: A Review, pp. 291–316. Springer, Heidelberg (2008)
9. Mocofan, M., Vasiu, R.: Dynamic textures indexing using the co-occurrence matrix features. In: Proceedings of 2012 7th IEEE International Symposium on Applied Computational Intelligence and Informatics (SACI), pp. 327–330, May 2012
10. Dubois, S., Peteri, R., Ménard, M.: A Comparison of Wavelet Based Spatio-temporal Decomposition Methods for Dynamic Texture Recognition, pp. 314–321. Springer, Heidelberg (2009)
11. Smith, J.R., Lin, C.Y., Naphade, M.: Video texture indexing using spatio-temporal wavelets. In: Proceedings of International Conference on Image Processing, vol. 2, pp. 437–440 (2002)

12. Bouthemy, P., Fablet, R.: Motion characterization from temporal co-occurrences of local motion-based measures for video indexing. In: Proceedings of Fourteenth International Conference on Pattern Recognition (Cat. No.98EX170), vol. 1, pp. 905–908, August 1998
13. Phillips, W., Shah, M., Lobo, N.D.V.: Flame recognition in video. In: Proceedings of Fifth IEEE Workshop on Applications of Computer Vision, pp. 224–229 (2000)
14. Günay, O.: Dynamic texture analysis in video with application to flame, smoke and volatile organic compound vapor detection. Ph.D. thesis, BIlkent university (2009)
15. Ma, Y., Cisar, P.: Event detection using local binary pattern based dynamic textures. In: IEEE Computer Society Conference on Computer Vision and Pattern Recognition Workshops 2009 (CVPR Workshops 2009), pp. 38–44. IEEE (2009)
16. Komulainen, J., Hadid, A., Pietikäinen, M.: Face spoofing detection using dynamic texture. In: Asian Conference on Computer Vision, pp. 146–157. Springer, Heidelberg (2012)
17. Hsu, W.L., Chen, T.H.: People gathering recognition based on dynamic texture detection. In: 2015 International Conference on Machine Learning and Cybernetics (ICMLC), vol. 1, pp. 334–339. IEEE (2015)
18. Li, J., Chen, L., Cai, Y.: Dynamic texture segmentation using 3-D fourier transform. In: Fifth International Conference on Image and Graphics ICIG 2009, pp. 293–298. IEEE (2009)
19. Chan, A.B., Vasconcelos, N.: Modeling, clustering, and segmenting video with mixtures of dynamic textures. IEEE Trans. Pattern Anal. Mach. Intell. **30**(5), 909–926 (2008)
20. Amiaz, T., Fazekas, S., Chetverikov, D., Kiryati, N.: Detecting regions of dynamic texture. In: International Conference on Scale Space and Variational Methods in Computer Vision, pp. 848–859. Springer, Heidelberg (2007)
21. Candes, E., Demanet, L., Donoho, D., Ying, L.: Fast discrete curvelet transforms. Multiscale Model. Simul. **5**(3), 861–899 (2006)
22. Lin, L., Xu, Y., Liang, X., Lai, J.: Complex background subtraction by pursuing dynamic spatio-temporal models. IEEE Trans. Image Process. **23**(7), 3191–3202 (2014)
23. Ali, I., Mille, J., Tougne, L.: Space-time spectral model for object detection in dynamic textured background. Pattern Recogn. Lett. **33**(13), 1710–1716 (2012)
24. Chan, A.B., Mahadevan, V., Vasconcelos, N.: Generalized stauffer-grimson background subtraction for dynamic scenes. Mach. Vis. Appl. **22**(5), 751–766 (2011)
25. Zhang, S., Yao, H., Liu, S.: Dynamic background modeling and subtraction using spatio-temporal local binary patterns. In: Proceedings of 15th IEEE International Conference on Image Processing 2008 (ICIP 2008), pp. 1556–1559. IEEE (2008)
26. Ramesh, V., et al.: Background modeling and subtraction of dynamic scenes. In: Proceedings of Ninth IEEE International Conference on Computer Vision, pp. 1305–1312. IEEE (2003)
27. Tesfaldet, M., Brubaker, M.A., Derpanis, K.G.: Two-stream convolutional networks for dynamic texture synthesis. arXiv preprint arXiv:1706.06982 (2017)
28. Funke, C.M., Gatys, L.A., Ecker, A.S., Bethge, M.: Synthesising dynamic textures using convolutional neural networks. arXiv preprint arXiv:1702.07006 (2017)
29. Zhu, Z., You, X., Yu, S., Zou, J., Zhao, H.: Dynamic texture modeling and synthesis using multi-kernel gaussian process dynamic model. Sig. Process. **124**, 63–71 (2016)
30. Lizarraga-Morales, R.A., Guo, Y., Zhao, G., Pietikäinen, M., Sanchez-Yanez, R.E.: Local spatiotemporal features for dynamic texture synthesis. EURASIP J. Image Video Process. **2014**(1), 17 (2014)

31. Sheikh, Y., Haering, N., Shah, M.: Shape from dynamic texture for planes. In: 2006 IEEE Computer Society Conference on Computer Vision and Pattern Recognition, vol. 2, pp. 2285–2292. IEEE (2006)
32. Polana, R., Nelson, R.: Temporal Texture and Activity Recognition (eds.), pp. 87–124. Springer, Dordrecht (1997)
33. Fable, R., Bouthemy, P.: Motion-Based Feature Extraction and Ascendant Hierarchical Classification for Video Indexing and Retrieval, pp. 221–229. Springer, Heidelberg (1999)
34. Fablet, R., Bouthemy, P., Perez, P.: Nonparametric motion characterization using causal probabilistic models for video indexing and retrieval. IEEE Trans. Image Process. **11**(4), 393–407 (2002)
35. Fablet, R., Bouthemy, P.: Motion recognition using spatio-temporal random walks in sequence of 2D motion-related measurements. In: Proceedings of 2001 International Conference on Image Processing (Cat. No.01CH37205), vol. 3, pp. 652–655 (2001)
36. Fablet, R., Bouthemy, P.: Motion recognition using nonparametric image motion models estimated from temporal and multiscale co-occurrence statistics. IEEE Trans. Pattern Anal. Mach. Intell. **25**(12), 1619–1624 (2003)
37. Lu, Z., Xie, W., Pei, J., Huang, J.: Dynamic texture recognition by spatio-temporal multiresolution histograms. In: Application of Computer Vision, WACV/MOTIONS 2005 Volume 1. Seventh IEEE Workshops on, vol. 2, pp. 241–246, January 2005
38. Rahman, A., Murshed, M.: Real-time temporal texture characterization using block-based motion co-occurrence statistics. In: 2004 International Conference on Image Processing ICIP 2004, vol. 3, pp. 1593–1596, October 2004
39. Rahman, A., Murshed, M., Dooley, L.S.: Feature weighting methods for abstract features applicable to motion based video indexing. In: Proceedings of International Conference on Information Technology: Coding and Computing ITCC 2004, vol. 1, pp. 676–680, April 2004
40. Rahman, A., Murshed, M.: A temporal texture characterization technique using block-based approximated motion measure. IEEE Trans. Circuits Syst. Video Technol. **17**(10), 1370–1382 (2007)
41. Rahman, A., Murshed, M.: Detection of multiple dynamic textures using feature space mapping. IEEE Trans. Circuits Syst. Video Technol. **19**(5), 766–771 (2009)
42. Fazekas, S., Chetverikov, D.: Dynamic texture recognition using optical flow features and temporal periodicity. In: 2007 International Workshop on Content-Based Multimedia Indexing, pp. 25–32, June 2007
43. Fazekas, S., Chetverikov, D.: A non-regular optical flow for dynamic textures. In: Fazekas, A., Hajdu, A. (eds.) KÉPAF 2007 6th conference of Hungarian Association for Image Processing and Pattern Recognition. Debrecen, KÉPAF Társ, pp. 157–164 (2007)
44. Fazekas, S., Chetverikov, D.: Analysis and performance evaluation of optical flow features for dynamic texture recognition. Sig. Process. Image Commun. **22**(7), 680–691 (2007)
45. Andrearczyk, V., Whelan, P.F.: Dynamic texture classification using combined co-occurrence matrices of optical flow. In: Irish Machine Vision & Image Processing Conference proceedings IMVIP, vol. 2015 (2015)
46. Péteri, R., Chetverikov, D.: Dynamic texture recognition using normal flow and texture regularity. In: Marques, J.S., de la Blanca, N.P., Pina, P. (eds.) Pattern Recognition and Image Analysis: Second Iberian Conference, pp. 223–230. Springer, Heidelberg (2005)

47. Péteri, R., Chetverikov, D.: Qualitative characterization of dynamic textures for video retrieval. In: Wojciechowski, K., Smolka, B., Palus, H., Kozera, R.S., Skarbek, W., Noakes, L. (eds.) Computer Vision and Graphics: International Conference, ICCVG 2004, Proceedings, Warsaw, Poland, September 2004, pp. 33–38. Springer, Dordrecht (2006)
48. Péteri, R.: Tracking dynamic textures using a particle filter driven by intrinsic motion information. Mach. Vis. Appl. **22**(5), 781–789 (2011)

$NH\infty$-SLAM Algorithm for Autonomous Underwater Vehicle

Fethi Demim[1(✉)], Abdelkrim Nemra[1], Hassen Abdelkadri[1], Kahina Louadj[2], Mustapha Hamerlain[3], and Abdelouahab Bazoula[1]

[1] Laboratoire Robotique et Productique, Ecole Militaire Polytechnique, EMP, Bordj El Bahri, Algiers, Algeria
demifethi@gmail.com, karim_nemra@yahoo.fr, imoteahst@outlook.com, abdelouahab.bazoula@gmail.com

[2] Laboratoire LIMPAF, Université de Bouira, Bouira, Algeria
louadj_kahina@yahoo.fr

[3] Centre de développement des technologies avancées, CDTA, Baba Hassen, Algiers, Algeria
mhamerlain@cdta.dz

Abstract. This paper describes an approach that combines the navigation data given by a Doppler Velocity Logs (DVL), the MTi Motion Reference Unit (MRU) and a Mechanically Scanned Imaging Sonar (MSIS) as a principal sensor to efficiently solve underwater Simultaneous Localization and Mapping (SLAM) problem in structured environments such as marine platforms, harbors, or dams, etc. The MSIS has been chosen of its capacity to produce a rich representation of the environment. In recent years, to solve the SLAM Autonomous Underwater Vehicle (AUV) problem, very few solutions have been proposed. Our contribution has introduced a method based on the Nonlinear H-infinity filter ($NH\infty$) to solve the SLAM-AUV problem. In this work, the $NH\infty$-SLAM algorithm is implemented to construct a map in partially structured environments and localize the AUV within this map. The data-set used in this paper are taken from the experiments carried out in a marina located in the Costa Brava (Spain) with the Ictineu AUV which is necessary to test different SLAM algorithms. The validation of the proposed algorithm through simulation in offline is presented and compared to the EKF-SLAM algorithm. The $NH\infty$-SLAM algorithm provides an accurate estimate than EKF-SLAM and good results were obtained.

Keywords: Autonomous navigation · Underwater vehicle
Localization · Map building

1 Introduction

Recently, efforts have been made to provide the vehicles with a greater degree of autonomy. This is achieved by equipping the submersible vehicles with their own power source and giving it the capacity to determine its actions based on the

O. Demigha et al. (Eds.): CSA 2018, LNNS 50, pp. 193–203, 2019.
https://doi.org/10.1007/978-3-319-98352-3_21

inputs from its own sensors and a pre-defined mission plan. These vehicles called Autonomous Underwater Vehicles (AUVs) have already succeeded in performing different types of tasks and have also presented new challenges. Simultaneous Localization and Mapping (SLAM) is an essential alternative for mobile robot navigation in unknown environments where global positioning system (GPS) is not available. Sometimes a GPS reading gathered at the surface is overlaid on the image, introducing an approximate location of the underwater vehicle. The problem of SLAM has attracted immense attention in the robotics literature. SLAM addresses the problem of the AUV moving through an environment in which no map is available a priori. When the underwater vehicle explores an unknown area, it must also be able to build a map of an unknown environment and localize itself in real time. The objective of developing navigation systems for AUV is to be used in industrial applications such as the inspection of hydroelectric dams, harbors, underwater cables, pipes and oceanography research.

This paper concentrates on autonomous underwater vehicle SLAM to build a map within an unknown environment while keeping track of its current position and it connected to platforms where structures made out of vertical planes are available in acoustic images [1]. The SLAM techniques use information from onboard robot sensors to provide feature relative localization give an priori map of the environment using the Extended Kalman Filter (EKF) which is presented by Chatila and Moutarlier [2]. Several algorithms of the EKF-SLAM have been implemented in different environments, such as indoors, underwater and outdoors [3]. There is another filter called Unscented Kalman Filter (UKF) that uses an only representation of a Gaussian random variable in (N) dimensions using sigma points ($2N+1$) [4]. There is another approach, known as Fast-SLAM utilizes Rao-Black wellised particle filter to solve the SLAM problem efficiently [5]. SLAM is one of the most active topics in robotics research and, although a large number of works have already been presented, there are still very few approaches applied to the field of underwater robotics.

As an alternative approach, there is another algorithm, known as the nonlinear H-infinity filter ($NH\infty$). The aim of our work is the implementation of a robust hybrid SLAM for a submarine robot. To do this, the AUV must merge the data from its proprioceptive sensor (DVL) and exteroceptive sensor (imaging sonar). Data fusion is based on two approaches. First, $H\infty$ filter is optimal in terms of minimizing the $H\infty$-norm of the gain between a set of disturbance inputs, and the estimation error [6]. Thus, the filter may be characterized by the fact that the worst-case gain is minimized. In contrast, the Kalman filter minimizes the mean square gain between the disturbances and the estimation error. The second difference is that the minimum mean square estimate of Kalman filter commutes with linear operations. However, the minimal $H\infty$-norm estimate does not possess this property and the $H\infty$ optimal estimator depends on the plant output being estimated. Our $H\infty$ filtering procedure uses a similar procedure as in Einicke and White [7]. An important contribution of this work is the development of a method for extracting acoustic features from underwater images acquired with a mechanically scanned imaging sonar (MSIS).

This method makes it possible to use a MSIS for SLAM as a low-cost alternative to digital sensors. The MSIS provides the acoustic imagery that feeds the feature extraction algorithm. The development of a new algorithm based on ($NH\infty$) is proposed for state and parameter estimation which is robust and stable to modeling uncertainties making it suitable for Autonomous Underwater Vehicle (AUV) localization and mapping problem. This new strategy retains the near optimal performance of the ($NH\infty$) when applied to an uncertain system, it has the added benefit of presenting a considerable improvement in the robustness of the estimation process. A data-set obtained during an experiment performed with the Ictineu AUV serves as a test for the proposed SLAM algorithm [8]. The vehicle, a low-cost research platform of reduced dimensions developed in the Underwater Robotics Laboratory of the University of Girona, played out 600 m the trajectory in an abandoned marina situated in the Spanish Costa Brava, near St. Pere Pescador [8].

The paper is organized as follows. Section 2 depicts the process model of the AUV and the observation model. The Sect. 3 describes the SLAM-AUV and also proposes our version of the implementation of the $NH\infty$-SLAM algorithm in detail. Results, discussion and conclusion are provided in Sect. 4.

2 Process Model of AUV

The AUV used in this work is an autonomous underwater vehicle. It is small ($0.8 \times 0.5 \times 0.5$ m), light (60 km in air), and very shallow-water vehicle (depth rating 30 m). The Ictineu vehicle (see Fig. 1) [8]. The vehicle's state vector X_V is represented as

$$X_V = (x, y, z, \psi, u, v, w, r, \psi_{L_0})^T \tag{1}$$

where, as defined in [17], $(x,\ y,\ z)^T$ represent the position, ψ represent the heading of the vehicle in the local reference frame L and $(u,\ v,\ w,\ r)^T$ are their corresponding linear and angular velocities on the vehicle's coordinate frame V. The term ψ_{L_0} represents the angle between the initial vehicle heading at the time step $T = 0$. From the [8,17] and as can be seen in Fig. 2, the discrete equation model of the AUV is

$$\begin{pmatrix} x \\ y \\ z \\ \psi \\ u \\ v \\ w \\ r \\ \psi_{L_0} \end{pmatrix}_{(k+1)} = \begin{pmatrix} x + (uT + n_u T^2/2)cos(\psi) - (vT + n_v T^2/2)sin(\psi) \\ y + (uT + n_u T^2/2)sin(\psi) + (vT + n_v T^2/2)cos(\psi) \\ z + wT + n_w T^2/2 \\ \psi + rT + +n_r T^2/2 \\ u + n_u T \\ v + n_v T \\ w + n_w T \\ r + n_r T \\ \psi_{L_0} \end{pmatrix}_{(k)} \tag{2}$$

Fig. 1. AUV Ictineu representation [15].

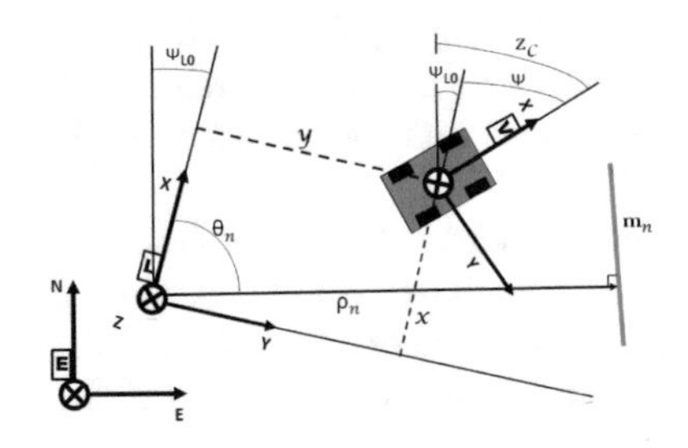

Fig. 2. AUV Kinematic model with different reference coordinate frames.

2.1 Direct Observation Line Based Model

We assume that we have m_i numbers of feature in the environment located at known points. It scans in a $2D$ plan by rotating a fanshaped sonar beam through a series of small-angle steps [8,13]. As imaging sonars operate at high frequency or very high frequency, typically from 100 kHz to 2 MHz. In the case of the representation of landmark by line, each line of the scan is considered a landmark and is represented by two parameters $[\rho_j^V;\ \theta_j^V]$. It contains the range and bearing of a line feature in the local frame respectively. The direct observation line based model Z_j^V is given by [8,14]:

$$Z_j^V(k+1) = h_i(X(k+1), N_{Z,j}) \tag{3}$$

$$Z_j^V(k+1) = \begin{bmatrix} \rho_j^V \\ \theta_j^V \end{bmatrix} = \begin{bmatrix} \rho_i - xcos\theta_i - ysin\theta_i \\ \theta_i - \psi \end{bmatrix} + N_{Z,j} \tag{4}$$

where h_i represents the nonlinear measurement function, $X(k+1)$ defines the state vector of the vehicle and i feature, $[\rho_j^V,\ \theta_j^V]^T$ are polar parameters in the vehicle reference frame and $N_{Z,j}$ represents the noise affecting the line feature observation, is a zero-mean white noise with covariance R_j.

2.2 Inverse Observation Line Based Model

Where the AUV state $(x,\ y,\ z)^T$ and noticing a feature m_{new} with coordinates $(x_{new}, y_{new}, z_{new})^T$ using a MSIS; let $Z_i = [\rho_{j,L}, \theta_{j,L}]^T$ is the observation of landmark m_{new} by the AUV. The feature mapping model is an inverse observation model $m_{new,j}$, knowing the state of the AUV and observation, it can be written as follows

$$m_{new,j} = \begin{bmatrix} \rho_{j,L} \\ \theta_{j,L} \end{bmatrix} = \begin{bmatrix} xcos(\psi + \theta_i) + ysin(\psi + \theta_i) + \rho_i \\ \psi + \theta_i \end{bmatrix} \quad (5)$$

3 SLAM-AUV

The autonomous underwater vehicles have the ability to go into inaccessible areas and do what other submarines could not do and go where they could not go. SLAM is the problem of constructing a model of the environment being traversed with onboard sensors [9], while at the same time maintaining an estimate of the vehicle location within the model [10,11]. As an alternative approach of AUV, there is a robust filter, known as the nonlinear H-infinity filter ($NH\infty$). This filter provides a robust and stable estimate to modeling uncertainties and errors [7]. Our $H\infty$ filtering procedure uses a similar procedure as in Einicke and White [7,12]. The strategy of ($NH\infty$) attempts to estimate the states given in the Eqs. '(6) and (7)' while satisfying the $H\infty$ filter performance criteria with respect to Δ_i and their norm bounds:

$$X(k+1) = f(X(k), U(k+1), W(k+1)) \quad (6)$$

$$Z(k+1) = h(X(k+1), V(k+1)) \quad (7)$$

The Eqs. 6 and 7 can be rewritten as:

$$\begin{aligned} X(k+1) &= \nabla F_V(k)X(k) + \nabla F_W(k)W(k) + \Omega(k) \\ &+ \Delta_1(\tilde{X}(k|k)) + \Delta_2(\tilde{X}(k|k))W(k) \end{aligned} \quad (8)$$

$$Z(k+1) = H(k)X(k+1) + V(k+1) + \Psi(k+1) + \Delta_3(\tilde{X}(k+1|k)) \quad (9)$$

$NH\infty$-SLAM Algorithm

- **Initialization**
  The state vector associated with the submarine robot and the initial covariance matrix are presented as follows:

$$\hat{X}(0) = \hat{X}_V(0) = \begin{bmatrix} 0\,0\,0\,0\,0\,0\,0\,0\,\hat{\psi}_{L_0} \end{bmatrix}^T \quad (10)$$

$$P(0) = P_V(0) = \begin{bmatrix} 0_{8\times 8} & 0_{8\times 1} \\ 0_{1\times 8} & \sigma^2_{\psi_{L_0}} \end{bmatrix} \quad (11)$$

  where: ψ_{L_0} takes its value from the available measurement of the attitude and a low-cost MRU sensor and $\sigma^2_{\psi_{L0}}$ is initialized accordingly with the precision of the same sensor.
- **Prediction Time Update**
  The prediction stage is a process, which deals with vehicle motion increases the uncertainty of the vehicle pose estimate. First, the state vector is augmented with a control input $U = (u,\ v,\ w,\ r)^T$.

– **Data association and update**

$$F_V = \left(\frac{\partial f}{\partial x} \; \frac{\partial f}{\partial y} \; \frac{\partial f}{\partial z} \; \frac{\partial f}{\partial \psi} \; \frac{\partial f}{\partial u} \; \frac{\partial f}{\partial v} \; \frac{\partial f}{\partial w} \; \frac{\partial f}{\partial r} \; \frac{\partial f}{\partial \psi_{L_0}} \right) \tag{12}$$

where F_V represents the Jacobian of $f(\hat{X}_V, U(k))$ with respect to $\hat{X}_V$. From the kinematic model, F_V is

$$F_V = \frac{\partial f}{\partial X_V}(\hat{X}_V(k), 0) \tag{13}$$

The prediction equations are presented as follows

$$\begin{cases} \hat{X}(k+1) = f(\hat{X}(k), U(k)) \\ \hat{P}(k+1|k) = \nabla F_V P(k|k) \nabla F_V^T + \nabla F_W Q \nabla F_W^T \end{cases} \tag{14}$$

where Q represents a noise with covariance matrix of input control U.

$$Q = \begin{bmatrix} \sigma_u^2 & 0 & 0 & 0 \\ 0 & \sigma_v^2 & 0 & 0 \\ 0 & 0 & \sigma_w^2 & 0 \\ 0 & 0 & 0 & \sigma_r^2 \end{bmatrix} \tag{15}$$

$\nabla F_V(k)$ and $\nabla F_W(k)$ are the Jacobian matrices of $f(\hat{X}(k), U(k))$ with respect to $\hat{X}(k)$ and $U(k)$ respectively such as:

$$\nabla F_V(k) = \begin{bmatrix} F_V(k) & 0_{9\times 2n} \\ 0_{9\times 2n} & I_{2n\times 2n} \end{bmatrix} \tag{16}$$

and

$$\nabla F_W(k) = \begin{bmatrix} F_W(k) \\ 0_{2n\times 4} \end{bmatrix} \tag{17}$$

The Jacobian necessary for this approximation is denoted by F_W, it is the derivative motion model f with respect to the input control $U = \left[u \; v \; w \; r\right]^T$.

$$F_W = \left[\frac{\partial f(\hat{X}(k), U(k))}{\partial U(k)}\right] = \left[\frac{\partial f}{\partial u} \; \frac{\partial f}{\partial v} \; \frac{\partial f}{\partial w} \; \frac{\partial f}{\partial r}\right] \tag{18}$$

The covariance matrix can be written as follows:

$$\begin{aligned} P(k+1|k) &= \begin{bmatrix} F_V(k) & 0_{9\times 2n} \\ 0_{9\times 2n} & I_{2n\times 2n} \end{bmatrix} P(k) \\ &\begin{bmatrix} F_V(k) & 0_{9\times 2n} \\ 0_{9\times 2n} & I_{2n\times 2n} \end{bmatrix}^T + \begin{bmatrix} F_W(k) \\ 0_{2n\times 4} \end{bmatrix} Q \begin{bmatrix} F_W(k) \\ 0_{2n\times 4} \end{bmatrix}^T \end{aligned} \tag{19}$$

To calculate the prediction error between the measure and its prediction, the innovation term $S_{j,i}$ is written as follows:

$$\nu_{j,i}(k+1) = Z_j^V(k+1) - h_i(\hat{X}(k+1|k)) \tag{20}$$

$$S_{j,i}(k+1) = H_i(k+1)P(k+1|k)H_i(k+1)^T + R \tag{21}$$

where $H_i(k+1)$ sets the Jacobian matrix of h_i with respect to the augmented state vector $\hat{X}(k+1|k)$

$$H_i(k+1) = \frac{\partial h_i(k)}{\partial X}(\hat{X}(k+1|k), 0) \tag{22}$$

such as:

$$H_V(k) = \frac{\partial h_i(k)}{\partial X_V(k)}, H_{m_i}(k) = \begin{bmatrix} 1 & xsin(\theta_i) - ycos(\theta_i) \\ 0 & 1 \end{bmatrix} \tag{23}$$

To verify the validity of the corresponding, an individual compatibility (IC) is implemented using the Mahalanobis distance [8, 18]:

$$D^2_{j,i} = \nu_{j,i}(k+1)^T S_{j,i}(k+1)^{-1} \nu_{j,i}(k+1) < \chi^2_{d,\alpha} \tag{24}$$

where $d = dim(h_i)$ and α represents the data association threshold.
For all the observers observed at time $(k+1)$, we get a real measure: $Z_i^V(k+1)$.
The predicted measure at time $(k+1)$ is obtained as follows

$$\hat{Z}_i^V(k+1) = h_i(\hat{X}(k+1)) \tag{25}$$

Since $h_i(k)$ depends on the position of the AUV $X_V(k)$ and the position of the ith feature (m_i) in the associated state vector with the extraction feature. For this, we rewrite it in the following form:

$$H_i^V(k+1) = H_i(k+1) \times \nabla G_i^V \tag{26}$$

where

$$\nabla G_i^V = \begin{pmatrix} 1\,0\,0\,0\,0_{1\times(2i+7)}\,0\,0\,0_{1\times(2n-(2i+8))} \\ 0\,1\,0\,0\,0_{1\times(2i+7)}\,0\,0\,0_{1\times(2n-(2i+8))} \\ 0\,0\,1\,0\,0_{1\times(2i+7)}\,0\,0\,0_{1\times(2n-(2i+8))} \\ 0\,0\,0\,1\,0_{1\times(2i+7)}\,0\,0\,0_{1\times(2n-(2i+8))} \\ 0\,0\,0\,0\,0_{1\times(2i+7)}\,1\,0\,0_{1\times(2n-(2i+8))} \\ 0\,0\,0\,0\,0_{1\times(2i+7)}\,0\,1\,0_{1\times(2n-(2i+8))} \end{pmatrix} \tag{27}$$

For all the associated landmarks between the actual measure and the predicted measure, at time $(k+1)$, we have the following correction equations:

$$S_{j,i}(k+1) = H_i^V(k+1)P(k+1|k)(H_i^V(k+1))^T + R_j \tag{28}$$

$$K_{j,i}(k+1) = P(k+1|k)(H_i^V(k+1))^T[S_{j,i}(k+1)]^{-1} \tag{29}$$

$$\begin{aligned} \hat{X}(k+1|k+1) = \hat{X}(k+1|k) + K_{j,i}(k+1) \\ Z_j^V(k+1) - h_i(\hat{X}(k+1|k)) \end{aligned} \tag{30}$$

$$\begin{aligned} \Sigma(k+1) = [&eye(11)P(k+1|k)eye(11)^T - \gamma^2 eye(11) \\ &- eye(11)P(k+1|k)JH^T; -JHP(k+1|k) \\ &eye(11)^T JHP(k+1|k)JH^T + R_j] \end{aligned} \tag{31}$$

$$P(k+1|k+1) = P(k+1|k) - P(k+1|k)[-eye(11)^T \\ JH^T](\Sigma(k+1))^{-1}[-eye(11); JH] \\ P(k+1|k) \tag{32}$$

where the covariance matrix on the measurement noise is represented as

$$R_j = \begin{bmatrix} \sigma^2_{\rho_i} & 0 \\ 0 & \sigma^2_{\theta_i} \end{bmatrix} \tag{33}$$

- **Map management**
  It is a step of increasing the state vector by the new detector feature [9,16,19].

4 Implementation, Simulations and Discussion

The sampling rates used for each filter and sensors used in this study are as follows: $f_{EKF} = f_{NH\infty} = 15$ Hz, $f_{MSIS} = 675$ KHz, $f_{DVL} = 1.5$ KHz

The simulation results provided in the following figures represent the estimated AUV position obtained using the EKF, and $NH\infty$ filters respectively with the kind of uncertainties which have been considered as follows:

$\sigma_x = \sigma_y = \sigma_z = 5 \times 10^{-6}$ m and $\sigma_\psi = 5 \times 10^{-3}$ rad

In order to test the reliability of the proposed algorithm, we used the experimental data on [7,15] by David Ribas. The Ictineu AUV accumulated an informational index along a 600 m worked direction which incorporated a small circle

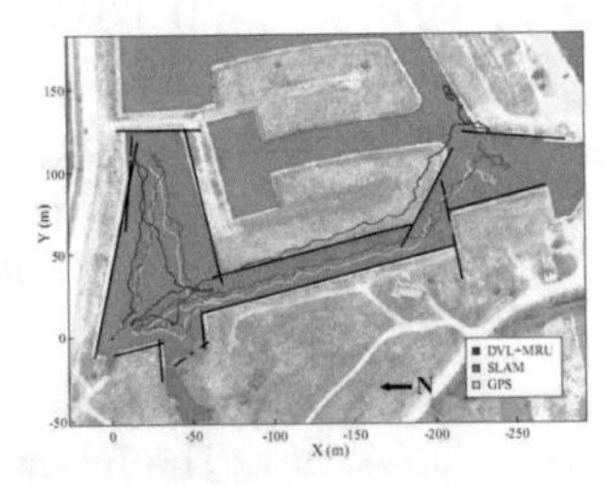

Fig. 3. Scenario satellite image of an abandoned marina represented by [8].

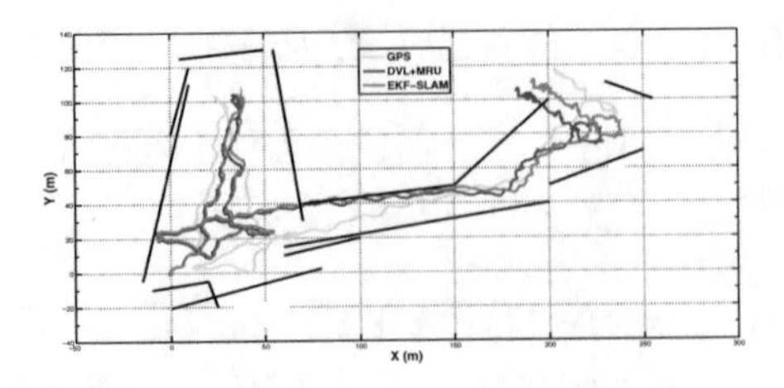

Fig. 4. Trajectory with EKF-SLAM algorithm.

around the principal water tank and a 200 m straight path through an outgoing canal (See Fig. 3). Figures 4 and 5 represent the trajectories obtained with the EKF and $NH\infty$ filters. It can be noticed that the map of $NH\infty$ filter provides the best result, in this case, as can be seen from Fig. 6, the $NH\infty$-SLAM algorithm gives the best results position and mapping than the EKF-SLAM algorithm. As it can be seen, the trajectory obtained by running the $NH\infty$ filter with performing the position updates with imaging sonar measurements is more accurate compared to the trajectory acquired by EKF. On the other hand, without using MSIS data, the trajectory obtained only by integrating the velocity measurements and does not update the estimate information, the process is inherently affected by drift. The evolution of vehicle position errors is shown in Fig. 7, were found that the $NH\infty$-SLAM performed significantly better than the EKF-SLAM, in terms of estimation error. It is clear, that the estimated uncertainty is very small of $NH\infty$-SLAM compared to the true uncertainty specifically for the X and Y states of the EKF-SLAM. In this case, the map generated by $NH\infty$-SLAM is not degraded in quality compared to EKF-SLAM.

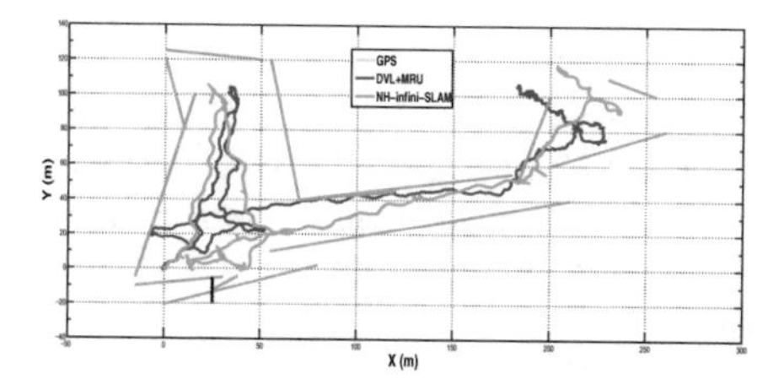

Fig. 5. Trajectory with $NH\infty$-SLAM algorithm.

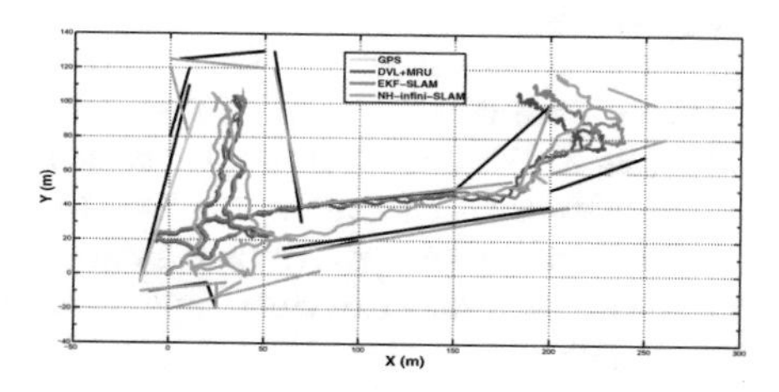

Fig. 6. Trajectories comparison for different filters.

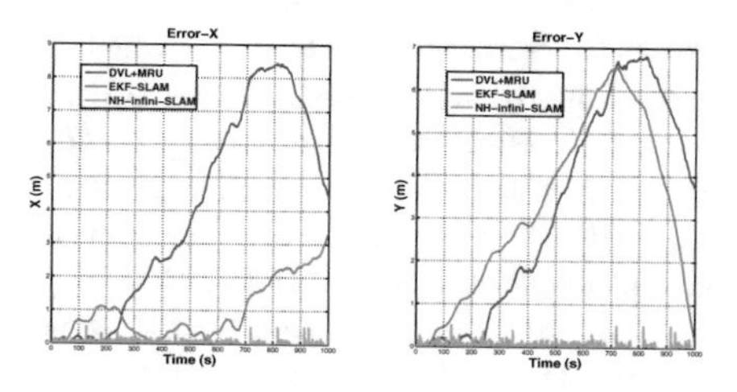

Fig. 7. Estimated position errors for different filter.

5 Conclusion

The aim of this paper is the application of SLAM techniques to the field of autonomous underwater navigation in structured underwater environments using an imaging sonar. The main contributions of this work consisted in implementing the $NH\infty$-SLAM algorithm. The proposed $NH\infty$-SLAM algorithm is validated through simulations and compared to the EKF-SLAM algorithm. A good result is obtained by $NH\infty$ which is more realistically than EKF of the same scene with the abandoned marina data-set.

Acknowledgment. A data-set obtained during an experiment performed with the Ictineu AUV serves as a test for our proposed SLAM algorithm in off-line. The authors are grateful to David Ribas, Pere Ridao, Juan Domingo Tardos and Jose Neira for their help with the experimental setup and data-set acquisition. The data-set of this work has been funded in part by projects DPI2005-09001-C03-01 and DPI2006-13578 of the Direction General Investigation of Spain.

References

1. Ribas, D., Ridao, P., Tardos, J.D., Neira, J.: Underwater SLAM in man-made structured environments. J Field Robot. **25**(8), 1–24 (2008)
2. Moutarlier, P., Chatila, R.: An experimental system for incremental environment modeling by an autonomous mobile robot. In: The 1st International Symposium on Experimental Robotics, Montreal (1989)
3. Davison, A.J., Reid, I.D., Molton, N.D., Stasse, O.: MonoSLAM: real-time single camera SLAM. In: Conference on Pattern Analysis and Machine Intelligence, vol. 29 (2007)
4. Holmes, S., Klein, G., Murray, D.W.: A square root unscented Kalman filter for visual mono SLAM. In: Proceedings of the IEEE International Conference on Robotics and Automation, Pasadena, CA, pp. 3710–3716, May 2008
5. Montemerlo, M., Thrun, S., Koller, D., Wegbreit, B.: FastSlam : a factored solution to the simultaneous localization and mapping problem. In: AAAI (2002)
6. Basar, T., Baernard, P.: $H\infty$ optimal control and related minimax design problems. In: A Dynamic Game Approach Systems and Control : Foundations and Applications. Birkhauser (1991)
7. Einicke, G., White, L.: Robust extended kalman filtering. IEEE Trans. Signal Process. **47**(9), 2596–2599 (1999)
8. Ribas, D., Ridao, P., Neira, J.: Underwater SLAM for structured environments using an imaging sonar. In: Springer Tracts in Advanced Robotics (2010)
9. Demim, F., Nemra, A., Louadj, K.: Robust SVSF-SLAM for unmanned vehicle in unknown environment. J. IFAC-Pap. OnLine Sci. Dir. **49**, 386–394 (2016)
10. Demim, F., et al.: Simultaneous localization and mapping algorithm for unmanned ground vehicle with SVSF filter. In: IEEE 8th International Conference on Modelling, Identification and Control, pp. 155–162 (2016)
11. Demim, F., et al.: A new adaptive smooth variable structure filter SLAM algorithm for unmanned vehicle. In: IEEE 6th International Conference on Systems and Control, pp. 6–13 (2017)

12. Nemra, A., Aouf, N.: Robust airborne 3D visual simultaneous localization and mapping with observability and consistency analysis. J Intell. Robotic Syst. **55**, 345–376 (2009)
13. Tena, I., Petillot, Y., Lane, D.M., Salson, C.: Feature extraction and data association for AUV concurrent mapping and localisation. In: Proceedings of the IEEE International Conference on Robotics and Automation, Seoul, Korea, pp. 2785–2790 (2001)
14. Ribas, D., Ridao, P., Neira, J., Tardos, J.D.: Line extraction from mechanically scanned imaging sonar. In: Martietal, J. (ed.) IbPRIA 2007, Part I, pp. 322–329. Springer, Heidelberg (2007)
15. Ribas, D.: Underwater SLAM for structured environments using an imaging sonar. Ph.D. Thesis, Girona university (2008)
16. Demim, F., Nemra, A., Louadj, K., Hamerlain, M., Bazoula, A.: Cooperative SLAM for multiple UGVs navigation using SVSF filter. J Control Measur. Electron. Comput. Commun. (Automatika) **58**(1), 119–129 (2017)
17. Fossen, T.I.: Marine Control Systems: Guidance, Navigation and Control of Ships, Rigs and Underwater Vehicles, 1st Edn. Marine Cybernetics AS, Trondheim (2002)
18. Demim, F., Nemra, A., Louadj, K., Hamerlain, M., Bazoula, A.: An adaptive SVSF-SLAM algorithm to improve the success and solving the UGVs cooperation problem. J. Exp. Theor. Artif. Intell. **30**(3), 389–414 (2017)
19. Demim, F., et al.: Simultaneous localization, mapping, and path planning for unmanned vehicle using optimal control. J Adv. Mech. Eng. **10**, 1–25 (2018)

3D Polynomial Interpolation Based Local Binary Descriptor

Elhaouari Kobzili[1](✉), Cherif Larbes[1], Ahmed Allam[1], and Fethi Demim[2]

[1] Department of Electrical Engineering, National Polytechnic School, 10 Avenue Hassen Badi BP 182, 16200 El Harrach, Algiers, Algeria
elhaouari.kobzili@g.enp.edu.dz, cherif.larbes@enp.edu.dz, ahmedallam900@gmail.com

[2] Laboratoire Robotique et Productique, Ecole Militaire Polytechnique, Bordj El Bahri, Algiers, Algeria
demifethi@gmail.com

Abstract. Many efforts are devoted to develop binary descriptors due to their low complexity, and flexibility in case of embedded systems. Almost all works on binary descriptor conception didn't exploit all information of a given patch; they just involved pixels intensities into binary test process. This kind of solution lack efficiency on patch description. In this paper, we propose to design a new descriptor based on 3D polynomial interpolation by used pixels intensities. We must take into account geometric positions of pixels. We suggest to divide the patch into equal grid cells (sub patches). Each sub patch undergoes a dimension augmentation. It becomes a 3-dimensional vector by considering intensities values as the third dimension. Based on 3D polynomial interpolation, we approximate the point cloud by a surface. This step is followed by a binary tests between all coefficients of polynomials situated in neighborhoods. Our method shows a considerable discrimination in case of high similarity. The results of our approach are evaluated on a well-known benchmark dataset exhibit a considerable robustness and reliability in front of severe changes. A computation costing is reported in the end of results section.

Keywords: Local Binary Descriptor · 3D polynomial interpolation · Recognition Matching · Computer vision · Pattern recognition

1 Introduction

In the last decade visual feature is become the base of many computer vision applications. We cite: object classes recognition and tracking [1], frame matching and stitching [2], augmented reality [3], and simultaneous localization and mapping [4, 5]. In order to deal with real time condition, the conception of a fast descriptor is crucial especially when we want to do a complex tasks as for the SLAM. Recently many works have been realized on binary descriptors, but almost approaches stay under expectation in term of accuracy, robustness, and efficiency against the well-known Scale Invariant Feature Transform (SIFT) [1]. By analyzing of the previous solutions of binary descriptors, and by exploiting the result of the work named as: a Moments based Local Binary Descriptor

O. Demigha et al. (Eds.): CSA 2018, LNNS 50, pp. 204–214, 2019.
https://doi.org/10.1007/978-3-319-98352-3_22

MOBIL [6], we propose a new scheme of a binary descriptor. However, instead of just exploited pixels intensities alone, we suggest to profit from pixels locations to generate a 3D model. This three dimension model has a considerable ability of distinctiveness between two very similar patches. It is a good manner of description in case of images matching. The main contributions of this work are:

- The elaboration of binary descriptor based on 3D interpolation.
- The studying of the tradeoff between accuracy and real time constraint.
- The including of the geometric modeling component for description instead of just intensities.

This paper is organized as follows: first, we will give the state of the art of our work. Second, we present an overview of our developed system. Third, we will present results of our contribution. Finally, we end this paper by a conclusion in which, we suggest some perspectives to improve our descriptor.

2 Related Work

Many vision tasks used features (ex: key points) with an efficient descriptor to maximize the score of correct match. However, the famous detector/descriptor is SIFT [1], it shows high robustness against severe changes. The SIFT presents a considerable complexity. It's not a suitable solution in case of real time applications. In order to speed up the matching process, Bay et al. suggest Speed-up Robust Feature (SURF) [7] as an improved solution near to SIFT in term of performance. The SURF presents a high reliability in case of real time. The majority of the previous approaches are considered under requirements in case of real time exigency. Recently, it has been appeared a faster binary descriptors. These descriptors allow to do a fast matching based on hamming distance by used the XOR logical operator. We mention Binary Robust Independent Elementary Features BRIEF [8]. This descriptor operates on a large intensities comparison, it is limited in term of rotational invariance. The binary descriptor efforts is followed by Binary Robust Invariant Scalable Key points (BRISK) [9]. It is qualified faster than SIFT and SURF. It is adapted with real time purpose. It is able to do the description in multilevel of scales. Recently, Ethan et al., propose an interested approach known as oriented fast and Rotated BRIEF (ORB) [10]. The ORB's authors overcame the lack of BRIEF descriptor by assuming the calculation of patch orientation based on patch moments [11]. In our work, we opt on the same method to order the patch into a referential orientation. An attractive solution was proposed by Alexandre et al., named as Fast Retina key point (FREAK) [12]. It is a bio inspired from human visual system (the retina). This approach is suggested to be faster and more robust than the previous descriptors. The author Xin et al. are proposed a descriptor known as: Ultra-Fast Feature for Scalable Augmented Reality on Mobile Devices (LDB) [13]. Authors didn't based just on the intensity test. They involved the first order gradient to provide a high description quality. This kind of descriptor suffers on description in case of severe viewpoints changes, and blur increase. In our work, we are based on very recent work of Bellarbi et al. (MOBIL) [6]. In his work, he mentions the advantage of exploited pixels intensities and theirs locations. The spatial dispatching of pixels plus its intensities make a synergy of discrimination between patches. The MOBIL's author improves the first versus of his descriptor, by used polar coordinate, its name is POLAR_MOBIL [14]. This work encourages us to explore deeply the geometric description to elaborate the best 3D representation. In this paper, we are based on the 3D polynomial interpolation. The objective is to find the curve which approximates the cloud of points in 3-dimensions of a given sub patch. We use polynomial of Lagrange [15] to model sub patches and by consequence, we get the 3D representation. The 3D model of cloud of points were addressed by many authors [16–18]. The elaboration of a geometric model were included in many fields as computer-aided design (CAD) [17], software as AutoCAD, infography, and reverse engineering.

3 An Overview of Our System

The conception of our system is divided into two steps: first, key points detection and second, key points description, as follow:

3.1 Detector

To do a description, first, we must detect the strongest points appeared in an image. In our work we adopt a fast detector known as Features from Accelerated Segment Test (FAST) [19]. Since, it do a quick scan of the image to find all probable key points with respect to a threshold. These key points are tested through a Non_Maxima suppression [19] to eliminate wrong points defining an edge and to keep just real corner. To be sure that all points are detected on multi-level of scale, we try to do the detection of points on 8 levels of scale reduction ($s = 1/\sqrt{2}$). To ensure rotation invariance, we opt to follow the same step of ORB and MOBIL. They are inspired theirs methods from work of Hu [11]. In this context, the patch orientation is got by the calculation of zeroth (m_{00}) and the first moments (m_{01}, m_{10}) based on (1, 2). The patch is reoriented to a referential angle, as in Fig. 2. In our work, we take an augmented patch around each key point of size (48×48) which is bigger than the suggested for description. In order to get a consistent orientation, we evaluate moments in a circular region of radius r.

$$m_{pq} = \sum_{x=-r}^{r} \sum_{y=-r}^{r} x^p y^p P(x, y) \tag{1}$$

The orientation calculation of the patch is given by (2)

$$\theta = a \tan 2\left(m_{01}, m_{10}\right) \tag{2}$$

3.2 Descriptor

After we get N key points of an image, we extract a larger (48×48) oriented patch, and we catch the centered area of size (32×32) from the bigger patch. We operate in this patch the description method. This patch is divided into (4 × 4) sub patches of size (8 × 8) pixels, as in Fig. 1. The sub patches are augmented on dimension by considered the *u*, and *v* as the first and the second dimension. The intensity is considered as the third dimension. We try to find the curve which interpolates the 64 points of a sub patch. We note, that each pixel of a sub patch is a point in three dimension of coordinate (*u*, *v*, *Intensity*). For our descriptor, we propose to use the polynomial of Lagrange [15] to find the 3D model of all points belongs a sub patch. To evaluate the polynomial coefficients, we base on the Vandermonde approach [15]. For more details about coefficients evaluations, we follow these steps. The explicit form of a surface which interpolates the cloud of points of a considered sub patch is given by (3).

1	2	3	4
5	6	7	8
9	10	11	12
13	14	15	16

Fig. 1. Patch divided into (4 × 4) sub patches.

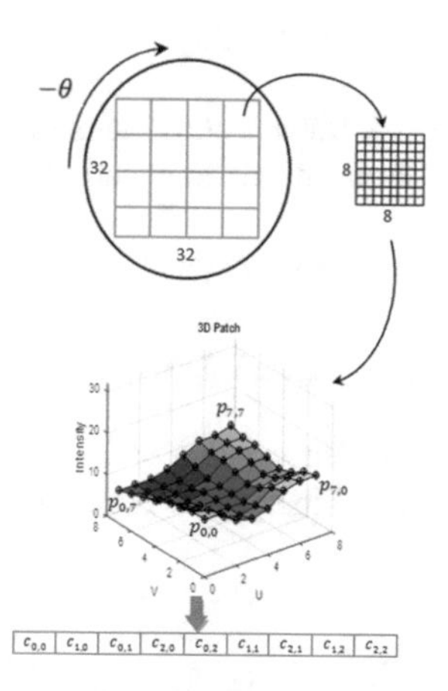

Fig. 2. 3D polynomial interpolation of a sub patch.

$$p_{u,v} = \sum_{i=0}^{n} \sum_{j=0}^{m} u^i v^j c_{i,j} \tag{3}$$

Where *u*, *v* are the 2D sub patches coordinates, and $c_{i,j}$ are the Vandermonde polynomial coefficients. By replacement, and with using coordinates of all the 3D points, we get a polynomial with unknown coefficients as in (4). With $p_{u,v}$ represents the pixel intensity at coordinate *u*, and *v*.

$$p_{u,v} = Poly\left(c_{i,j}\right) \tag{4}$$

Case 1: polynomial of degree $n = 2, m = 2$

$$p_{u,v} = c_{0,0} + c_{1,0}u + c_{0,1}v + c_{2,0}u^2 + c_{0,2}v^2 + \\ c_{1,1}uv + c_{2,1}u^2v + c_{1,2}uv^2 + c_{2,2}u^2v^2 \tag{5}$$

The formulation of (5) in matrix form is given by (6). With M is a matrix of size (64×9) given by (7). And "***e***" is a column vector of values equal to one (8).

$$M \cdot c_{i,j} = p_{u,v} \tag{6}$$

$$M = \left[e \ u \ v \ u^2 \ v^2 \ uv \ u^2v \ uv^2 \ u^2v^2\right] \tag{7}$$

$$e\left(size_{sub\_patch}, 1\right) = 1 \tag{8}$$

The solution of (6) provides the 9 polynomial coefficients $c_{i,j}$ given by (9). As defined by Matlab form, it is a rear division given by (10).

$$c_{i,j} = \left(M^T M\right)^{-1} M^T p_{u,v} \tag{9}$$

$$c_{i,j} = M \backslash p_{u,v} \tag{10}$$

We suppose the equality (11). Since coordinates of a sub patch u, v are invariants (12), it is possible to calculate T on off line to speed up the description process.

$$T = \left(M^T M\right)^{-1} M^T \tag{11}$$

$$u, v \in \{0, 1, 2, 3, 4, 5, 6, 7\} \tag{12}$$

In case of a polynomial of degree $n = 2, m = 2$, we get 9 coefficients for each sub patch. For each patch we perform a binary tests between all the nearest neighbors of sub

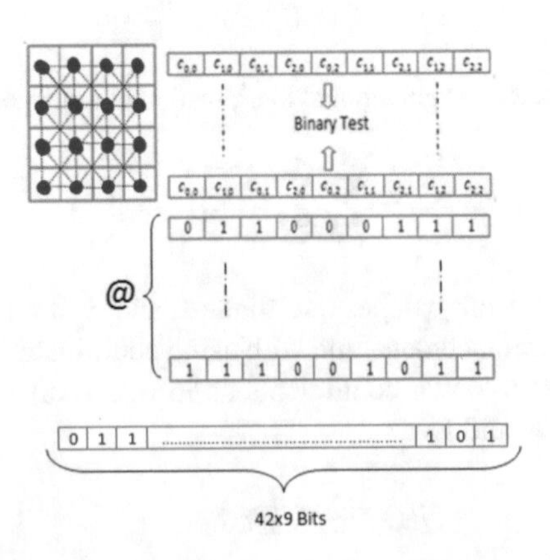

Fig. 3. Descriptor generation of a patch.

patches. Finally, and after the concatenation of all binary tests results, we get a binary descriptor of size (42 × 9) bits, as defined by Fig. 3.

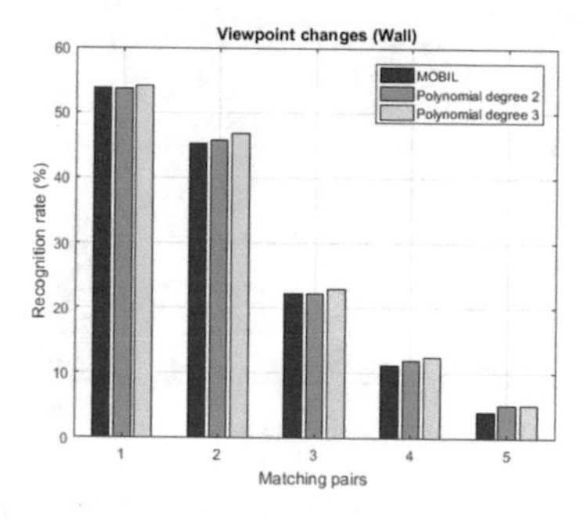

Fig. 4. Recognition rate with view point changes.

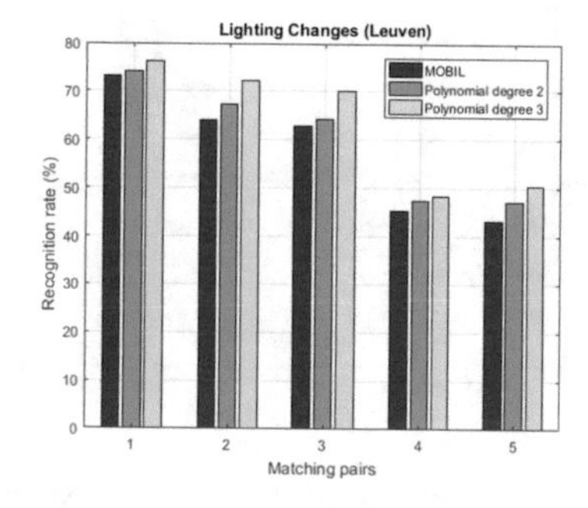

Fig. 5. Recognition rate with lighting changes.

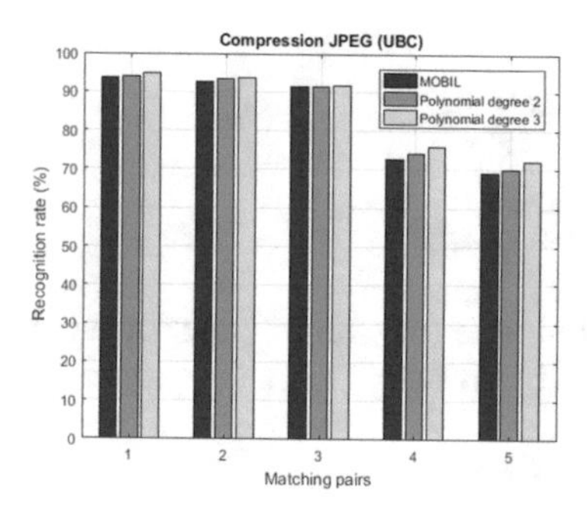

Fig. 6. Recognition rate with JPEG compression.

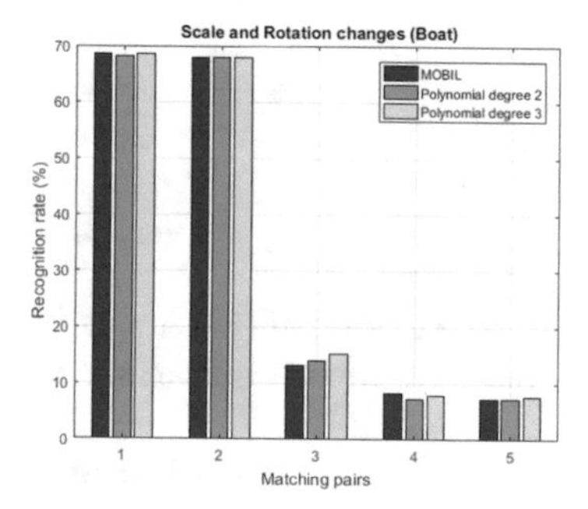

Fig. 7. Recognition rate with scale and rotation.

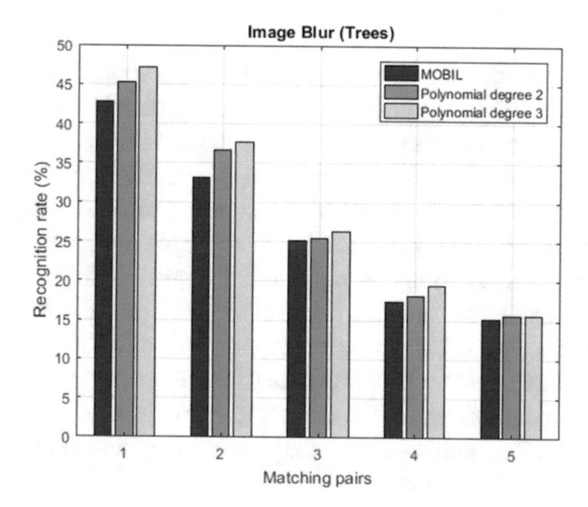

Fig. 8. Recognition rate with image blur.

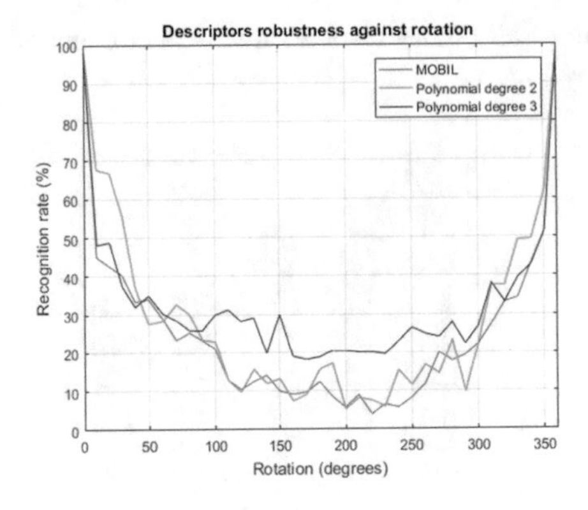

Fig. 9. Robustness of MOBIL, and polynomial descriptors against synthetic rotations.

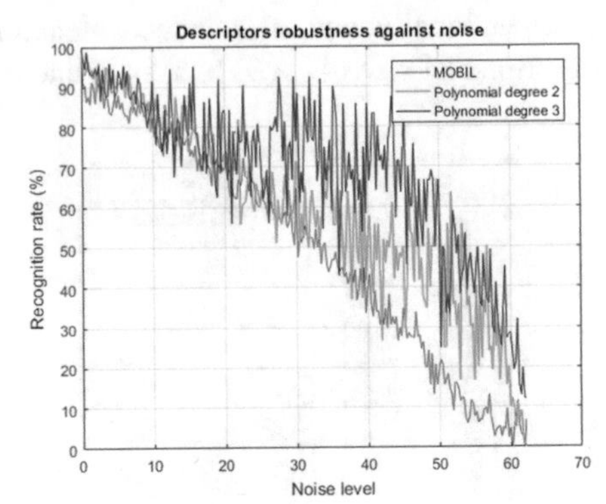

Fig. 10. Robustness of MOBIL, and polynomial descriptor against Gaussian noise.

Case 2: polynomial of degree $n = 3, m = 3$

$$\begin{aligned} p_{u,v} = c_{0,0} + c_{1,0}u + c_{0,1}v + c_{2,0}u^2 + \\ c_{0,2}v^2 + c_{1,1}uv + c_{2,1}u^2v + c_{1,2}uv^2 + \\ c_{2,2}u^2v^2 + c_{3,0}u^3 + c_{0,3}v^3 + c_{1,3}uv^3 + \\ c_{3,1}u^3v + c_{2,3}u^2v^3 + c_{3,2}u^3v^2 + c_{3,3}u^3v^3 \end{aligned} \tag{13}$$

The formulation of (13) in matrix form is similar to (6) with M larger than the first case. We notice that M is a matrix of size (64×16) given by (14).

$$\begin{aligned} A &= \left[e\, u\, v\, u^2\, v^2\, uv\, u^2v\, uv^2\, u^2v^2\right] \\ B &= \left[u^3\, v^3\, uv^3\, u^3v\, u^2v^3\, u^3v^2\, u^3v^3\right] \\ M &= [A\, B] \end{aligned} \tag{14}$$

The solution of this equations system provides 16 polynomial coefficients using (9). By a quick scan of the 3D interpolation field, a tradeoff between accuracy and real time leads us to use a polynomial of degree $n = 3, m = 3$. But in this situation, we get a longer descriptor of size (42×16), this last increase the processing time of the hamming distance. This polynomial degree induces on increasing of the calculation time for each coefficient. If the degree of the polynomial is more augmented, we violate the real time constraint. Moreover, increasing the polynomial degree didn't mean systematically improvement in interpolation quality, because we falls into the Runge problem [20]. This phenomenon is happened, when we have a cloud of points of high dynamic variation on the one side, and on the other side presence of points with slow dynamic. This situation carries the interpolation process to enter into numerical instability. The binary test for each two sub patches generates a binary number, which reflects the polynomial degree as defined in the following Eq. (15).

$$\tau(p,x,y) = \begin{cases} 1{:}c_{i,j}(x) \geq c_{i,j}(y) \\ 0{:}c_{i,j}(x) < c_{i,j}(y) \end{cases} \quad (15)$$

With $c_{i,j}(x)$, $c_{i,j}(y)$ represent coefficients for two sub patches in neighborhoods. In this paper, we select just the comparison between the nearest sub patches, because the geometric behavior has a significant meaning especially for sub patches of the nearest neighborhoods. In this optic, it rests just 42 possibilities of binary test between the (4×4) sub patches of Fig. 1. Finally, the descriptor generation is got by Eq. (16).

$$d_n(patch) = \sum_{i=1}^{n} 2^{i-1}\tau(p,x,y) \quad (16)$$

4 Results

The environment of development is Matlab, installed in a computer of a characteristic; Intel(R) Core (TM), processor i5 of frequency 2.40 GHz. The descriptor evaluation is realized by used Mikolajczyk dataset [21] as given by Figs. 11, 12, 13, 14 and 15.

Fig. 11. Matching with view point changes.

Fig. 12. Matching with lighting changes.

Fig. 13. Matching with compression.

Fig. 14. Matching with scale and rotation.

Fig. 15. Matching with blur.

4.1 Accuracy and Robustness Evaluation

Our descriptor shown a high performance compared to MOBIL descriptor in term of robustness against severe changes, especially view point in Fig. 4, the illumination change in Fig. 5, and the image blur in Fig. 8. Our descriptor is similar to MOBIL in case of Compression in Fig. 6, and rotation with zoom as given by Fig. 7. Our descriptor is favorable in case of place recognition for loop closing of the SLAM task due to the high recognition rate marked (Recognition rate is the number of inliers divided by the total number of matching). In Fig. 9, we remark that our descriptor resists more than MOBIL against synthetic in plane rotations. For a polynomial of degree 2, we notice, that it gives the same performance compared to MOBIL descriptor. But for degree 3, we have a good response in matching rate especially between 50 and 300 degrees. In Fig. 10, we notice that both descriptors fall down dramatically with noise level increase (Gaussian noise of mean 0 and a noise standard deviation increased gradually). But our descriptor (degree 2, 3) resists more than MOBIL and presents a remarkable robustness against noise. The chattering presented in our two graphs is due to the sensitivity to noise, and also to the inexactitude of rotation calculation. This fluctuation is defined into a limited band, and our descriptor always keeps on the recognition quality compared to MOBIL.

4.2 Real Time Evaluation

The efficiency of our descriptor increases remarkably with polynomial degree increase, but after the degree three, we can't get a considerable improvement. It is not interested when we wish to deal with real time purpose. The time needed to do one description in case of our approach is not preferable (Table 1), but it is possible to be decreased by choosing just five coefficients rather than 9 or 16. These coefficients selection permits to stay near computation time of MOBIL. It must balance between accuracy and computation time to deal with the same performance given by the complete descriptor. To go further and to reach real time condition, we can base on the look up table (LUT), to reorient the patch to a referenced angle. Anyway, it is possible to improve the computation time by used another software as Visual studio by profited from the power of OpenCV library. Designing a hardware architecture of our descriptor, and embedded it on the field programmable gate array (FPGA) can be an adapted solution for real time purpose. It's also possible for the other mobile devices.

Table 1. Time per description for MOBIL and our descriptor.

Descriptors	Time per description (ms)
MOBIL	0.994
Polynomial degree of $n = 2, m = 2$	1.75
Polynomial degree of $n = 3, m = 3$	2.8

5 Conclusion

This paper presents a timely subject, which is the binary descriptors. We are based on the geometric modeling to provide for a patch an intrinsic representation. The different results permit to judge that our solution can be a good alternative for patch description. This is due to their efficiency, and robustness compared to MOBIL. Our descriptor can be embedded easily in any numerical device due to their low computation time, and its binary form. It can easily be involved in system with high complexity as visual SLAM. As a future work, we suggest to do a deep learning to select the best binary test between polynomial coefficients. The goal is to eliminate the redundancy and select just the best comparison response in term of discrimination as in ORB and MOBIL. We plan to find another alternative of rotation invariance to improve our descriptor quality and minimize rotation sensitivity. We notice, that our descriptor is an appropriate solution to be adopted in our future designed visual Mono-SLAM.

References

1. Lowe, G.: Distinctive image features from scale-invariant keypoint. Int. J. Comput. Vis **60**(2), 91–110 (2004)
2. Snavely, N., Seitz, S.M., Szeliski, R.: Skeletal sets for efficient structure from motion. In: Proceedings of Computer Vision and Pattern Recognition (CVPR), vol. 1, p. 2. Publisher (2008)
3. Marchand, E., Uchiyama, H., Spindler, F.: Pose estimation for augmented reality: a hands-on survey. IEEE Trans. Vis. Comput. Graph. **22**(12), 2633–2651 (2016)
4. Klein, G.G., Murray, D.: Parallel tracking and mapping for small AR workspaces. In: Proceedings of IEEE ACM International Symposium Mixed Augmented Reality. Japan, pp. 225–234 (2007)
5. Mur-Artal, R., Montiel, J.M.M., Tardos, J.D.: ORB-SLAM: a versatile and accurate monocular slam system. IEEE Trans. Robot. **31**(5), 1147–1163 (2015)
6. Bellarbi, A., Otmane, S., Zenati, N., Benbelkacem, S.: MOBIL: a moments based local binary descriptor. In: International Symposium on Mixed and Augmented Reality (ISMAR 2014), pp. 251–252. IEEE, Munich (2014)
7. Bay, H., Ess, A., Tuytelaars, T., van Gool, L.: Speeded-up robust features (SURF). Comput. Vis. Image Underst. **110**(3), 346–359 (2008)
8. Calonder, M., Lepetit, V., Strecha, C., Fua, P.: BRIEF binary robust independent elementary features. In: Daniilidis, K., Maragos, P., Paragios, N. (eds.) Computer Vision, ECCV 2010, LNCS, vol. 6314, pp. 778–792. Springer, Germany (2010)
9. Leutenegger, S., Chli, M., Siegwart, R.Y.: BRISK: binary robust invariant scalable keypoints. In: Proceedings of the IEEE International Conference on Computer Vision (ICCV 2011), pp. 2548–2555. IEEE, Spain (2011)
10. Rublee, E., Rabaud, V., Konolige, K., Bradski, G.: ORB: an efficient alternative to SIFT or SURF. In: Proceedings of the IEEE International Conference on Computer Vision (ICCV 2011), pp. 2564–2571. IEEE, Spain (2011)
11. Hu, M.K.: Visual pattern recognition by moment invariants. IRE Trans. Inf. Theory IT **8**, 179–187 (1962)
12. Alahi, A., Ortiz, R., Vandergheynst, P.: FREAK: fast retinal keypoint. In: Proceedings of Computer Vision and Pattern Recognition (CVPR), pp. 510–517. IEEE (2012)
13. Yang, X., Cheng, K.T.: LDB: an ultra-fast feature for scalable augmented reality on mobile devices. In: International Symposium on Mixed and Augmented Reality (ISMAR), pp. 49–57. IEEE (2012)
14. Bellarbi, A., Zenati, N., Otmane, S., Belghit, H.: Learning moment-based fast local binary descriptor. J. Electron. Imaging **26**(2), 1017–9909 (2017)
15. Farin, G.: Courbes et surfaces pour la CGAO. Masson, Paris (1992)
16. Besl, P.J.: Geometric modeling and computer vision. Proc. IEEE **76**(8), 936–958 (1988)

17. Moron, V.: Mise en correspondance de données 3D avec un model CAO: application à l'inspection automatique. D. thesis, Dept. Auto Ind, Lyon Univ., France, (1996)
18. Faux, I.D., Pratt, M.: Computational Geometry for Design and Manufacture. Ellis Harwood Series in Mathematics and its applications. Halsted Press, Chichester (1981)
19. Rosten, E., Drummond, T.: Machine learning for high speed corner detection. In: European Conference on Computer Vision (ECCV), vol. 1 (2006)
20. Demailly, J.P.: Analyse numérique et équations différentielles. EDP sciencies, Grenoble (2006)
21. Mikolajczyk, K., Schmid, C.: A performance evaluation of local descriptors. IEEE Trans. Pattern Anal. Mach. Intell. **27**(10), 1615–1630 (2005)

A Novel Hybrid Approach for 3D Face Recognition Based on Higher Order Tensor

Mohcene Bessaoudi[1(✉)], Mebarka Belahcene[1], Abdelmalik Ouamane[1], Ammar Chouchane[1], and Salah Bourennane[2]

[1] LI3C Laboratory, University of Biskra, Biskra, Algeria
bessaoudi.mohcene@gmail.com
[2] Institut Fresnel, Université de Marseille, Marseille, France

Abstract. This paper presents a new hybrid approach for 3D face verification based on tensor representation in the presence of illuminations, expressions and occlusion variations. Depth face images are divided into sub-region and the Multi-Scale Local Binarised Statistical Image Features (MSBSIF) histogram are extracted from each sub-region and arranged as a third order tensor. Furthermore, to reduce the dimensionality of this tensor data, we use a novel hybrid approach based on two steps of dimensionality reduction multilinear and non-linear. Firstly, Multilinear Principal Component Analysis (MPCA) is used. MPCA projects the input tensor in a new lower subspace in which the dimension of each tensor mode is reduced. After that, the non-linear Exponential Discriminant Analysis (EDA) is used to discriminate the faces of different persons. Finally, the matching is performed using distance measurement. The proposed approach (MPCA+EDA) has been tested on the challenging face database Bosporus 3D and the experimental results show that our method achieves a high verification performance compared with the state of the art.

Keywords: 3D face verification · Depth image · Tensor representation · Histograms local features

1 Introduction

Face recognition is an important research topic in computer vision area due to their various applications such as, surveillance systems, criminal identification, and human robot-interaction, etc. During the last decades, most research works have been interesting in two-dimensional images and only a few of them have been utilizing the depth images converted from the 3D scans [1, 2]. 2D face verification and identification are still challenging tasks due to the facial appearance changes caused by many factors in uncontrolled environments such as illuminations, expressions, pose variations, and occlusions. More recently, with the progress of 3D digital sensors and scanners [3], using 3D face information can offer a solution to the previous problems and challenges, and many experiments show that the performance on the 3D shape channel is better than on 2D texture alone [4, 5].

O. Demigha et al. (Eds.): CSA 2018, LNNS 50, pp. 215–224, 2019.
https://doi.org/10.1007/978-3-319-98352-3_23

In face recognition systems, facial representation plays the main role in the choice of algorithms and mathematical tools, in linear and nonlinear transformations based on vectors and data representation in matrices have been largely used because of their computational and conceptual simplicity. Among the most used methods, there are Principal Component Analysis (PCA) [6], linear Discriminant Analysis (LDA) [7], Kernel Principal Component Analysis (KPCA) [8], and Exponential Discriminant Analysis (EDA) [9]. These algorithms usually handle the input face image as a 1D- feature vector. Unfortunately, they are not a natural way to represent the facial data, since the image vectorization process implies losing the pixels' location information. Thus, these methods induce the small sample size problem [10].

On the other hand, multilinear transformations which are extensions of the traditional linear transformations to high order tensors representation, offer a powerful and a natural way to represent the set of face images without breaking the original structure of data [11].

In our framework, a set of face images is represented as a third-order tensor based on the local descriptor histograms: Multi-Scale Local Binarised Statistical Image Features (MSBSIF) [12] extracted from 3D face images. The face tensor is projected into new lower dimensional and more discriminative power is done.

The main contributions of this paper are:

- We propose a robust and automatic 3D face verification system, based on high order tensor representation of face data.
- We propose a new approach for tensor dimensionality reduction based on two subspaces transformation methods: (1) multilinear transformation using MPCA and (2) another nonlinear transformation based on the advantage of discrimination between classes given by EDA algorithm. MPCA is used to reduce the dimension of each tensor modes and avoid the small sample size (SSS) problem when the number of features is greater than the number of samples.
- We empirically evaluate our approach against the State-of-The-Art methods on the challenging face dataset Bosporus 3D.

2 Notations of Tensor Algebra

The variables and mathematical notations used in this paper are defined as follows: tensors are denoted by bold uppercase letters, e.g., **A**, **B**, **X**; matrices by italic uppercase letters, e.g., *U*, *T*, *V*; vectors by lowercase bold letters, e.g., **z**, **u**, **w**; and scalars by lowercase and uppercase letters, e.g., i, j, K, L;

A tensor is a multidimensional array [10, 13]. It is the higher order generalization of scalar (zero-order tensor), vector (1st-order tensor), and matrix (2nd-order tensor). An Nth-order tensor A is represented as $\mathbf{A} \in \mathbb{R}^{I_1 \times I_2 \times \cdots \times I_N}$, the elements of the tensor **A** are noted by $a_{i_1 i_2 \ldots i_N}$, where I_k represents the dimension of mode k and $1 \le i_k \le I_k$, $1 \le k \le N$.

The k-mode unfolding of the nth-order tensor $\mathbf{A} \in \mathbb{R}^{I_1 \times I_2 \times \cdots \times I_N}$ into matrix $A^{(k)} \in \mathbb{R}^{I_k \times \prod_{i \neq k} I_i}$ i.e. $\mathbf{A} \Rightarrow {}_k A^{(k)}$ is defined by:

$$A^{(k)}_{i_k j} = \mathbf{A}_{i_1 \cdots i_N}, \quad j = 1 + \sum\nolimits_{q=1,\ q \neq k}^{N} (i_l - 1) \prod\nolimits_{p=l+1, p \neq k}^{N} I_p \tag{1}$$

The unfolding operation on a 3rd-order tensor is illustrated in Fig. 1

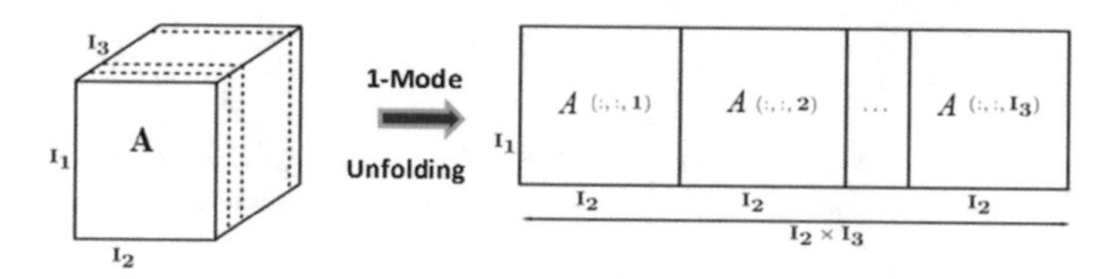

Fig. 1. Example of tensor unfolding.

The k-mode product of a tensor $\mathbf{A} \in \mathbb{R}^{I_1 \times I_2 \times \cdots \times I_N}$ with a matrix $U \in \mathbb{R}^{I_k \times J_k}$($k = 1, 2, \ldots,$ N) is an $I_1 \times I_2 \times \cdots I_{k-1} \times I^{'}_k \times I_{k+1} \times \ldots .. \times I_N$ tensor denoted by:

$$\mathbf{B} = \mathbf{A} \times {}_k U \tag{2}$$

Figure 2 illustrates example 1-mode product of third-order tensor $X \in \mathbb{R}^{11 \times 5 \times 4}$ with matrix $G^T \in \mathbb{R}^{4 \times 11}$.

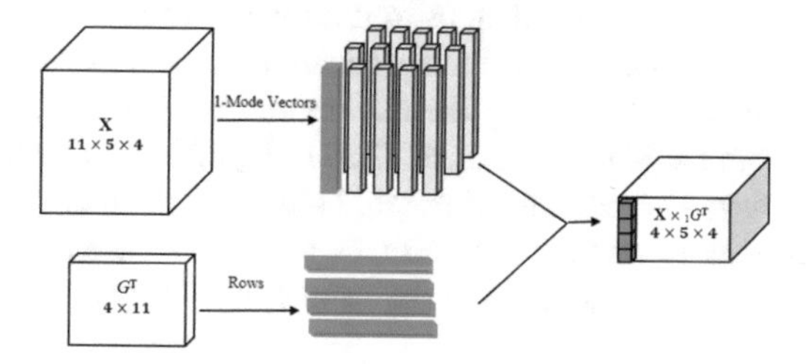

Fig. 2. Illustration of 1-mode product of third-order tensor $X \in \mathbb{R}^{11 \times 5 \times 4}$ with matrix $G^T \in \mathbb{R}^{4 \times 11}$.

3 Overview of the Proposed 3D Face Verification System

In this section, we explain the details of the proposed 3D face verification system based on tensor representation. As depicted in Fig. 3. Depth face images are preprocessed and described by BSIF features that are used to form the tensor representation. Afterward, the tensor data is projected into a lower subspace using our approach MPCA+EDA. In the test phase, face matching is performed in the new subspace using cosine similarity measurement. In the following, we present the details of our method.

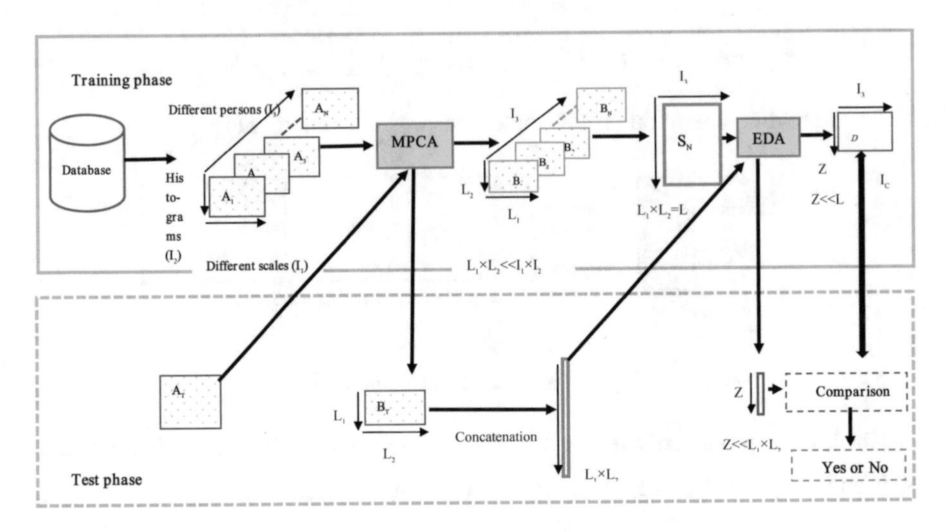

Fig. 3. Overview of the proposed approach

3.1 Preprocessing

Face preprocessing includes several steps to improve the quality of 3D face data. The face area is cropped using an elliptical mask centered in the nose tip point. The coordinates of nose tip points are located by the maximum intensity method [14]. The facial region is cropped from the 3D depth image. Generally, the data obtained from 3D scanning process are noisy; also contains different holes in some parts. For that, a median filter is applied in order to remove the noise, as well as, we use cubic interpolation of the neighboring pixels to fill the holes in the input depth image.

3.2 Feature Extraction

For feature extraction, depth image are described using local descriptor Binarised Statistical Image Features (BSIF) [15]. We use this descriptor because it achieved the best performances in previous works [16, 17]. In order to keep the spatial face structure, the faces' image is subdivided into P non-overlapping faces regions and the histograms of these rectangular blocks are concatenated forming a feature vector v of size $n = P \times 256$. The feature vectors of different faces are finally assembled to create the 3rd order tensor data. The feature extraction and tensor design steps of our system are illustrated in Fig. 4. In our case, the three tensor modes (I_1, I_2, I_3) are defined as follows:

I_1: The feature vectors (concatenated histograms).
I_2: The different scales used for 3D face images.
I_3: The different persons of the database.

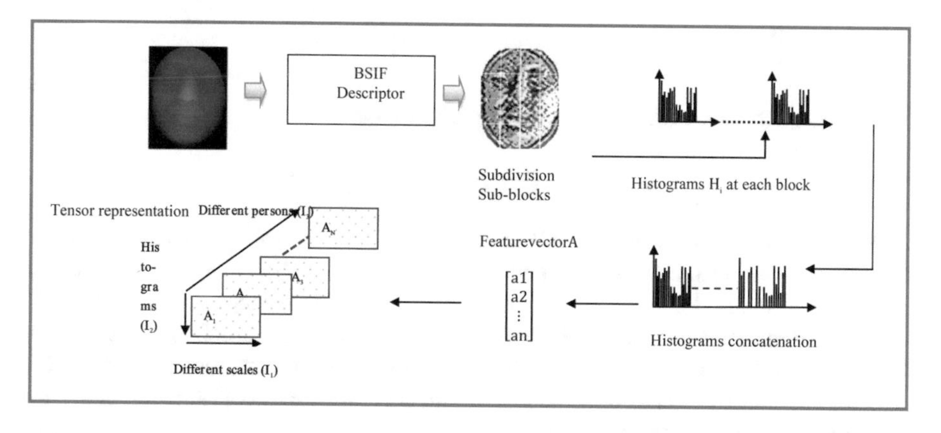

Fig. 4. Feature extraction and tensor design.

3.3 Dimensionality Reduction and Classification

MPCA+EDA

During this step, tensor data is transformed into a lower discriminative subspace. For this crucial operation, we use the multilinear PCA algorithm [11]. Let the training samples represented as the mth-order tensor $\mathbf{A} \in \mathbb{R}^{I_1 \times \cdots \times I_m}$(In our case m = 2). Therefore, the sample set of all training data can be characterized as an 3rd -order tensor $\tilde{\mathbf{A}} \in \mathbb{R}^{I_1 \times I_2 \times \cdots \times I_m \times N}$. MPCA aim to find the m projection matrices $\{G^{(n)} \in R^{I_n \times P_n}, n = 1, \ldots, m\}$ in order to map the original tensor A into a novel tensor set subspace $\{\mathbf{Y} \in \mathbb{R}^{P_1 \times P_2 \times \cdots P_m}\}$ with $p_n < I_n$. The multilinear PCA algorithm is given as follows:

1. Compute the main tensor $\bar{\mathbf{A}}$ of all training samples $\mathbf{A}_i$

$$\bar{\mathbf{A}} = \frac{1}{N} \sum\nolimits_i^N \mathbf{A_i} \tag{3}$$

2. Center the input training samples

$$\tilde{\mathbf{A}}_\mathbf{i} = \mathbf{A}_\mathbf{i} - \bar{\mathbf{A}}, i = 1 \ldots, m \tag{4}$$

3. Unfold the tensor of the training samples into a matrix $\tilde{A}_{i(k)}$, and find the eigenvector decomposition of the covariance matrix for each k-mode

$$C_{(k)} = \sum\nolimits_i^m \tilde{A}_{i(k)} . \tilde{A}_{i(k)}^T = S_{(k)}^T \Lambda_{(k)} G_{(k)} \tag{5}$$

Where $S_{(k)}$ is the eigenvectors matrix and $\Lambda_{(k)}$ is the eigenvalues matrix. Then, the matrix $G_{(k)}$ comprise containing the chosen eigenvectors that match the largest $n_{(k)}$ eigenvalues is given by

$$G_{(k)} = \left[g_1, g_2, \cdots, g_{n_{(k)}}\right] \quad (6)$$

4. The reduced tensor after the projection is given by

$$\mathbf{B} = \mathbf{A} \times_1 G^{(1)^T} \times_2 G^{(2)^T} \ldots \times_N G^{(N)^T}$$

MPCA ensures the projection of the tensor to a lower multilinear subspace. Therefore, we get a reduced core tensor in which ($L_1 \times L_2 << I_1 \times I_2$). However, MPCA is an unsupervised method that does not take the label classes into account. In our case, the discrimination between data belonging to different classes is the crucial goal. To handle this issue, we integrate a nonlinear discrimination algorithm to the original MPCA process. We employ the Exponential Discriminant Analysis (EDA) [9] to maximize the discrimination between the classes while reducing the tensor dimensionality at the same time. First, we unfold the reduced tensor **B** according to the 1st-mode (N = 1) to get the unfolding matrix S_N of size $L \times I_3$ ($L = L_1 \times L_2$). Then, we apply EDA to these vectors. The EDA algorithm is given as follows:

1. Compute the intra-class (S_w) and the inter-class (S_b) scatter matrices of B_N:

$$S_w = \sum_{i=1}^{C} \frac{1}{n_i} \sum_{j=1}^{n_i} \left(\mathbf{b}_j^i - \mathbf{m}^i\right)\left(\mathbf{b}_j^i - \mathbf{m}^i\right)^T \quad (7)$$

Where C is the number of classes of matrix B_N, and ni is the number of samples of class i.

$$S_b = \sum_{i=1}^{C} \left(\mathbf{m}^i - \bar{\mathbf{m}}\right)\left(\mathbf{m}^i - \bar{\mathbf{m}}\right)^T \quad (8)$$

Where m^i is the mean of class C_i and $\bar{m}$ is the mean of all data samples.

2. Find a projection matrix H that maximizes the inter-class distances while minimizing the intra-class distances, defined as:

$$H = \arg\max_H \frac{\left|H^T \exp(S_b) H\right|}{\left|H^T \exp(S_w) H\right|} = \arg\max_H \frac{\left|H^T (V_b^T \exp(\Lambda_b) V_b) H\right|}{\left|H^T \left(V_w^T \exp(\Lambda_w)\, V_w\right) H\right|} \quad (9)$$

Where V_b, Λ_b are the eigenvectors of and eigenvalues of S_b. V_w and Λ_w are the eigenvectors and the eigenvalues of S_w, respectively. The projection matrix H is composed of the leading eigenvectors of $(\exp(S_b))^{-1}\exp(S_w)$.

After using the EDA algorithm, we get a reduced training matrix D of size $Z \times I_3$ ($Z << L$).

In the test phase, the test image A_T is processed in the same manner as the training images and projected into MPCA subspace, then concatenated as a vector of size $L_1 \times L_2$. This vector is projected into EDA subspace. Then, we get a reduced feature vector of size where $Z << L_1 \times L_2$, for the test candidate. This test vector is matched with the training features (claimed identity Ic) of size $Z \times D_3$ using the cosine similarity. The cosine similarity between two vectors **x** and **y** is defined as follows:

$$\cos(\mathbf{x}, \mathbf{y}) = \frac{\mathbf{x}^{T}.\mathbf{y}}{\|\mathbf{x}\|.\|\mathbf{y}\|} \tag{10}$$

Where $\|.\|$ is the Euclidean norm.

4 Experiments and Results

This section provides the details about the experimental results of the proposed system. First, we describe the challenging face database, Bosphorus 3D [18] used to carry out the experiments as well as the parameters setting. Then, we proceed with the evaluation of three variants in our system: the multilinear approach using MPCA, the nonlinear approach using EDA and our hybrid approach (MPCA+EDA). Finally, we compare our best results with the State-of-The-Art.

4.1 Bosphorus 3D Database

Bosphorus 3D face database comprises 4666 scans from 105 different subjects in various expressions, poses and occlusion conditions. Bosphorus database is collected by a structured-light 3D scanner. In our experiments, we evaluate the proposed approach on the same protocol as defined by Erdogmus and Dugelay [19]. An example of pose variation from this database is shown in Fig. 5.

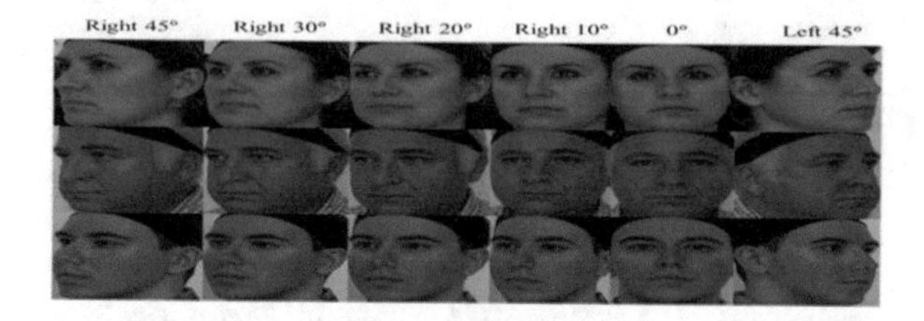

Fig. 5. Example of pose variation from Bosphorus 3D face database.

4.2 Parameter Settings

We report the parameters selected in the experiments as follows. Depth face images are resized to 100×120 pixels. Each face image is subdivided into 20 blocks, where the size of each bloc is 30×40 pixels. For MSBSIF descriptor, we used four scales in which the filter size is $l = \{7, 9, 11, 13\}$. The histograms of all the 20 blocks are concatenated to a vector of dimension 20×256.

The training data is used for estimating the subspace projection matrices. For projections, the final lower dimension is calculated automatically by retaining 99% energy of the eigenvalues. In order to assess the verification performance of different approaches, we report the verification rate (VR) at 0.001 FAR (False Accept Rate).

4.3 Results and Discussion

In this subsection, we report on a variety of evaluations to show the effectiveness of the proposed approach MPCA+EDA. The obtained results for each approach are provided in Table 1. The comparison of all experimental results in terms of ROCs curves is shown in Fig. 6. By analyzing the obtained results, we make the following observations:

Table 1. Face verification rates VR at 0.001 FAR

Method	VR (%)
Multilinear: MPCA	69.78
Nonlinear : EDA	85.53
Combination: MPCA+EDA	**92.12**

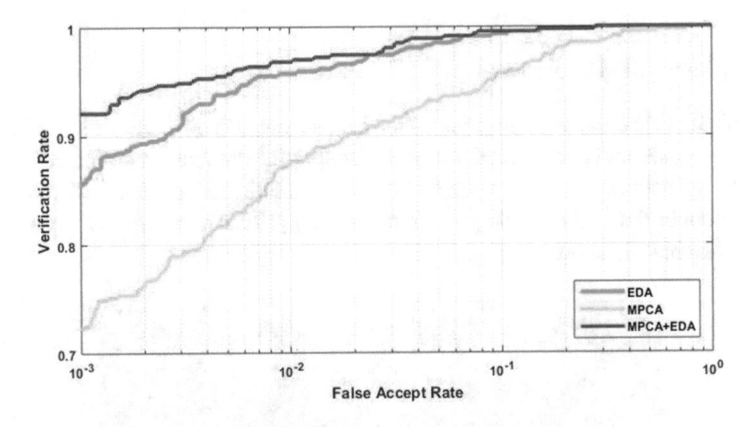

Fig. 6. ROC curves of the three methods.

As can be seen in Table 1, MPCA+EDA gives the best performance with a verification rate (VR) of 92.12% compared to MPCA and EDA, that achieve a verification rate of 69.78% and 85.53%, respectively. Therefore, our hybrid approach MPCA+EDA based on a multilinear step using MPCA followed by the nonlinear step based on EDA, gives the best performance (reaches a gain of 6.59% in the verification rate), compared to each method independently.

The comparison of the three methods in terms of ROCs curves is shown in Fig. 5. The ROC depicts the probability of a correct match to the false acceptance rate. The ROC curve of our approach MPCA+EDA has the better performances and it is the more stable one, which means that the verification rate of our algorithm is greater than the verification rate of the other methods at each false acceptance rate.

Comparison with State-of–The-Art Methods

Table 2 compares the verification performance achieved by our proposed approach with the State-of-The-Art methods on Bosphorus 3D database. The best verification rate of our approach attained to 92.12%. Our results are competitive compared to the best results presented in recent work by Ouamane et al. [17] using a 3rd order tensor face representation. Their work used a hybrid approach that has yielded the currently best verification rate, obtained by the fusion of two multiscale descriptors (MSLPQ +MSBSIF) and the application of the covariance normalization. In contrast, in our work we used only one descriptor MSBSIF and achieved the second rank.

Table 2. Comparison of verification rates with state of the art.

Method	Data representation	Method	VR (%)
Liu et al. [20]	Matrix	SHF	81.4
Erdogmus et al. [19]	Matrix	LBP	69.09
Elaiwat et al. [5]	Matrix	CLF	77.50
Ming [5]	Matrix	OSR	93.95
Ouamane et al. [17]	3rd order tensor	TEDA+WCCN	92.97
Our approach	**3rd order tensor**	**MPCA+EDA**	**92.12**

5 Conclusion

In this paper, we presented an effective approach for 3D face verification in real environments using depth images. The proposed approach is based on high order tensor representation. Histograms of multi-scale local descriptor BSIF are used for characterizing 3D face images and different face variations are modeled as a multilinear algebra problem, where face images with these challenges are represented as a 3rd-order tensor. This latter is reduced and projected using hybrid subspace transformation (MPCA +EDA). The results of experiments conducted on the Bosphorus 3D face database confirm the robustness of the proposed approach to the problems of illumination, expression and occlusion variations. As future work, we aim to generalize our approach to other 3D databases and investigate alternative face tensor designs.

References

1. Kakadiaris, I.A., Toderici, G., Evangelopoulos, G., Passalis, G., Chu, D., Zhao, X., Shah, S.K., Theoharis, T.: 3D-2D face recognition with pose and illumination normalization. Comput. Vis. Image Underst. **154**, 137–151 (2017)
2. Soltanpour, S., Boufama, B., Jonathan Wu, Q.M.: A survey of local feature methods for 3D face recognition. Pattern Recogn. **72**, 391–406 (2017)
3. Elaiwat, S., Bennamoun, M., Boussaïd, F., et al.: A curvelet-based approach for textured 3D face recognition. Pattern Recogn. **48**(4), 1235–1246 (2015)

4. Wang, X., Ruan, Q., Ming, Y.: 3D face recognition using corresponding point direction measure and depth local features. In: 2010 IEEE 10th International Conference on Signal Processing (ICSP), pp. 86–89. IEEE (2010)
5. Ming, Y.: Rigid-area orthogonal spectral regression for efficient 3D face recognition. Neurocomputing **129**, 445–457 (2014)
6. Abdi, H., Williams, L.J.: Principal component analysis. Wiley Interdisc. Rev. Comput. Stat. **2**(4), 433–459 (2010)
7. Pang, Y., Wang, S., Yuan, Y.: Learning regularized LDA by clustering. IEEE Trans. Neural Networks **25**(12), 2191–2201 (2014)
8. Welling, M.: Kernel principal components analysis. Adv. Neura L Inf. Process. Syst. **15**, 70–72 (2003)
9. Zhang, T., Fang, B., Tang, Y.Y., et al.: Generalized discriminant analysis: a matrix exponential approach. IEEE Trans. Syst. Man Cybern. Part B (Cybernetics) **40**(1), 186–197 (2010)
10. Lai, Z., Xu, Y., Yang, J., et al.: Sparse tensor discriminant analysis. IEEE Trans. Image Process. **22**(10), 3904–3915 (2013)
11. Lu, H., Plataniotis, K.N., Venetsanopoulos, A.N.: MPCA: multilinear principal component analysis of tensor objects. IEEE Trans. Neural Networks **19**(1), 18–39 (2008)
12. Arashloo, S.R.: Multiscale binarised statistical image features for symmetric unconstrained face matching. In: 2014 22nd Iranian Conference on Electrical Engineering (ICEE), pp. 1377–1382. IEEE (2014)
13. Liu, Y., Liu, Y., Zhong, S.: Tensor distance based multilinear globality preserving embedding: a unified tensor based dimensionality reduction framework for image and video classification. Expert Syst. Appl. **39**(12), 10500–10511 (2012)
14. Ouamane, A., Belahcene, M., Benakcha, A., et al.: Robust multimodal 2D and 3D face authentication using local feature fusion. SIViP **10**(1), 129–137 (2016)
15. Kannala, J., Rahtu, E., BSIF: binarized statistical image features. In: 2012 21st International Conference on Pattern Recognition (ICPR), pp. 1363–1366. IEEE (2012)
16. Ammar, C., Mebarka, B., Abdelmalik, O., et al.: Evaluation of histograms local features and dimensionality reduction for 3D face verification. J. Inf. Process. Syst. **12**(3), 468–488 (2016)
17. Ouamane, A., Chouchane, A., Boutellaa, E., et al.: Efficient tensor-based 2d + 3d face verification. IEEE Trans. Inf. Forensics Secur. **12**(11), 2751–2762 (2017)
18. Savran, A., Alyüz, N., Dibeklioğlu, H., et al.: Bosphorus database for 3D face analysis. In: Biometrics and Identity Management, pp. 47–56 (2008)
19. Erdogmus, N., Dugelay, J.-L.: 3D assisted face recognition: dealing with expression variations. IEEE Trans. Inf. Forensics Secur. **9**(5), 826–838 (2014)
20. Liu, P., Wang, Y., Huang, D., et al.: Learning the spherical harmonic features for 3-D face recognition. IEEE Trans. Image Process. **22**(3), 914–925 (2013)

Subjective and Objective Evaluation of Noisy Multimodal Medical Image Fusion Using 2D-DTCWT and 2D-SMCWT

Abdallah Bengueddoudj(✉) and Zoubeida Messali

Department of Electrical Engineering, University of Bordj Bou Arreridj, El Anasser, Algeria
Beng.abdallah@hotmail.com

Abstract. This paper focuses on the evaluation of noisy image fusion for medical images obtained from different modalities. In general, medical images suffer from poor contrast and are corrupted by blur and noise due to the imperfection of image capturing devices. In order to improve the visual and quantitative quality of the fused image, we compare two algorithms with other fusion techniques. The first algorithm is based on Dual Tree Complex Wavelet Transform (DTCWT) while the second is based on Scale Mixing Complex Wavelet Transform (SMCWT). The tested algorithms are using different fusion rules in each one, which leads to a perfect reconstruction of the output (fused image), this combination will create a new method which exploits the advantages of each method separately. DTCWT presents a good directionality since it considers the edge information in six directions and provide approximate shift invariant as well as SM-CWT, the goal of PCA is to extract the most significant features (wavelet coefficients in our case) to improve the spatial resolution. We compared the tested methods visually and quantitatively to recent fusion methods presented in the literature over several sets of medical images at multiple levels of noise. Further, the tested fusion algorithms have been tested up to the important level of Gaussian, salt & pepper and speckle noise (350 test). For the quantitative quality, we used several well-known fusion metrics. The results show that the tested methods outperform each method individually and other algorithms proposed in the literature.

Keywords: Multimodal medical images fusion
Dual Tree Complex Wavelet Transform
Scale Mixing Complex Wavelet Transform · Fusion metrics
Noisy image fusion · PCA

1 Introduction

Currently, medical image fusion has become a useful and essential tool for clinical therapy and diagnosis. The idea of image fusion is based on the combination of information acquired using different imaging sensors into a single image [1]. The primary objective of image fusion is to preserve the relevant information of each source images and generate the fused image without introducing artefacts or distortions. The source

O. Demigha et al. (Eds.): CSA 2018, LNNS 50, pp. 225–234, 2019.
https://doi.org/10.1007/978-3-319-98352-3_24

images are captured using different biomedical sensors to extract complementary information about human tissues. Magnetic Resonance Image (MRI), Computed Tomography (CT), Positron Emission Tomography (PET) and Single Photon Emission Tomography (SPECT) are a few examples of biomedical sensors. Each one of these imaging techniques has their own characteristics and limitations. For example, CT images take excellent information about bones and other hard tissue whereas MR images are perfectly describing information about soft tissues. Hence, we can conclude that these complementarities may be combined in order to create a new image which exploits the advantages of each image source separately.

In the present work, we have evaluated two fusion algorithms using several sets of noisy medical images at multiple levels of noise. The first algorithm called Bayes-SMCWT [2] is based on 2d-SMCWT and integrates a Bayesian approach in the fusion rule, while the second one (called FSPCA-DTCWT) [3] is based on 2d-DTCWT and uses a feature selection strategy as fusion rule for the detail coefficients. Both 2d-SMCWT and 2d-DTCWT incorporate PCA to merge the approximation coefficients as will be described later in Sect. 3. The tested fusion methods are compared to recent fusion techniques presented in the literature in terms of visual quality and quantitatively using well known fusion performance metrics (four metrics). The rest of this paper is organized as follows: Image fusion literature is discussed in Sect. 2. A brief description of the tested fusion algorithms is given in Sect. 3. Experimental results and performance evaluations are given in Sect. 4. Finally, a conclusion of the work is given in Sect. 5.

2 Background and Literature

Numerous image fusion methods have been proposed in the literature, these methods can be divided into three different levels according to the stage at which the fusion mechanism takes place, namely: pixel level, feature level and decision level [4]. In the first level, the fusion is performed directly on pixel values of the source images to generate a single fused image. Pixel-level fusion algorithms became the most popular methods for multimodal image fusion, they can be classified into three categories [5]: substitution techniques, domain transform techniques and optimization approaches.

Fusion methods belong to the first category suffer from contrast degradation and artefacts in the fused images. Many image fusion algorithms based on multiresolution techniques can be used in order to overcome the limitations of substitution methods. Pyramid and Wavelet transforms are the most popular multiresolution approaches and they are widely used in image fusion. In general, wavelet transform methods perform better than pyramid transform methods, since it has a good directional information and provides a better representation of the decomposed components and better results than the Pyramid Transform [6]. Discrete Wavelet Transform (DWT) is one of the popular Wavelet Transform used in image fusion. DWT provides a good spectral information and better directional information along three spatial orientations (vertical, horizontal, and diagonal) as compared to pyramid representations.

Multiple researches have proved that DWT suffers from shift sensitivity, absence of phase information and have poor directionality [6]. To overcome these limitations, DTCWT and SMCWT [7, 8] inherit all the advantages of DWT and provide an approximate shift invariant with better directionality, and provide perfect reconstruction using short linear-phase filters. For these reasons, we have chosen DTCWT and SMCWT and incorporate different fusion rules for the developed algorithms.

Generally, the images generated by medical instruments suffers from various problems such as poor contrast, blur and noise due to applications of various quantization, reconstruction and enhancement algorithms. Therefore, medical image fusion needs a special attention. The objective evaluation of medical image fusion methods has become a challenging research area in recent years, as no reference exist to prove fusion quality. Beside the visual analysis, the quantitative analysis plays an important role in performance evaluation. Moreover, the most existing image fusion techniques ignore the problem caused by noise effects. Therefore, in this paper, we have subjectively and objectively evaluated the performance of medical image fusion at different noise levels using two image fusion algorithms. The objective evaluation of the tested fusion schemes is performed with various fusion performance metrics. The visual results are shown for two sets of noisy multimodal medical images, and comparison has been done with various wavelet transform-based fusion methods. Further, fusion algorithms have been tested against different levels of Gaussian, salt & pepper and speckle noise and image fusion performance metric plots are shown for Gaussian, salt & pepper and speckle noise.

3 The Tested Fusion Algorithms

3.1 Pixel Level Image Fusion Based on SMCWT Using Bayesian Estimation and PCA

In [2], we proposed a multimodal medical image fusion algorithm (called Bayes-SMCWT). The 2d-SMCWT presents an efficient tool in image decomposition since it preserves the energy that avoids contrast distortions. Moreover, it has a better representation of the input images with several hierarchies of detail coefficients (see Fig. 1) and guarantees the orthogonality and the inverse transform can be performed in a straightforward manner.

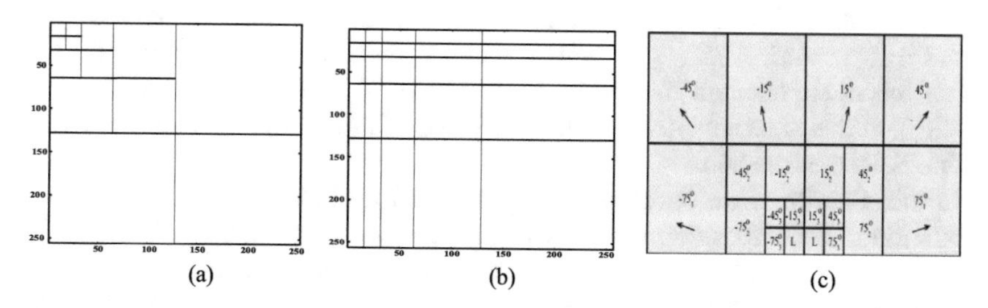

Fig. 1. 2D wavelet decomposition: (a) DWT, (b) SMCWT, (c) DTCWT.

Bayes-SMCWT algorithm involves two different fusion rules for the detail and approximation wavelet coefficients. For the detailed sub bands coefficients, we used Bayesian MAP estimation technique, this fusion rule takes into account the mutual correlation between the source images and the fused image, while a MAX-PCA fusion rule is adopted to merge the approximate coefficients.

The overall schematic diagram illustrating Bayes-SMCWT fusion algorithm is shown in Fig. 2(a). The reader can refers to [2] for more details about the different steps of the Bayes-SMCWT algorithm.

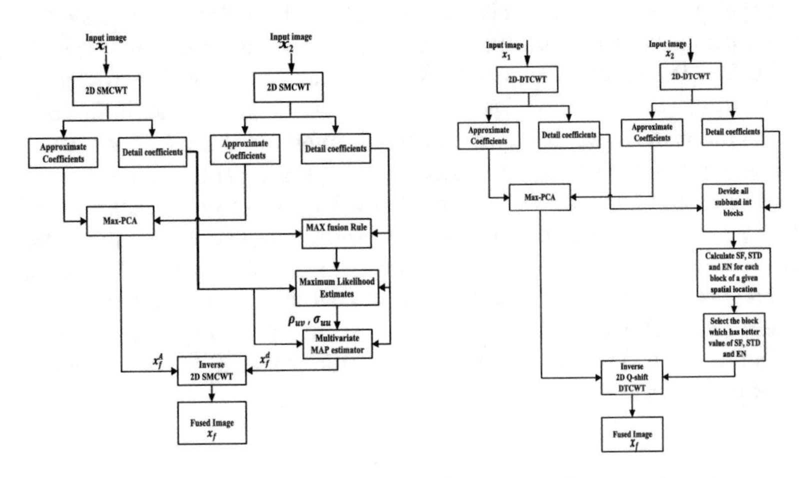

Fig. 2. Schematic diagram of the tested algorithms: (a) Bayes-SMCWT algorithm [2], (b) FSPCA-DTCWT algorithm [3].

3.2 Feature Level Image Fusion Based on DTCWT Using Feature Selection and PCA

The Second algorithm (FSPCA-DTCWT) [3] uses DTCWT, which is an extension of the DWT. The DTCWT, as the name implies, consists of two trees of real filters and provides six pairs of subbands (for both the real and imaginary wavelet coefficients) using complex scaling and wavelet functions; where the 2D DWT only separates information into horizontal, vertical and diagonal information, the 2D DTCWT separates the same information into six directional subbands, with the angles centered around ±15°, ±45°, ±75°, and their negative equivalents as shown in Fig. 1. The overall schematic diagram illustrating FSPCA-DTCWT fusion algorithm is shown in Fig. 2(b). In this algorithm, the detailed wavelet coefficients have been merged using a Feature Selection method.

In this algorithm, we developed a region-based method to merge the detail coefficients in the decomposition levels, which involve the computation of statistical features such as the standard deviation, spatial frequency and the entropy of the detailed coefficients within a local neighboring region for the decomposed source images A and B. This is mainly used as a measure of the activity of the pixel centered at that region. The approximate coefficients have been fused by applying the MAX-PCA fusion Rule. The reader can refer to [3] for more detail about FSPCA-DTCWT algorithm.

4 Experimental Results

In this section, the performance of Bayes-SMCWT and FSPCA-DTCWT algorithms is presented. The tested fusion algorithms were applied to two different noisy medical image data sets of different modalities, namely: MRI and CT images (Fig. 3). As opposed to CT images, which provide excellent hard tissue imaging such as for bones. Clear and detailed images of soft tissue can be obtained from MR images. The fusion of MR and CT images integrate all possible relevant and complementary information from both MR/CT into one single image, which will be more useful for a better medical diagnosis. The experiments are performed for Bayes-SMCWT as well as for FSPCA-DTCWT for salt and pepper noise at density level d = 0.05. The results of the tested algorithms are compared with different existing multiscale transform-based methods including: Discrete Wavelet Transform (DWT) [9], Shift Invariant Discrete Wavelet Transform with Haar Wavelets (SIDWT) [10], DTCWT with Mean-Max rules [11] and Discrete Cosine Harmonic Wavelet (DCHWT) [12]. Experiments have been conducted using Matlab 2013b with three levels of decomposition for all multiscale transform methods as well as the tested algorithms.

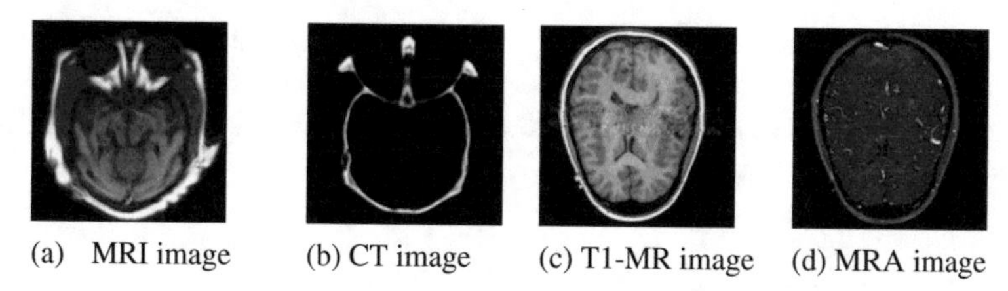

Fig. 3. Medical input images: (a) and (b) the first set, (c) and (d) the second set.

Figures 4 and 5 show the results for the two groups of noisy medical images. It is clear that the fused images obtained from the tested methods depict the best visual quality with the highest contrast compared to the other ones. Hence, the tested algorithms successfully handle noise and generate fused images with good visual quality.

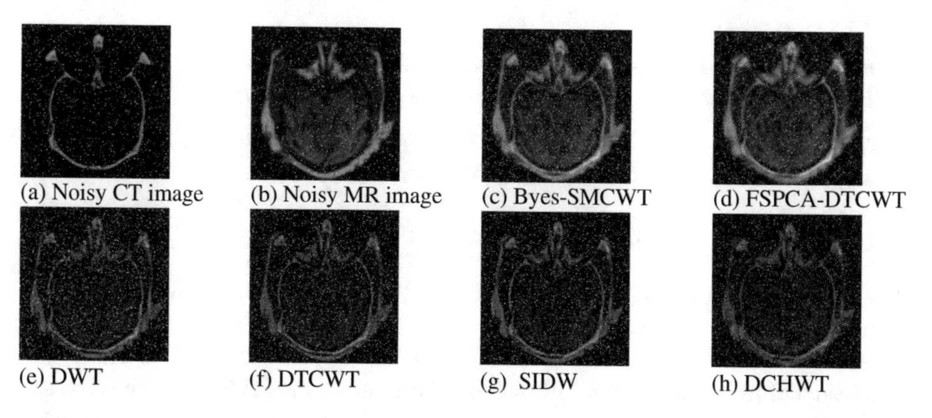

Fig. 4. Fusion result for the first set of noisy medical images at noise density $d = 0.05$.

However, due to the random nature of noise, the subjective analysis is not sufficient for the evaluation of the tested fusion methods on noisy images. Therefore, the robustness of the tested fusion algorithms is tested against noise with fusion metrics including: Standard Deviation (STD) [6], Entropy (EN) [6], Visual information fidelity (VIF) [13] and Correlation Coefficient (CC) [14].

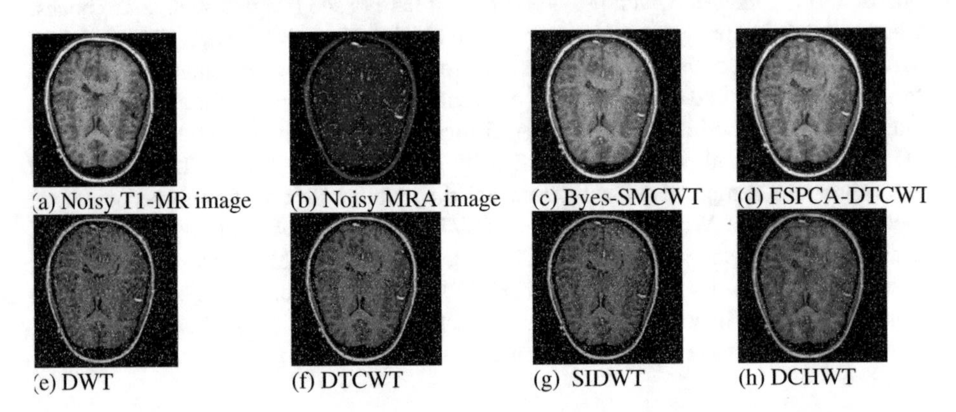

Fig. 5. Fusion results for the second set of noisy medical images at noise density d = 0.05.

- Standard Deviation (STD) calculates the information about contrast, The STD of an $M \times N$ image is given by:

$$\sigma = \left(\frac{1}{M \times N} \sum_{m=1}^{M} \sum_{n=1}^{N} (f(m,n) - \mu)^2 \right)^2, \tag{1}$$

where $f(m,n)$ is the pixel intensity value at (m,n) and μ is the mean value of the image.

- Entropy (EN) [6] is used to characterize the texture and to measure the information quantity contained in an image. It is defined as:

$$EN = -\sum_{l=1}^{L} P_l log_2 P_l, \tag{2}$$

where L represents the number of grey levels and P_l is the ratio of the number of the pixels with grey value is equal to l over the total number of pixels.

- The Correlation Coefficient (CC) [14] measures the similarity between the source images A and B and the fused one F, CC is given by:

$$CC(A,B,F) = \text{Corr}(A,F) + \text{Corr}(B,F), \tag{3}$$

- Visual information fidelity (VIF) [13] is a full reference image quality assessment metric. It is based on natural scene statistics and the notion of image information extracted by the human visual system (HVS). The quantity of visual information is considered as the amount of information extracted by the HVS from the reference image after it has been passed through the distortion channel. More details about the mathematical formulation of VIF can be found in [13, 15].

The values of fusion results for the same groups of the input images (Figs. 3 and 4) are illustrated in Tables 1 and 2, respectively. Before discussing these values of fusion metrics, we note that better medical image fusion results should have the higher value of STD, EN, VIF and CC. From the data values of Tables 1 and 2, we notice that the tested fusion algorithms give the best values of the objective indicators in most of the cases. Bold values indicate the best results obtained with respect to the different metrics when comparing among the various methods.

Table 1. Fusion metrics for first set of images corrupted with salt and pepper noise of density $d = 0.05$.

Methods	Entropy	STD	VIF	CC
Bayes-SMCWT	**6.6098**	52.5145	**0.4692**	**0.6997**
FSPCA-DTCWT	**6.7030**	47.6936	**0.4150**	0.6391
DWT	5.4620	52.2208	0.2671	0.6698
SIDWT	4.8715	51.9755	0.2872	0.6840
DTCWT	4.6552	**55.6657**	0.3856	0.6180
DCHWT	5.6604	40.0790	0.2273	0.6755

Table 2. Fusion metrics for the second set of medical images corrupted with salt and pepper noise of density $d = 0.05$.

Methods	Entropy	STD	VIF	CC
Bayes-SMCWT	**7.2196**	**73.1342**	0.6338	0.7885
FSPCA-DTCWT	**7.2073**	**70.4036**	0.5912	0.7701
DWT	6.2040	63.3976	0.4436	0.7716
SIDWT	6.0018	63.0579	0.4990	0.7867
DTCWT	5.8728	63.3648	**0.6560**	0.7748
DCHWT	6.7043	54.0458	0.4319	**0.7894**

Further, the tested algorithms are rigorously tested against different levels (350 tests) of Gaussian, salt and pepper and speckle noise. We performed experiments by adding different values of noise variances and noise density to input images. The plots are shown in Fig. 6 for Standard Deviation (STD) and Entropy (EN). By observing the plots of the fusion metrics, for the different types of noise, one can easily conclude that the values of STD and EN obtained by the tested algorithms are not affected by noise i.e. EN and STD are continuously increasing up to variance and density 0.2 in the different cases. These are desirable features for a better fusion process. Therefore, one can conclude that the tested noisy fusion algorithms are robust against the three types of noise. The plots of STD against the three different types of noise using DTCWT-, SIDWT, and DWT based methods show good fusion results. However, the plots of EN against salt and pepper and speckle noise using these methods in Figs. 6(b) and (c) respectively, show poor robustness against noise.

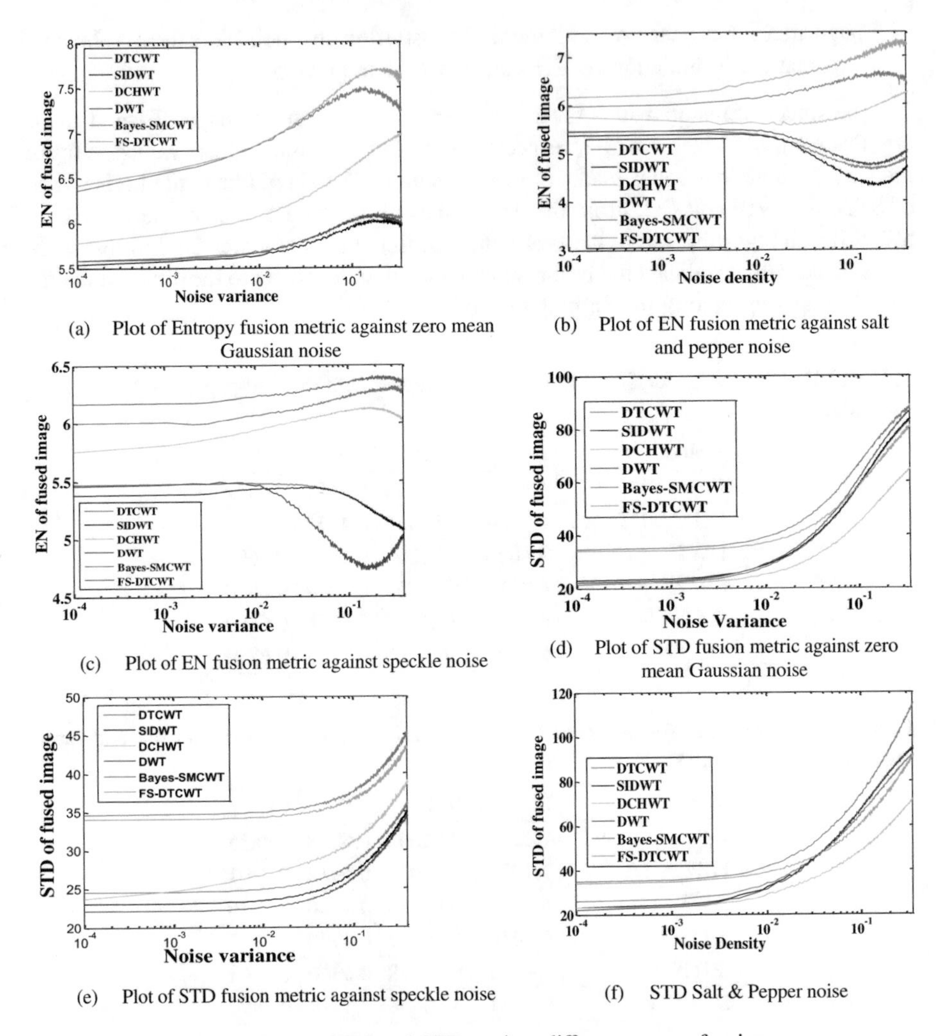

Fig. 6. Plots of EN and STD against different types of noise.

In order to evaluate the tested algorithms for the best choice of window size, which defines the number of wavelet coefficients over a local neighborhood, we conducted several tests against salt and pepper noise at density d = 0.05 by varying window sizes for the two tested algorithms. The visual and quantitative results are shown in Fig. 7 and Table 3, respectively. The values of STD, EN and VIF in Table 3 as well as the visual results in Fig. 7 for Bayes-SMCWT show that an increase of the window's size in this case does not provide any significant change in fusion performance except an increasing computation time. However, the size of the window in FSPCA-DTCWT has a remarkable effect, for example, a loss of information is clearly observed for a 32×32 window size which is also indicated by a decreased value of STD (79.89). This is mainly due to the large amount of noise appear in the increased window size.

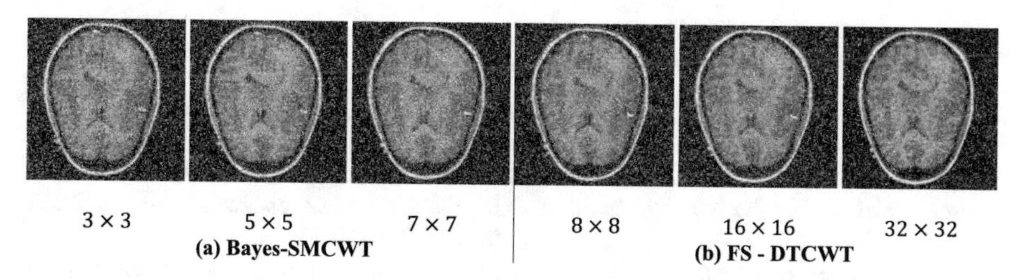

Fig. 7. Fusion results for the second set of noisy medical images at noise density $d = 0.15$ using different window sizes.

Table 3. Fusion metrics for the second set of noisy medical images with noise density $d = 0.15$ using different window sizes.

Window sizes	STD	EN	VIF
FS-DTCWT			
08 × 08	81.43	7.35	0.49
16 × 16	80.79	7.35	0.49
32 × 32	79.89	7.38	0.49
Bayes-SMCWT			
03 × 03	82.00	7.41	0.54
05 × 05	81.98	7.41	0.54
07 × 07	81.96	7.41	0.54

From the above discussion, one can conclude that the quantitative results of the tested method are consistent with the visual results.

Analyzing the above results of subjective and objective evaluations, we can see that the tested fusion algorithms are better than other wavelet transform based fusion methods. Also, the tested methods are able to preserve detail information such as edges and boundaries in a better way.

5 Conclusion

In this paper, we subjectively and objectively evaluated the performance of noisy multimodal medical image fusion using two fusion algorithms based on DTCWT and SMCWT. The results were compared visually and quantitatively with several wavelet transform based fusion techniques. The robustness of Bayes-SMCWT and FSPCA-DTCWT algorithms is proved by performing both objective and subjective evaluation using STD, EN, VIF and CC metrics. The tested fusion algorithms have been extensively tested against different types of noise (Gaussian, salt & pepper and speckle) and plots of STD and EN demonstrate the superiority of Bayes-SMCWT and FSPCA-DTCWT algorithms over DTCWT, SIDWT, DWT and DCHWT noisy medical image fusion techniques.

References

1. James, A.P., Dasarathy, B.V.: Medical image fusion: a survey of the state of the art. Inf. Fusion **19**, 4–19 (2014)
2. Bengueddoudj, A., Messali, Z., Mosorov, V.: A novel image fusion algorithm based on 2D scale-mixing complex wavelet transform and Bayesian MAP estimation for multimodal medical images. J. Innov. Opt. Health Sci. **10**, 1750001 (2017)
3. Bengueddoudj, A., Messali, A.: An efficient algorithm for multimodal medical image fusion based on feature selection and PCA using DTCWT (FSPCA-DTCWT). Med. Technol. J. **2** (1), 179–192 (2018)
4. Ganasala, P., Kumar, V.: CT and MR image fusion scheme in nonsubsampled contourlet transform domain. J. Digit. Imag. **27**, 407–418 (2014)
5. Daneshvar, S., Ghassemian, H.: MRI and PET image fusion by combining IHS and retina-inspired models. Inf. Fusion **11**, 114–123 (2010)
6. Singh, R., Khare, A.: Fusion of multimodal medical images using Daubechies complex wavelet transform – a multiresolution approach. Inf. Fusion **19**, 49–60 (2014)
7. Selesnick, I.W., Baraniuk, R.G., Kingsbury, N.C.: The dual-tree complex wavelet transform. IEEE Sig. Process. Mag. **22**, 123–151 (2005)
8. Remenyi, N., Nicolis, O., Nason, G., Vidakovic, B.: Image denoising with 2D scale-mixing complex wavelet transforms. IEEE Trans. Image Process. **23**, 5165–5174 (2014)
9. Chiorean, L., Vaida, M.-F.: Medical image fusion based on discrete wavelet transform using Java technology. In: Proceedings of the ITI 2009 31st International Conference on Information Technology Interfaces, pp. 55–60 (2009)
10. Rockinger, O.: Image sequence fusion using a shift-invariant wavelet transform. Proc. Int. Conf. Image Process. **3**, 288–291 (1997)
11. Singh, R., Srivastava, R., Prakash, O., Khare, A.: Multimodal medical image fusion in dual tree complex wavelet transform domain using maximum and average fusion rules. J. Med. Imag. Health Inform. **2**, 168–173 (2012)
12. Shreyamsha Kumar, B.K.: Multifocus and multispectral image fusion based on pixel significance using discrete cosine harmonic wavelet transform. Sig. Image Video Process. **7**, 1125–1143 (2013)
13. Sheikh, H.R., Bovik, A.C.: Image information and visual quality. IEEE Trans. Image Process. **15**, 430–444 (2006)
14. Crowley, J.L., Martin, J.: Experimental comparison of correlation techniques. In: IAS-4, International Conference on Intelligent Autonomous Systems, Karlsruhe (1995)
15. Han, Y., Cai, Y., Cao, Y., Xu, X.: A new image fusion performance metric based on visual information fidelity. Inf. Fusion **14**, 127–135 (2013)

A Benchmark of Motion Detection Algorithms for Static Camera: Application on CDnet 2012 Dataset

Kamal Sehairi[1(✉)], Chouireb Fatima[1], and Jean Meunier[2]

[1] TSS Laboratory, University Amar Telidji of Laghouat, Laghouat, Algeria
k.sehairi@lagh-univ.dz
[2] DIRO, University of Montreal, Montreal, Canada

Abstract. The main aim behind this study relates to comparing a variety of motion detection methods for a mono static camera and it endeavors to identify the best method for different environments and complex backgrounds whether in indoor or outdoor scenes. For this reason, we used the CDnet 2012 video dataset as a benchmark that comprises numerous challenging problems, ranging from basic simple scenes to complex scenes affected by shadows and dynamic backgrounds. Eleven ranging from simple to complex motion detection methods are tested and several performance metrics are used to achieve a precise evaluation of the results. This study sets its objective to enable any user to identify the best method that suits his/her need.

Keywords: Motion detection · Background modelling
Background subtraction · Object detection · Video surveillance

List of symbols

$I_t(x, y)$	Pixel coordinates (x, y) in the current image I_t.
Th, T	Threshold values.
U	The energy function.
U_m	Energy that ensures spatio-temporal homogeneity.
U_a	The adequacy energy that ensures good coherence of the solution compared to the observed data.
$c(s, r)$	Denotes a set of binary cliques associated with the chosen neighbourhood system.
$B_t(x, y)$	Pixel coordinates (x, y) in the background image B_t.
α, ρ	Learning rates.
(μ, σ)	Mean and standard deviation of Gaussian distribution.
$\omega_{k,t}$	The estimated weight associated with the kth Gaussian at time t.
D	Deviation threshold.
$H_{x,y,q}$	Spatio-temporal histogram for each pixel.
$P_{x,y,q}$	Probability density functions for each pixel.
$E_{x,y}$	The entropy of the pixel (x, y).

O. Demigha et al. (Eds.): CSA 2018, LNNS 50, pp. 235–245, 2019.
https://doi.org/10.1007/978-3-319-98352-3_25

Φ	Quantization function quantizes the 256 grey level values into Q grey levels.
ES	Eigen-space formed using N reshaped frames.
C	Covariance matrix.
V, V_M	The eigenvector matrix. The M eigenvectors in V that correspond to the M largest eigenvalues.

1 Introduction

Motion detection is the first and important step in intelligent visual surveillance system; it aims to find moving target objects in an input video sequence. The resulting information from motion detection task provides support to higher-level tasks such as automated object classification and behavior recognition. Though, detection of moving object is a difficult task due to several reasons such as noise in video sequence, complex backgrounds, sudden illumination changes, and shadows of static and dynamic objects. Different approaches have been proposed by researchers for moving object detection. These approaches can be further classified [1–3] into three types: background subtraction, [4, 5] temporal differencing, [6, 7] and optical flow [8, 9]. Temporal differencing technique is computationally simple and fast. It is adaptive to dynamic environment, but generally it is not effective in extracting all the relevant pixels of a target object. There may be holes left inside moving objects.

Background subtraction is a popular method to detect stationary foreground objects; but it may suffer from dynamic background changes such as the entrance of new background object or sudden illumination change. Many background modelling approaches have been proposed to overcome these problems. Bouwmans [10] classified these approaches into seven categories: basic background modelling, statistical background modelling, fuzzy background modelling, background clustering, neural network background modelling, wavelet background modelling, and background estimation. Further, Cheung and Kamath [11] classified background modelling techniques into two broad categories: non-recursive and recursive. A non-recursive technique uses a sliding-window approach for background estimation. It stores a buffer containing the previous L video frames and estimates the background image on the basis of the temporal variation of each pixel within the buffer. Non-recursive techniques are highly adaptive, as they do not depend on previous data beyond the frames stored in the buffer [12]. Optical-flow-based motion segmentation can be used to detect moving objects for both the case of static and moving camera. Though, most optical flow methods are computationally expensive and cannot be implemented in real-time without specialized hardware [13]. In [14], motion detection methods have been classified into matching methods, energy-based methods, statistical methods, gradient methods, and hybrid methods.

Several recent studies have tried to compare different motion detection methods. Piccardi [4] examined seven motion detection methods and performed a comparative performance analysis based on speed, memory requirements, and accuracy; however, he did not conduct any quantitative evaluation. Toyama et al. [15] presented a comparative study between eight motion detection methods and a method which they proposed.

This method combines three algorithms (pixel-level, region, and frame-level). The result of this comparison is that their method outperforms the other methods. Unfortunately, they did not present a detailed performance evaluation; only the number of false negative and the number of false positive were considered as evaluation criteria. Though, they did not conduct a comparison between these methods. In [16], Benezeth et al. performed a comparison between seven background subtraction methods in the case of different types of videos (camera-captured, semi-synthetic, and synthetic); two evaluation criteria (precision and recall) were used to classify these methods. This comparison is also based on the computational complexity as well as the memory requirements of each method. However, their video dataset did not present a wide variety of challenges due to factors including bad weather and camera jitter. In addition, semi-synthetic and synthetic videos are not real and not all the ground truths of the dataset are provided.

Bouwmans [17] presented an overview of traditional and recent approaches, and classified them in terms of the mathematical models used. In addition, he presented the available resources, datasets and libraries which can be used to evaluate and compare these approaches. But, he did not perform a quantitative evaluation.

The main objective of this work is to compare different motion detection methods and determine for which type of scenes and situations each method suits best. We should note that this work differs from our previous work in [18], since we apply the tested methods on the CDnet 2012 dataset [19], which includes a total of 35 videos of indoor and outdoor scenes totaling nearly 108000 images categorized into 6 challenges (baseline, dynamic background, camera jitter, shadow, intermittent object motion and thermal). Each scene represents diverse moving objects, such as boats, cars, trucks, cyclists, and pedestrians. For each image, a ground truth is provided to allow a precise comparison and ranking of motion detection methods. Additionally, seven distinct metrics have been used for evaluation: recall, specificity, false positive rate, false negative rate, percentage of wrong classification, precision, and F-measure.

2 Motion Detection Methods

An effective video surveillance system depends closely on the technique used for motion detection in the video sequence. Detection of moving object can be complex due to several reasons such as complex backgrounds (e.g. tree leaf movements), shadows, weather conditions (e.g. rain or snow-fall), and sudden illumination changes, as well as problems related to the moving object itself, such as the similarity of its color to the background color, its size, and its distance from the camera. Hence, recently, several methods have been proposed to overcome these problems. Table 1 summarizes the motion detection algorithms used in this comparative study, in which we group them in three categories, the methods are: Euclidean distance (delta frame DF) [7, 20, 21], three-frame difference (3FD) [22–24], adaptive background (running average filter RAF) [12, 15, 25, 26], forgetting morphological temporal gradient (FMTG) [27], background estimation [28, 29], Markov Random Field (MRF)-based detection algorithm [30–32], running Gaussian average (RGA) [33–35], mixture of Gaussians (MoG) [36–38], spatio-temporal entropy image (STEI) [39, 40], difference-based spatio-temporal entropy image (DSTEI) [40, 41], and eigen-background (Eig-Bg) [42]. More details can be found in [18].

Table 1. Motion detection algorithms used in this comparative study

Category	Type of method	Method	The background	The foreground
Temporal differencing methods	Basic	DF	$I_{t-1}(x,y)$	$\|I_t(x,y) - I_{t-1}(x,y)\| \geq Th$
	Basic	3FD	$I_{t-2}(x,y), I_{t-1}(x,y)$	$\zeta_t(x,y) = Min(\|I_{t-2}(x,y) - I_{t-1}(x,y)\| \geq Th_{t_1},$ $\|I_t(x,y) - I_{t+1}(x,y)\| \geq Th_{t_2})$
	Statistical - Spatio-temporal histogram	MRF	$I_{t-1}(x,y)$	$U(e,o) = U_m(e) + U_a(o,e)$ $U_a(o,e) = \frac{1}{2\sigma^2}[o - \psi(e)]^2$ $\psi(e) = \begin{cases} 0 & \text{if } \|I_t(x,y) - I_{t-1}(x,y)\| < Th \\ \alpha & \text{else} \end{cases}$ $U_m(e) = \sum_{c \in C} V_c(e_s, e_r) = V_s(e_s, e_r) + V_p(e_s^t, e_s^{t-1}) + V_p(e_s^t, e_s^{t+1})$ $\begin{cases} V_s(e_s, e_r) = \begin{cases} -\beta_s & \text{if } e_s = e_r \\ +\beta_s & \text{if } e_s \neq e_r \end{cases} \\ V_p(e_s^t, e_s^{t-1}) = \begin{cases} -\beta_p & \text{if } e_s^t = e_s^{t-1} \\ +\beta_p & \text{if } e_s^t \neq e_s^{t-1} \end{cases} \\ V_p(e_s^t, e_s^{t+1}) = \begin{cases} -\beta_f & \text{if } e_s^t = e_s^{t+1} \\ +\beta_f & \text{if } e_s^t \neq e_s^{t+1} \end{cases} \end{cases}$
Background subtraction methods	Basic	RAF	$B_t(x,y) = (1-\alpha)B_{t-1}(x,y) + \alpha I_{t-1}(x,y)$	$\|I_t(x,y) - B_t(x,y)\| \geq Th$
	Basic	FMTG	$M_t(x,y) = \alpha I_t(x,y) + (1-\alpha)\max\{I_t(x,y), M_{t-1}(x,y)\}$ $m_t(x,y) = \alpha I_t(x,y) + (1-\alpha)\min\{I_t(x,y), m_{t-1}(x,y)\}$	$M_t(x,y) - m_t(x,y) \geq Th$
	Basic	$\sum\Delta$	(1) Computation of $\Sigma\Delta$ mean $\left.\begin{array}{l} M_0(x,y) = I_0(x,y) \\ M_t(x,y) = M_{t-1}(x,y) + \text{sgn}(I_t(x,y) - M_{t-1}(x,y)) \end{array}\right\}$ (2) Computation of difference $\Delta_t(x,y) = \|M_t(x,y) - I_t(x,y)\|$ (3) Computation of $\Sigma\Delta$ variance $\left.\begin{array}{l} V_0(x,y) = \Delta_0(x,y) \\ \text{if } \Delta_t(x,y) \neq 0, \\ V_t(x,y) = V_{t-1}(x,y) + sgn(N \times \Delta_t(x,y) - V_{t-1}(x,y)) \end{array}\right\}$	$\Delta_t(x,y) - V_t(x,y) \geq 0$
	Statistical - temporal histogram	RGA	$\mu_t = \alpha I_t + (1-\alpha)\mu_{t-1}$ $\sigma_t^2 = \alpha(I_t - \mu_t)^2 + (1-\alpha)\sigma_{t-1}^2$	$\|I_t - \mu_t\| > D\sigma_t$
	Statistical - temporal histogram	MoG	(1) Initialisation of : $\mu_k, \sigma_k, \omega_k$ k = 1,…, M (2) Match test : $\|\mu_k - X_t\| \leq D\sigma_k$ k = 1,…, M if match is found Update the parameters of the matched gaussian as follow: $\rho = \frac{\alpha}{\omega}$ $\omega_t = (1-\alpha)\omega_{t-1} + \alpha$ $\mu_t = \rho X_t + (1-\rho)\mu_{t-1}$ $\sigma_t^2 = \rho(X_t - \mu_t)^2 + (1-\rho)\sigma_{t-1}^2$ for the other distributions update: $\omega_t = (1-\alpha)\omega_{t-1}$ if no match found: create a new distribution with: $\mu_t = X_t, \sigma_t^2 = \sigma_0^2, \omega_t = \min(\omega_t)$ (3) order the distributions according to ω_k/σ_k (4) the first B distributions in the ranking order satisfying $\sum_{k=1}^{B} \omega_k > T$ are considered as background	(5) perform test match with these selected background to determine the foreground $\|\mu_k - X_t\| > D\sigma_k$ k = 1,…, M
	Matrix decomposition	Eig-Bg	(1) eigen-space construction $\mu = \sum_{k=1}^{N} I_k$ $C = Cov(ES) = ES \cdot ES^T = \frac{1}{N}\sum_{k=1}^{N}[I_k - \mu] \cdot [I_k - \mu]^T$ $with\, ES = [I_1\, I_2 \ldots I_N]$ $C = U \Lambda U^T$ (2) projection of I on eigen-space $I' = U_M^T(I - \mu)$ U_M consists of the M eigenvectors in U that correspond to the M largest eigenvalues (3) Background construction by back projection $B = U_M I' + \mu$	$\|I - B\| > Th$
Energy based methods	Spatio-temporal histogram	STEI	N/A	$E_{x,y} = -\sum_{q=1}^{Q} P_{x,y,q}\log(P_{x,y,q})$ where $P_{x,y,q} = \frac{H_{x,y,q}}{N}$ and $N = L \times w \times w$
	Spatio-temporal histogram	DSTEI	N/A	$D_t = \Phi(\|I_t - I_{t-1}\|)$ $H_{i,j,q}(N) = \frac{1}{L}\sum_{k=1}^{L} h_{i,j,q}(k)$ $H_{i,j,q}(k+1) = \alpha H_{i,j,q}(k) + (1-\alpha)h_{i,j,q}(k+1)$ $E_{x,y} = -\sum_{q=1}^{Q} P_{x,y,q}\log(P_{x,y,q})$ where $P_{x,y,q} = \frac{H_{x,y,q}}{N}$ and $N = L \times w \times w$

3 Evaluation Metrics and Performance Analysis

To provide detailed comparison of these methods, we use the same seven metrics as those used for CDnet2012: recall (Re), specificity (Sp), false positive rate (FPR), false negative rate (FNR), percentage of wrong classification (PWC) [12], precision [12], and F-measure [43]. For comparison purpose, we have adopted the same approach used on CDnet [18, 19] to generate the results. First, we compute all the metrics for each video in each category; then a category average metric was computed:

$$M_c = \frac{1}{N_c}\sum_{v} M_{v,c} \tag{1}$$

Where M_c represents one of the seven metrics (Re, Sp, FPR, FNR, PWC, Pr and F1), N_c is the number of videos in each category, and v is a particular video in category c. We also define an overall average metric (OAM), which is the simple average of the category averages:

$$OAM = \frac{1}{C}\sum_{c=1}^{C} M_c \tag{2}$$

Where C is the number of categories. In order to rank all the methods, first, for each category c, we compute the rank of each method for metric M. Then, we compute the average rank of this method across all the metrics:

$$RM_c = \frac{1}{7}\sum_{n=1}^{7} Rank(M_c) \tag{3}$$

Subsequently, we compute the average over all the categories to obtain the average rank across categories, RC, for each method:

$$RC = \frac{1}{C}\sum_{c=1}^{C} RM_c \tag{4}$$

In addition, we compute the average rank across the OAM for each method:

$$R = \frac{1}{7}\sum_{n=1}^{7} rank\,(OAM) \tag{5}$$

4 Results and Discussion

We apply all the motion detection methods described in Sect. 2 on the CDnet dataset, which includes different scenarios and challenges (Baseline, Camera jitter, Dynamic background, Intermittent object, Shadow and Thermal) [19]. All the motion detection methods are followed by an automatic thresholding operation in order to determine region changes and remove small changes in luminosity, except for the RGA and MoG methods, in which the threshold is fixed to 2.5σ, where σ denotes the standard deviation. We select Otsu's thresholding method based on a previous study [13].

For the eigen-background method, we set the number of training images as N = 28. These training images are equally spaced by 10 frames, and the number of eigen-background vectors is set at M = 3.

For the mixture of Gaussians method, the parameters used for this evaluation are selected in accordance with the work of Nikolov *et al.* [43], who have measured the accuracy of the algorithm as a function of each variable parameter. Further, they have proposed a set of optimal parameters to improve the performance of the MoG algorithm. Accordingly, we select the number of Gaussians as K = 3, the learning rate as $\alpha = 0.01$, the foreground threshold as $T = 0.25$, the deviation threshold as D = 2.5, and the initial standard deviation as $\sigma_{init} = 20$.

For the running Gaussian average method, the learning rate is set as $\alpha = 0.01$ and the deviation threshold is set to D = 2.5. We note that MoG and RGA are applied to grey level images in all our tests. For the STEI and DSTEI methods, to construct the spatio-temporal histogram, we select a 3×3 window with 5 images, and the number of grey levels is set as Q = 100.

For the MRF-based motion detection algorithm, according to the work of Caplier [31], we set the four parameters as $\beta_s = 20, \beta_p = 10, \beta_f = 30, \alpha = 10$. For the $\Sigma\Delta$ method, the only parameter to be set was N; we selected N = 3 (typically, N = 2, 3, or 4) [29].

Finally, for the forgetting morphological temporal gradient method and the adaptive background detection method, the parameter α should take values in the interval. In all our tests for these last two methods, we chose $\alpha = 0.1$. All these parameters are typical or recommended values.

The overall results of testing these methods using the CDnet dataset are reported in Table 2, where the entries are sorted by average rank across categories (RC). It is clear from Table 2 that the STEI method generate poor results compared to the other methods; the use of entropy alone as a metric to detect moving objects does not yield good results because of the spatio-temporal accumulation window object edges, which can lead to high diversity (high entropy), and thus alters the segmentation result. Moreover, this error will spread to the entire edge region. Adding the delta frame to this method (DSTEI) increases the precision and decreases the percentage of wrong classification considerably. However, from the overall results in Table 2, we note that the DSTEI does not achieve significant improvement over the FD method, owing to the drawbacks of using the spatio-temporal accumulation window and the tails caused by using inappropriate values of α to compute the spatio-temporal histogram recursively.

Table 2. Overall results across all categories

	Recall	Specificity	FPR	FNR	PWC	Precision	F-Measure	R	RC
RGA	0,26505	0,99497	**0,00503**	0,73495	**4,19859**	**0,59907**	0,33006	**2,85714**	**4,47619**
Eig-Bg	**0,62747**	0,94214	0,05786	**0,37253**	7,16652	0,51787	**0,49452**	5,28571	4,80952
GMM	0,16616	**0,99693**	0,00307	0,83384	4,17175	0,63707	0,23899	3,28571	5,45238
FMTG	0,33262	0,95680	0,04320	0,66738	7,53384	0,50232	0,27755	6,42857	5,64286
RAF	0,27558	0,96552	0,03448	0,72442	6,90696	0,50048	0,25658	6,00000	5,88095
DSTEI	0,24499	0,96359	0,03641	0,75501	7,18776	0,52441	0,21885	6,85714	5,95238
FD	0,15265	0,97515	0,02485	0,84735	6,41522	0,55956	0,15622	6,42857	6,04762
3FD	0,10307	0,98558	0,01442	0,89693	5,58560	0,57070	0,12242	6,85714	6,16667
$\Sigma - \Delta$	0,11619	0,99272	0,00728	0,88381	4,84171	0,46134	0,15251	6,57143	6,20557
MRFMD	0,07506	0,98936	0,01064	0,92494	5,35660	0,51765	0,09631	7,42857	7,21429
STEI	0,47427	0,76004	0,23996	0,52573	25,51530	0,13410	0,15525	8,00000	7,76190

The MRF-based method also shows poor results, but unlike the STEI and DSTEI methods, this method has a very low false positive rate, low percentage of wrong classification (PWC), and acceptable precision. The poor results are due to the initialization step based on the delta frame method, leading to incompletely segmented moving objects containing holes. Nevertheless, this method performs well by enhancing the image difference, especially in the case of noise or complex backgrounds.

As expected, the frame difference and three-frame difference methods do not show good results, except for high precision, mainly because of the incomplete segmentation of the entire shape of the moving object while preserving only the edges. Moreover, overlap in the case of slow moving objects or objects that are far from the camera is another drawback of this method. To overcome the problem of incomplete segmentation, the threshold operation in the delta frame method is usually followed by morphological operations to link the edges of the moving objects. Then, regions and holes in the image are filled. Another solution is to combine frame difference methods with a background subtraction method.

From Table 2, we can see that the $\sum\Delta$ method achieves some improvement over MRFMD (both methods are based on the delta frame method). The $\sum\Delta$ method is characterized by a low false positive rate, low percentage of wrong classification, and high specificity, i.e. a large number of background pixels are correctly classified. The forgetting morphological temporal gradient method (FMTG) yields acceptable results, especially good segmentation in the case of slow moving or small moving objects. However, it is characterized by high PWC caused by artificial tails due to inappropriate choice of α. Moreover, this method is conducive to temporal noise. As with FMTG, the running average filter (RAF) method produced acceptable results. This was expected because FMTG is based on recursive operation of the running average filter. The second-best method according to the evaluation results (Table 2) is the eigen-background method, which yields good detection results by extracting accurately all the moving regions. We note that it had a high recall, high F1-score, and low false positive rate. Furthermore, its results are strongly dependent on the images that form the eigen-space; the presence of moving objects in this space could alter the detection results.

Table 3. Top three methods for all categories

	1st	2nd	3rd
Baseline	Eig-Bg	$\Sigma - \Delta$	RGA
Camera jitter	Eig-Bg	RGA	MoG
Dynamic background	RGA	GMM	DSTEI
Intermittent object motion	$\Sigma - \Delta$	DSTEI	RGA
Shadow	RGA	GMM	3FD
Thermal	RAF	Eig-Bg	FD

Finally, the methods based on Gaussian distribution show best performance among all the methods, with high precision, high F1-score, low false positive rate, and low percentage of wrong classification. Remarkably, the RGA method outperforms the MoG method (Table 2), because of the larger number of parameters required to be set accurately for the MoG algorithm $(K, \alpha, T, D, \sigma)$, which, must be set according to the challenging conditions presented by a video (daylight/night, indoor/outdoor, complex/simple background, with/without noise). Thus, in most cases, the RGA method seems to be sufficient (Table 3). Moreover, its computational complexity is lower than that of the MoG method, which is in agreement with the findings of Piccardi [4] and Benezeth et al. [16]. In addition, we note that the Eig-Bg method performs better than the RGA method in the case of easy scenes such as those in the "baseline" category, and its execution time depends on the number of eigenvectors. In our case, the execution time seems to be acceptable; however, the memory requirements make it unsuitable for real-time applications.

5 Conclusion

In this paper, we presented a review and comparison of different motion detection methods using one of the most recent, complete, and challenging datasets, i.e., CDnet. Detailed pixel evaluation has been performed using different metrics to enable a user to determine the appropriate method for his/her need.

From the results reported, we can conclude that there is no ideal method for all situations; each method performs well in some cases and fails in others. However, if we have to choose one method among all the tested methods, it would undoubtedly be the running Gaussian average (RGA) method, which ranks among the three best methods in nearly all the categories and has an acceptable execution time. In the future, we will test other methods in order to expand the scope of this study and provide users with a complete benchmark of motion detection methods.

References

1. Kim, I., Choi, H., Yi, K., Choi, J., Kong, S.: Intelligent visual surveillance — a survey. Int. J. Control Autom. Syst. **8**(5), 926–939 (2010)
2. Paul, M., Haque, S., Chakraborty, S.: Human detection in surveillance videos and its applications - a review. EURASIP J. Adv. Sig. Process. **176**(1), 1–16 (2013)
3. Ko T.: A survey on behavior analysis in video surveillance for homeland security applications. In: AIPR 2008, Washington, DC, USA (2008)
4. Piccardi, M.: Background subtraction techniques: a review. In: 2004 IEEE International Conference on Systems, Man and Cybernetics, vol. 4, pp. 3099–3104 (2004)
5. Migliore, D.A., Matteucci, M., Naccari, M.: A revaluation of frame difference in fast and robust motion detection. In: Proceedings of 4th ACM International Workshop on Video Surveillance and Sensor Networks (VSSN 2006), pp. 215–218. ACM, New York (2006)
6. Yalamanchili, S., Martin, W., Aggarwal, J.: Extraction of moving object descriptions via differencing. Comput. Vis. Graph. **18**(2), 188–201 (1982)
7. Jain, R., Martin, W., Aggarwal, J.: Segmentation through the detection of changes due to motion. Comput. Vis. Graph. **11**(1), 13–34 (1979)
8. Lucas, B.D., Kanade, T.: An iterative image registration technique with an application to stereo vision. In: Proceedings of 7th International Joint Conference on Artificial Intelligence, vol. 2, pp. 674–679. Morgan Kaufmann Publishers Inc., San Francisco (1981)
9. Horn, B., Schunck, B.: Determining optical flow: a retrospective. Artif. Intell. **59**(1–2), 81–87 (1993)
10. Bouwmans, T.: Recent advanced statistical background modeling for foreground detection – a systematic survey. Recent Pat. Comput. Sci. **4**(3), 147–176 (2011)
11. Cheung, S., Kamath, C.: Robust background subtraction with foreground validation for urban traffic video. EURASIP J. Appl Sig. Process. **2005**(14), 2330–2340 (2005)
12. Elhabian, S., El-Sayed, K., Ahmed, S.: Moving object detection in spatial domain using background removal techniques: state-of-art. Recent Pat. Comput. Sci. **1**(1), 32–54 (2010)
13. Sehairi, K., Chouireb, F., Meunier, J.: Comparison study between different automatic threshold algorithms for motion detection. In: 2015 4th International Conference on Electrical Engineering (ICEE), Boumerdes, December 2015, pp. 1–8 (2015)
14. Hammami, M., Jarraya, S.K., Ben-Abdallah, H.: A comparative study of proposed moving object detection methods. J. Next Gener. Inf. Technol. **2**(2), 56–68 (2011)
15. Toyama, K., Krumm, J., Brumitt, B., et al.: Wallflower: principles and practice of background maintenance. In: The Proceedings of the Seventh IEEE International Conference on Computer Vision, Kerkyra, September 1999, pp. 255–261 (1999)
16. Benezeth, Y., Jodoin, P.-M., Emile, B., et al.: Comparative study of background subtraction algorithms. J. Electron. Imag. **19**(3), 033003 (2010)
17. Bouwmans, T.: Traditional and recent approaches in background modeling for foreground detection: an overview. Comput. Sci. Rev. **11–12**, 31–66 (2014)
18. Sehairi, K., Chouireb, F., Meunier, J.: Comparative study of motion detection methods for video surveillance systems. J. Electron. Imag. **26**(2), 023025 (2017). https://doi.org/10.1117/1.JEI.26.2.023025
19. Goyette, N., Jodoin, P.-M., Porikli, F., et al.: Changedetection.net: a new change detection benchmark dataset. In: 2012 IEEE Computer Society Conference on Computer Vision and Pattern Recognition Workshops, Providence, RI, USA, June 2012, pp. 1–8 (2012)
20. Wang, Z., Xiong, J., Zhang, Q.: Motion saliency detection based on temporal difference. J. Electron. Imag. **24**(3), 033022-10 (2015)

21. Radzi, S.S.M., Yaakob, S.N., Kadim, Z., et al.: Extraction of moving objects using frame differencing, ghost and shadow removal. In: 2014 5th International Conference on Intelligence Systems, Modelling and Simulation, Langkawi, January 2014, pp. 229–234 (2014)
22. Chen, C., Zhang, X.: Moving vehicle detection based on union of three-frame difference. In: International Conference on Advances in Electronic Engineering, Communication and Management, vol. 140, pp. 459–64. Springer (2012)
23. Wang, Z.: Hardware implementation for a hand recognition system on FPGA. In: 2015 IEEE 5th International Conference on Electronics Information and Emergency Communication, Beijing, China, May 2015, pp. 34–38 (2015)
24. Zhang, Y., Wang, X., Qu, B.: Three-frame difference algorithm research based on mathematical morphology. Proc. Eng. **29**, 2705–2709 (2012)
25. Heikkilä, J., Silvén, O.: A real-time system for monitoring of cyclists and pedestrians. Image Vis. Comput. **22**(7), 563–570 (2004)
26. Tan, X., Li, J., Liu, C.: A video-based real-time vehicle detection method by classified background learning. World Trans. Eng. Technol. Educ. **6**(1), 189–192 (2007)
27. Richefeu, J.C., Manzanera, A.: A new hybrid differential filter for motion detection. In: Computer Vision and Graphic. Computational Imaging and Vision, vol. 32, pp. 727–732. Springer, Dordrecht (2006)
28. Manzanera, A., Richefeu, J.: A new motion detection algorithm based on Σ-Δ background estimation. Pattern Recogn. Lett. **28**(3), 320–328 (2007)
29. Lacassagne, L., Manzanera, A., Denoulet, J.: High performance motion detection: some trends toward new embedded architectures for vision systems. J. Real-Time Image Proc. **4** (2), 127–146 (2008)
30. Bouthemy, P., Lalande, P.: Recovery of moving object masks in an image sequence using local spatiotemporal contextual information. Opt. Eng. **32**(6), 1205 (1993)
31. Caplier, A., Luthon, F., Dumontier, C.: Real-time implementations of an MRF-based motion detection algorithm. Real-Time Imag. **4**(1), 41–54 (1998)
32. Denoulet, J., Mostafaoui, G., Lacassagne, L., et al.: Implementing motion Markov detection on general purpose processor and associative mesh. In: 7th International Workshop on Computer Architecture Machine Perception (CAMP 2005), pp. 288–293 (2005)
33. Wren, C., Azarbayejani, A., Darrell, T., et al.: Pfinder: real-time tracking of the human body. IEEE Trans. Pattern Anal. Mach. Intell. **19**(7), 780–785 (1997)
34. Tang, Z., Miao, Z., Wan, Y.: Background subtraction using running Gaussian average and frame difference. In: Ma, L., Rauterberg, M., Nakatsu, R. (eds.) Proceedings of the 6th International Conference on Entertainment Computing (ICEC 2007), pp. 411–414. Springer, Berlin (2007)
35. Jabri, S., Duric, Z., Wechsler, H., et al.: Detection and location of people in video images using adaptive fusion of color and edge information. In: Proceedings of 15th International Conference on Pattern Recognition, Barcelona, vol. 4, pp. 627–630 (2000)
36. Stauffer, C., Grimson, E.: Adaptive background mixture models for real-time tracking. In: Proceedings of Computer Vision and Pattern Recognition, pp. 246–252 (1999)
37. Bouwmans, T., El Baf, F., Vachon, B.: Background modeling using mixture of Gaussians for foreground detection - a survey. Recent Pat. Comput. Sci. **1**(3), 219–237 (2008)
38. Lumentut, J., Gunawan, F., Diana: Evaluation of recursive background subtraction algorithms for real-time passenger counting at bus rapid transit system. Proc. Comput. Sci. **59**, 445–453 (2015)
39. Ma, Y-F., Zhang, H-J.: Detecting motion object by spatio-temporal entropy. In: IEEE International Conference on Multimedia and Expo, ICME 2001, Tokyo, Japan, August 2001, pp. 265–268 (2001)

40. Jing, G., Siong, C.E., Rajan, D.: Foreground motion detection by difference-based spatial temporal entropy image. In: Proceedings of 2004 IEEE Region 10 Conference (TenCon 2004), vol. 1, pp. 379–382. IEEE CS Press (2004)
41. Chang, M.C., Cheng, Y.J.: Motion detection by using entropy image and adaptive state-labeling technique. In: 2007 IEEE International Symposium on Circuits & Systems, New Orleans, LA, pp. 3667–3670 (2007)
42. Oliver, N., Rosario, B., Pentland, A.: A Bayesian computer vision system for modeling human interactions. IEEE Trans. Pattern Anal. Mach. Intell. **22**(8), 831–843 (2000)
43. Nikolov, B., Kostov, N., Yordanova, S.: Investigation of mixture of Gaussians method for background subtraction in traffic surveillance. Int. J. Reason. Based Intell. Syst. **5**(3), 161–168 (2013)

Machine Learning and Data Science

Potentials of Computational Intelligence for Big Multi-sensor Data Management

Allel Hadjali(✉)

Ecole Nationale Supérieure de Mécanique et D'Aérotechnique, Poitiers, France
allel.hadjali@ensma.fr

Abstract. Due to the important development in Hardware technologies, sensors are everywhere. So, many real-world applications in different domains (such as Defense, Industry, Transport, Energy, Surveillance, Climate and Weather, Healthcare) use multi-sensors to collect tremendous amounts of data about their states and environments. Such data are inherently uncertain, erroneous and noisy on the one hand, and voluminous, distributed and continuous on the other hand. One of the major challenges the Governments, Industry, Companies and Organizations have to face today is how to manage and make sense of Big sensor data for the purpose of decision making. Recent advancements in science and technologies (like Computational Intelligence and Machine Learning) are opening the road to more advanced analytics techniques that can allow for the sensor data characteristics and extract useful insights. This allows building solutions that provide fast time responses and less resources consuming. In this talk, we show how techniques stemming from the recent Computational Intelligence field can contribute to the above solutions to manage and handle Sensor data. Some examples from the aeronautic/space domain are used to motivate our propositions.

O. Demigha et al. (Eds.): CSA 2018, LNNS 50, p. 249, 2019.
https://doi.org/10.1007/978-3-319-98352-3_26

Clustering Approach for Data Lake Based on Medoid's Ranking Strategy

Redha Benaissa[1,2,3](✉), Farid Benhammadi[1](✉), Omar Boussaid[2](✉), and Aicha Mokhtari[3](✉)

[1] DBE Laboratory, Ecole Militaire Polytechnique, Bordj el Bahri, Algiers, Algeria
benaissa.redha@gmail.com, fbenhammadi2008@gmail.com

[2] University Lumiere Lyon 2, 5, avenue Pierre Mends, 69676 Bron, France
omar.boussaid@univ-lyon2.fr

[3] RIIMA Laboratory, USTHB University, Algiers, Algeria
aicha_mokhtari@yahoo.fr

Abstract. A number of conventional clustering algorithms suffer from poor scalability, especially for data lake. Thus many modified clustering algorithms have been proposed to speed up these conventional algorithms based on the employment of data sampling techniques. However, these representations require the number of clusters to proceed to centroid selection for final data clustering. To address this limitation, this paper develops a two-phase clustering-based methodology. In the first phase, rather than attempting to construct a random sampling, we define a novel approach that computes plausible sample points, uses them as centroids for the final clusters. To speedup our clustering algorithm in the second phase we propose a parallelization scheme in conjunction with a Spark parallel processing implementation. Computational experiments reveal that the Global sampling method is more effective in terms of both quality and stability compared to the most popular K-means algorithm for the same parameter settings.

1 Introduction

Everyday a large amount of data is generated from various and heterogeneous sources. Conventional clustering systems cannot efficiently handle such a huge volume of data due to their poor scalability [1]. Parallel clustering systems are techniques that can reduce this huge volume by grouping similar data together through the assignment of data splits to the parallel nodes. The use of parallel clustering model has been recognized by many of big data handlers based on the sampling and filtering data [2]. These models divide the big data into clusters, with the data having the same characteristics on a given cluster. With the evolution of cloud computing, clustering parallel processing techniques have gained popularity and several researchers have proposed modifications to the existing clustering techniques in order to handle large datasets. These clustering techniques can be broadly categorized into three basic classes: sampling methods, streaming methods and parallel methods [3]. However, this clustering requires

O. Demigha et al. (Eds.): CSA 2018, LNNS 50, pp. 250–260, 2019.
https://doi.org/10.1007/978-3-319-98352-3_27

two factors that are influencing the future of data lake processing: (i) the specification of the number of clusters to construct the representative centroids or core points and (ii) the adequate sampling process to speed up any clustering algorithm with high accuracy results.

The purpose of this paper is to provide a parallel clustering based on mediods ranking method in the light of the aforementioned two factors that are influencing the future of data lake processing. Using a global sampling scheme, first our method constructs plausible centroids of final clustering. This construction is based on point ranking expressed in term of the number of minimal intra-distances (minimal similarities of a pair of data points). Note that in our method the plausible centroids are real data points. Second, we address the parallelization issue of our clustering method in conjunction with a Spark platform parallel processing implementation [16] to reduce the computational time of our clustering algorithm. Using some well-known large benchmarking data sets, experimental results show that the proposed parallel clustering method results in a higher accuracy and speedup than most popular k-means large scale clustering algorithms.

The remainder of this paper is organized as follows. Section 2 is dedicated to a general discussion of the parallel clustering models with a brief description of their performance on big data. The concept of data lake constitutes the topic of Sect. 3. Our global sampling technique in conjunction with a Spark parallel processing implementation is treated in Sect. 4. Computation evaluation results are discussed in Sect. 5. Finally, Sect. 6 provides final conclusions on our findings and future research directions.

2 Related Work

Many conventional clustering algorithms have been modified for big data. Based on the clustering processing these algorithms can be classified into three different categories such as sampling methods, streaming methods and parallel methods [3]. Sampling methods permit to reduce the clustering computation time by first choosing a subset of the big data set and then using this subset to find the final clusters. The sampling process picks a subset of the given input and makes inference on the original dataset. The key idea behind all sampling-based clustering methods is to obtain the cluster representatives, using only the sampled subset, and then assign the remaining data points to the closest representative. The popular methods to efficiently cluster big data sets of this category employ random sampling [4–6]. Other techniques are based on intelligent sampling scheme such as corset sampling [7–9] or deterministic sampling [2]. This latter approach uses many sampling stages that can be used to speed up any clustering algorithm. Empirical results show that this technique results in a speedup of more than an order of magnitude over conventional hierarchical clustering algorithms. Moreover, the obtained accuracy is better. The second clustering category uses a single pass over an arbitrary sized data set such as streaming clustering [10–12]. For instance, Guha et al. [11] proposed the most popular and widely applied streaming clustering. They first summarize the data stream into a larger number of clusters than desired. Then, they cluster the obtained centroids in the first step. Finally, the third clustering category represents the distribution or parallelism of the conventional clustering process using parallel, distributed, grid or cloud computing models. Parallel algorithms for hierarchical clustering have been given in [13] for several models of computing. The basic idea behind this parallel algorithm is to reduce the clustering problem to those of finding a minimum spanning tree in a complete graph and finding connected components in a forest. Other techniques speed up the clustering process by first dividing the task into a number of independent sub-tasks on the same data that can be processed simultaneously, and then efficiently merging these solutions into the final solution [14,15]. For instance, in [14], the MapReduce framework is employed to speed up the k-means and the k-medians clustering algorithms. The data set is split among many processors and a small representative data sample is obtained from each of the processors. These representative data points are then clustered to obtain the cluster centers. According to these centers, the remaining data points are clustered. The approach proposed in [15] uses the parallel tabu search clustering algorithm on Spark platform. Based on the same idea, the authors utilize the centroid-driven orientation of the k-means algorithm under the guidance of a simple version of tabu search. This strategy facilitates the parallel implementation of the k-means in the Spark environment. Computational experiments disclose that the proposed approach can generate better solutions than the k-means algorithm in terms of both quality and stability, while achieving a similar accelerating ratio when running on multiple computing nodes.

3 Data Lake Concept

Data lake is an emerging concept that has gained increasing popularity to organize and build the next generation of systems to address new big data challenges. Data Lake has been introduced as a flat architecture that supports broader analysis on various types of data from different sources [17]. Figure 1 shows the simple architecture that we propose for Data Lakes based on spark environment. The data lake contains data in their raw format (unstructured, semi-structured, and structured data) at scale and support data transformations by integrating Big Data processing frameworks such as Apache Hadoop and Apache Spark. This concept is accepted as a way to describe any large data pool in which the schema and data requirements are not defined until the data is queried. So the data lake loads all the data and defines the structure of the data at the time it is used with a powerful programming framework, such as MapReduce. For this reason, the data lake has some capabilities such as [18]

- To capture and store raw data at scale for a low cost.
- To store many types of data in the same repository.
- To perform transformations on the new data processing.
- To define the structure of the data at the time it is used.
- To perform single subject analytics based on very specific use case.

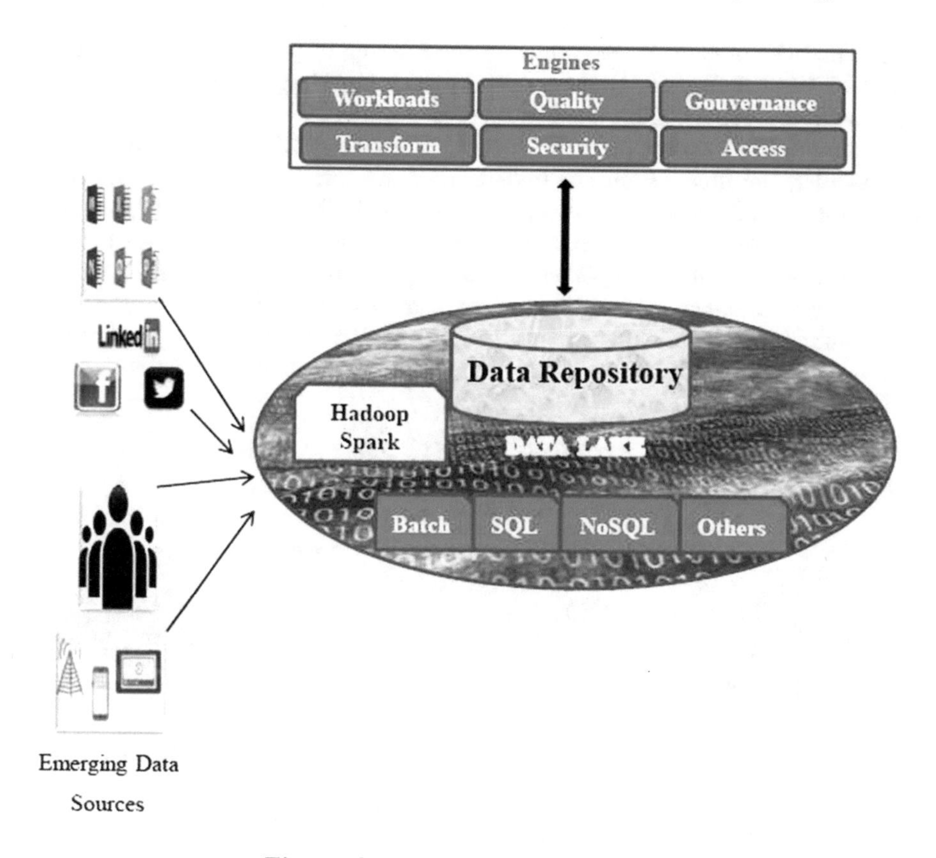

Fig. 1. Overall data lake architecture.

4 A New Parallel Clustering Model

4.1 Global Sampling Strategy

An important factor that influences clustering algorithms for Big data is the simplest and intelligent sampling strategies employed to construct the centroids or core points for final clustering of the remaining points of the datasets. These strategies are particularly attractive but they require the number of clusters and levels to construct the final centroids. Also, they depend on the local split clustering solutions only without the global sample points of the dataset. To address these limitations, we propose a new sampling strategy based on a sample points ranking method. This strategy involves selecting plausible medoids sequentially, to ensure global coverage of the chosen data points. So that the total intra-distance of sampled data points according to the chosen medoids is minimized.

The idea of our sampling is to process the global given input data in the first step and make inferences to construct the sample points ranking of the original dataset. Then each rank identified k plausible medoids (k is the plausible target number of clusters according to the ranking of points where accuracy is minimal. See the formula 1) for clustering process. When this happens, all sample points different of the plausible medoid points are clustered into k clusters according to their closest plausible medoid. Figure 2 provides an overview of our methodology for global sampling to determine the number of clusters. The global sampling strategy consists of two phases. Phase 1, the plausible medoid points construction, makes the minimal necessary adjustments to create plausible clustering centers. Phase 2, the plausible medoids combination, searches for possible clustering results that have decreased intra/inter distances.

In the first stage of our sampling strategy, it divides the collection of objects into a number of parts such that each part contains some center points.

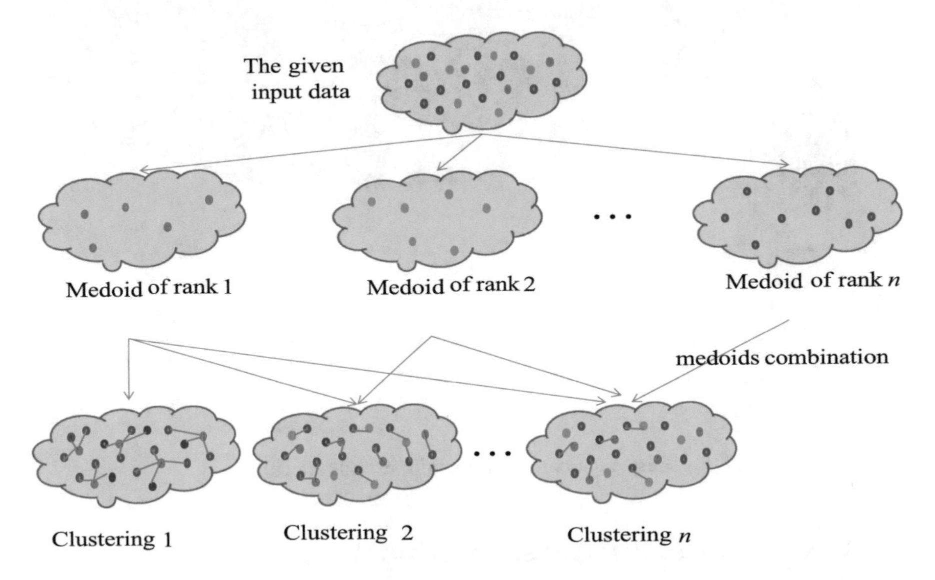

Fig. 2. Methodology for our global sampling strategy for clustering.

Let $\mathfrak{D} = \{x_1, x_2, \ldots, x_n\}$ be a given set of n objects where each object (point) $x_i \in \mathcal{X}$ and $\mathcal{X} \subseteq \mathfrak{R}^d$ for some dimension d. Let $\mathcal{C} = \{C_1, C_2, \ldots, C_k\}$ be a set of plausible clusters which reflect the possible grouping of the objects. In our global sampling, we first compute the ranking score $\mathcal{R}_j(x_i)$ of each point $x_i \in \mathcal{X}$ for $i \neq j$. This score is computed according to the following formula:

$$\mathcal{R}(x_i) = \sum_{j \neq i} \mathcal{SC}_j(x_i) \tag{1}$$

To find the score $\mathcal{SC}_j(x_i)$ for each data point x_i, we compute the similarity of a pair of data points (x_i, x_j) for all $i \neq j$. This similarity represents the distance between x_i and all the points x_j for $i \neq j$ as follows:

$$\mathcal{SC}_j(x_i) = \begin{cases} 1 \text{ if } d^2(x_i, x_j)_{i \neq j} = Min\{d^2(x_l, x_j)\}_{l \neq j \text{ and } l \neq i} \\ 0 \text{ if } \text{ otherwise.} \end{cases} \tag{2}$$

where $d^2(x_i, x_j)_{i \neq j}$ represents the distance between x_i and x_j points defined by:

$$d^2(x_i, x_j)_{i \neq j} = \|x_i - x_j\|^2 \tag{3}$$

Now we construct plausible medoid sets according to the ranking of points. Let $\mathfrak{M} = \{\mathcal{M}_1, \mathcal{M}_2, \ldots, \mathcal{M}_r\}$ be a set of all plausible medoid points on $\mathfrak{D}$ where r is the number of ranks. Therefore, we define $\mathcal{M}_r$ set, which is the plausible medoid set as below:

$$\mathcal{M}_r = \{x_i | x_i \in \mathfrak{D}, \mathcal{R}(x_i) = r\} \tag{4}$$

To illustrate the first phase of our global sampling, let consider a set of objects $\mathfrak{D} = \{x_1, x_2, \ldots, x_8\}$ and assume that the resulting distances between x_i points are reported by the tabular representations in Table 1. Applying Eqs. 1–4, we obtain the following two plausible medoid sets according to their ranks:

$$\mathcal{M}_1 = \{x_1, x_2, x_3, x_4, x_7, x_8\}$$
$$\mathcal{M}_2 = \{x_5, x_6\}$$

In the second stage, we combine the collection of the resulting plausible medoid points into a number of sets such that each set can be considered as cluster centers for the clustering process. This combination of medoids points of each rank produces several plausible clustering results according to each combined medoid points. Finally, the clustering with high result is retained. Note that we combine only medoid points with prior ranks in our clustering approach based on Clustering accuracy. To measure this accuracy we used distances which are summed over all the points. So given $\mathfrak{C} = \{C_1, C_2, \ldots C_{k_c}\}$ a set of clusters corresponding to a given input objects set, we first identify the medoid m_k of each cluster C_k (center). Then we calculate the distance $d^2(m_k, x_i)$ of each point $x_i \in C_k$ to the medoid m_k of the cluster it belongs to. Then we define the clustering accuracy of our algorithm by the following equation:

$$A_{\mathfrak{C}} = \sum_{k=1}^{k_c} \sum_{x_i \in C_k} d^2(m_k, x_i) \tag{5}$$

Table 1. Distance.

$\mathcal{X}$	x_1	x_2	x_3	x_4	x_5	x_6	x_7	x_8
x_1	0.000	0.021	1.282	1.308	0.940	0.955	1.075	1.492
x_2		0.000	1.277	1.300	1.092	1.085	1.263	1.611
x_3			0.000	0.001	0.454	0.293	0.679	1.186
x_4				0.000	0.477	0.313	0.715	1.205
x_5					0.000	0.042	0.202	0.502
x_6						0.000	0.133	1.269
x_7							0.000	1.269
x_8								0.000

Finally, the retained clustering that we seek is the clustering which minimize all accuracies and it can be expressed as

$$Min\{A_{\mathfrak{C}_{pc}}\} \tag{6}$$

where pc is the number of plausible clustering sets. A sequential pseudo code for our global sampling strategy of our medoid clustering is supplied by the following algorithm.

Algorithm 1. Sequential Global Sampling-based Medoid Clustering

---

Input: A set of n data points.
Output: The best k clusters.

1. Calculate all pair-wise point distances and place them in a $n \times n$ matrix. This matrix is called the dissimilarity matrix.
2. Calculate all medoid ranks using the dissimilarity matrix.
3. Find the plausible medoid sets with the medoid ranking. Start with r ranks labeled $1, 2,$
4. **for** (i=1:r) **do**
5. Construct the combined medoid sets from
$$m_c = \{m | m \in \mathcal{M}_j \text{ and } j \leq i\}$$
6. Start with $|m_c|$ clusters (nodes) labeled $1, 2, 3, \ldots, |m_c|$ where each cluster has one medoid (center) point.
7. Calculate accuracy of $|m_c|$ clusters.
8. **end for**
9. Retain the $|m_c|$ clusters whose accuracy is minimal.

4.2 Parallel Clustering Algorithm

When we apply our clustering model for Big data extirped from the data lake, parallelism becomes inevitable. A few parallel models of data lake have been introduced in the literature based on spark environment. In this section, we present parallelizations of our global sampling based clustering paradigm using this environment. We have parallelized two stages of our global sampling-based medoid clustering. The first stage is the process of plausible medoid sets construction. Since the data points can be treated independently in this process, during the pair dissimilarity matrix calculation (map process) each map computes the similarities between data points of two splits in parallel. In this stage, the reduce process relies on the output of map to compute the global dissimilarity matrix and the plausible medoid sets construction. The second phase is the process of clustering task according to the combined medoid sets. Since the data points can be treated independently, during the clustering each data split points can be treated independently (map operation) each map clustered each points into the closest medoid in parallel. Such a medoid is given a unique integer as ID. Finally, the reduce process of this stage allows to calculate the minimal accuracy to select the best clustering results. Algorithm 2 presents a flowchart of our parallel global sampling-based medoid clustering.

Algorithm 2. Parallel Global Sampling-based Medoid Clustering

Input: A set of n data points; integer p.
Output: The best k clusters.

1. Split the data points into p equal sized splits: $S_1, S_2, \ldots, S_p$ using data.mapPartitions (Spark method).
   STAGE 1:
2. Assign each pair of splits (S_i, S_j) to a single processor.
3. Calculate the pair dissimilarity matrix.
4. Repeat $\frac{p\times(p-1)}{2}$ times steps 3 through 4.
5. Put all of the $\frac{p\times(p-1)}{2}$ dissimilarity matrix together.
6. Calculate medoid ranks
7. Determine Medoid sets according to their ranks.
   STAGE 2:
8. Calculate the combined medoid set $|m_r|$.
9. Assign combined medoid set $|m_r|$ to all processors.
10. Assign each split S_i to a single processor.
11. Cluster S_i into $|m_r|$ clusters and find the center of each of these final clusters.
12. Repeat $\frac{p\times(p-1)}{2}$ times steps 4 through 5.
13. Calculate clustering accuracy
14. Repeat $\frac{r\times(r-1)}{2}$ times steps 2 through 7.
15. Sort the clustering results based on their accuracies. This completes the parallel clustering task.

5 Computational Results

This section describes the experimental results of our global sampling-based clustering method. In order to measure the internal validity of our clustering, we first compare the sequential results obtained by our method and the conventional K-means algorithm for the same parameter settings (the same number of clusters) for all data split data sets. There are several internal validation measures in the literature. In this paper, we will use Davies-Bouldin index (DB) [19]. Using the dataset Glass downloaded from the UCI open dataset depository[1], Fig. 3 compares the minimal, maximal and mean DB index values, respectively, of the proposed method with Kmeans when we have the same parameter settings. Note that clustering results for Kmeans algorithm are computed as the average DB indices taken to assign each data point to the clusters 20 times. As expected, the proposed algorithm was better than the maximal and mean k-means algorithm results. Our DB index values are comparable to those of the minimal Kmeans, showing that they yielded similar clustering results. Clearly, Kmeans performs

[1] https://archive.ics.uci.edu/ml/datasets.

the worst while our clustering method is sometimes the best. The proposed algorithm achieves higher clustering quality, in terms of DB index, although it takes slightly longer to assign objects to clusters. This is due to the fact that our clustering algorithm needs more combined clusters in the second stage. This limitation can be treated using the parallelization process for our algorithm. As a conclusion, experimental results show that the accuracy obtained by the proposed method is very competitive with that of Kmeans algorithm in terms of clustering quality.

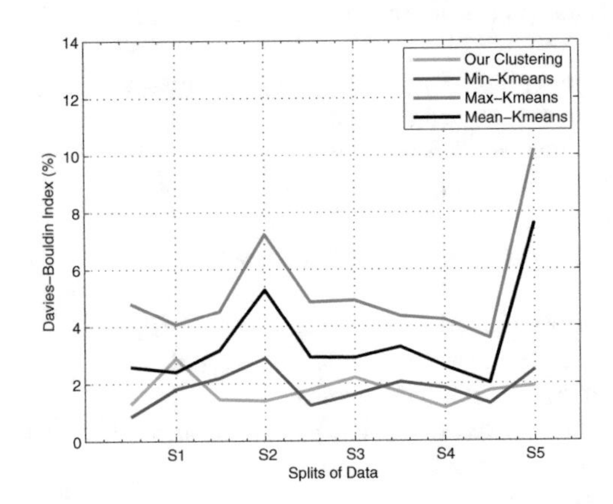

Fig. 3. The DB index values comparison results for Glass dataset.

For data lake, we first divide the data points into p equal sized splits and hence each split has a size equal to $\ell = n/p$ where n is the total number of data points. For instance, we assume that we have m virtual machines.

In the first stage, our global sampling-based clustering maps each split's pair (p_i, p_j) to each virtual machine processing. The most computationally intensive operations in the first stage are the computation of the dissimilarity matrix for n points with the complexity $\mathcal{O}(n^2)$. According to the number of machines, this process requires $\frac{(p\times(p-1))}{2m}$ map operations to calculate all dissimilarity submatrix and partial medoid points' ranks. To obtain the final ranks with the correspondings plausible medoid sets, we use reduce operations to group the partial medoid points for the same rank. Thus the map and reduce operations generate plausible medoid points. The runtime complexity of this stage is $\mathcal{O}(\frac{p\times(p-1)}{2m}\cdot \ell^2)$.

However, in the second stage, our parallel clustering algorithm requires the plausible medoid data points to be replicated in all the machines and randomly split the remaining data points into p splits of ℓ size. To assign n points to that cluster whose center is closest among k cluster centers, the complexity can be done in $\mathcal{O}(nk)$. Proceeding in a similar manner, we need $\mathcal{O}(lk)$ time to assign each split in map operations and hence the total time spent in the second stage is $\mathcal{O}(\frac{r(r-1)}{2m}\cdot lk)$. In summary, the overall running time complexity of the proposed sampling-based clustering is: $\mathcal{O}(\frac{p\times(p-1)}{2m}\cdot \ell^2) + \mathcal{O}(\frac{r(r-1)}{2m}\cdot lk)$

6 Conclusion

In this paper, we worked on a clustering problem related to data lake analytics. we proposed a parallel global sampling-based clustering method that employs medoid point ranking technique and the simple medoids combination. As a comparison, computational experiments disclose that our method can generate clustering quality similar than the Kmeans algorithm for the same parameter settings. We thus feel that this strategy is a very effective sampling technique to achieve our objective of clustering data lake sets efficiently and accurately.

The future direction to explore concerns the medoids combination strategies based on the medoid distribution because we believe that this is suitable to cluster Big data extirped from data lake sets more efficiency.

References

1. Saha, S.: Novel Algorithms for Big Data Analytics, Ph.D. thesis University of Connecticut (2017)
2. Rajasekaran, S., Saha, S.: A novel deterministic sampling technique to speedup clustering algorithms. In: 9th International Conference on Advanced Data Mining and Applications (ADMA), pp. 34–46 (2013)
3. Chitta, R., Jain, A.K., Jin, R.: Sparse kernel clustering of massive high-dimensional data sets with large number of clusters. In: Proceedings of the Ph.D. Workshop at the International Conference on Information and Knowledge Management, pp. 11–18 (2015)
4. Zhang, T., Ramakrishnan, R., Linvy, M.: BIRCH: an efficient data clustering method for very large data sets. Data Min. Knowl. Disc. **1**(2), 141–182 (1997)
5. Guha, S., Rastogi, R., Shim, K.: CURE: an efficient clustering algorithm for large databases. Inf. Syst. **26**(1), 35–58 (2001)
6. Achlioptas, D.: Database-friendly random projections: Johnson-Lindenstrauss with binary coins. J. Comput. Syst. Sci. **66**(4), 671–687 (2003)
7. Kaufman, L., Rousseeuw, P.J.: Finding Groups in Data: An Introduction to Cluster Analysis. Wiley Blackwell, New York (2005)
8. Har-Peled, S., Mazumdar, S.: On coresets for k-means and k-median clustering. In: Proceedings of the ACM Symposium on Theory of Computing, pp. 291–300 (2004)
9. Wang, L., Leckie, C., Kotagiri, R., Bezdek, J.: Approximate pairwise clustering for large data sets via sampling plus extension. Pattern Recogn. **44**(2), 222–235 (2011)
10. Aggarwal, C.C.: A survey of stream clustering algorithms. In: Data Clustering: Algorithms and Applications, pp. 231–258 (2013)
11. Guha, S., Meyerson, A., Mishra, N., Motwani, R., O'Callaghan, L.: Clustering data streams: theory and practice. IEEE Trans. Knowl. Data Eng. **5**, 515–528 (2003)
12. Kranen, P., Assent, I., Baldauf, C., Seidl, T.: The ClusTree: indexing micro-clusters for anytime stream mining. Knowl. Inf. Syst. **29**(2), 249–272 (2011)
13. Rajasekaran, S.: Efficient parallel hierarchical clustering algorithms. IEEE Trans. Parallel Distrib. Syst. **16**(6), 497–502 (2005)
14. Ene, A., Im, S., Moseley, B.: Fast clustering using MapReduce. In: Proceedings of the International Conference on Knowledge Discovery and Data Mining, pp. 681–689 (2011)

15. Lu, Y., Cao, B., Rego, C., Glover, F.: A Tabu search based clustering algorithm and its parallel implementation on spark. arXiv preprint arXiv:1702.01396 (2017)
16. Zaharia, M., Chowdhury, M., Franklin, M.J., et al.: Spark: cluster computing with working sets. HotCloud **10**(10), 95 (2010)
17. Terrizzano, I., Schwarz, P.M., Roth, M., Colino, J.E.: Data wrangling: the challenging journey from the wild to the lake. In: 7th Biennial Conference on Innovative Data Systems Research (CIDR 15), Asilomar, California, USA, pp 4–7 (2015)
18. Fang, H.: Managing data lakes in big data era. In: The 5th Annual IEEE International Conference on Cyber Technology in Automation, Control and Intelligent Systems, pp. 820–824 (2015)
19. Davies, D., Bouldin, D.: A cluster separation measure. IEEE PAMI **1**(2), 224–227 (1979)

A Data Clustering Approach Using Bees Algorithm with a Memory Scheme

Mohamed Amine Nemmich[1(✉)], Fatima Debbat[1(✉)], and Mohamed Slimane[2(✉)]

[1] Department of Computer Science, Mascara University, Mascara, Algeria
amine.nemmich@gmail.com, debbat_fati@yahoo.fr
[2] Laboratoire Informatique (EA6300), Université François Rabelais Tours, 64, Avenue Jean Portalis, 37200 Tours, France
mohamed.slimane@univ-tours.fr

Abstract. The Bees Algorithm (BA) is one of the most recent swarm-based meta-heuristic algorithms that mimic the natural foraging behavior of honey bees in order to solve optimization problems and find the optimal solution. Clustering analysis, used in various science fields and applications, is an important tool and a descriptive process attempting to identify similar classes of objects based on the values of their attributes. To solve clustering problems there are diverse ways, including machine learning techniques, statistics, and metaheuristic methods. In this work, an improved Bees Algorithm with memory scheme (BAMS), which is a modified version of the BA algorithm, is used for data clustering. In the BAMS algorithm, a simple memory scheme is introduced to prevent visiting sites which are close to previously visited sites and to avoid visiting sites with the same fitness or worse. Four real-life data sets are applied to validate the proposed algorithm, and results of this study are compared to BA and others state-of-the-art methods. The experimental results show that the proposed algorithm outperforms other methods.

Keywords: Clustering data · Metaheuristic algorithm · Bees algorithm
Memory scheme · Optimization

1 Introduction

Clustering is a machine learning task that consists of grouping similar instances. It is an unsupervised classification technique that aims to discover natural clustering of instances according to the similarities of their main characteristics. Clustering is a procedure that clusters a set of instances so that the instances in the same clusters have more similarities than the other clusters, based on previously defined criteria [1].

Clustering is a difficult problem and has been worked on by many researchers from different kinds of disciplines and fields to solve the large range of problems such as math programming, data mining, scientific analysis, pattern recognition, image quantization, spatial data analysis, bio-engineering, medical analysis, stock market [2]. Broadly, the clustering techniques are classified into two categories: partitioning and hierarchical techniques. The partitional clustering determines all clusters at once, whereas, in

O. Demigha et al. (Eds.): CSA 2018, LNNS 50, pp. 261–270, 2019.
https://doi.org/10.1007/978-3-319-98352-3_28

hierarchy-based approaches, the successive clusters found using previously established clusters [3].

Many clustering methods have been proposed. K-means algorithm is the most well-known and broadly applied partitional clustering method. It classifies and groups the objects of the whole dataset S based on features into K clusters (i.e. $C_1, C_2, \ldots, C_k$) by choosing K points of data in a random manner as a set of cluster centers in order to minimize the Euclidean distances sum between clusters and points of data close to them. Total Euclidean distance is shown in Eq. (1) [4].

$$E = \sum_{j=1}^{K} \sum_{X_i \in C_j}^{N_j} \left\| X_i - Z_j \right\| \tag{1}$$

where K is the number of clusters, X_i is i^{th} data point and belongs to the cluster C_j, N_j is the number of data points in C_j and Z_j is the center of C_j.

The K-means algorithm is powerful, efficient, and simple and performs well for processing a large amount of data. It has a linear time complexity [5]. However, the algorithm has several shortcomings such as sensitivity to noises, dependence upon its initial selection of cluster centers and dead unit problem. Another limitation of K-means is that as the dataset becomes large, it can gradually converge to the suboptimal local minimum, and even after repetitive application the global optimum cannot be attained [4].

To avoid k-means drawbacks several heuristic clustering applications have been brought in literature such as neural network, tabu search (TS), simulated annealing (SA), Genetic Algorithm (GA) and fuzzy clustering method [6]. Nowadays, many swarm intelligence algorithms are successfully applied to clustering such as Ant Colony Optimization (ACO), Particle Swarm Optimization (PSO) and Artificial Bee Colony (ABC) [6], etc. The Bees Algorithm (BA) is a recent population-based search metaheuristic inspired by the food foraging behavior of bees, it is introduced in 2005 by Pham et al. [7]. BA has enjoyed an important attention of researchers because it has been demonstrated as an efficient and robust optimization tool. The BA has been widely and successfully used to solve various engineering problems in many domains such as production scheduling problem, function optimization, robotic swarm coordination, supply chain optimization, image thresholding, pattern recognition, etc. [8].

In this paper, a new variant of BA called Bees Algorithm with memory scheme (BAMS) is presented to overcome local optima and get the global solution to the clustering problem. In this algorithm, a simple memory scheme is introduced to prevent visiting sites which are close to previously visited sites and to avoid visiting sites with the same fitness or worse.

The rest of this article is arranged as follows: In Sect. 2, a brief description of the Bees Algorithm (BA) in its standard form is done, following by the introduction of an optimization algorithm named BA with memory scheme (BAMS) algorithm for solving clustering problems in Sect. 3. We present the results for optimization on several real datasets in Sect. 4. Comparison and discussion with other well-known algorithms are also summarized in this section such as k-means, Genetic Algorithm (GA), Ant Colony Optimization (ACO), Particle Swarm Optimization (PSO) and Artificial Bee Colony (ABC). Finally, Sect. 5 presents the concluding remarks.

1. Initialize population with random solutions.
2. Evaluate fitness of the population.
3. **While** (stopping criterion not met)
4. //Forming new population.
5. Select sites for neighborhood search.
6. Recruit bees for selected sites (more bees for best e sites) and evaluate fitnesses.
7. Select the fittest bee from each patch.
8. Assign remaining bees to search randomly and evaluate their fitnesses.
9. **End While**.

Fig. 1. Basic steps of the Bees-algorithm [7].

2 The Bees Algorithm

In the field of science, a recent intelligent optimization method based on natural environment of the bees, is known as Bees Algorithm (BA), belongs in the group of population-based metaheuristics and it is proposed by Pham et al. [7]. Its inspiration has its roots in the natural foraging behavior of the bees to find the optimal solution to certain optimization problems. In the BA, bees in charge of domain search correspond to function evaluation at the corresponding points of the function space. Each food source refers to a point in the solution space. Food quality corresponds to function fitness, and the waggle dance can be thought as a ranking of evaluated solutions [7].

The BA starts out by initializing the population of scouts randomly on the space of search. Then, the fitness of the points inspected by the scouts is evaluated. The scouts with the highest fitness are selected for neighborhood search (i.e. local search) as "selected bees" [9]. To avoid duplication, a neighborhood (called a 'flower patch') is created around every best solution; furthermore, the forager bees are affected and recruited. This step is called local search or exploitation [9]. The remaining of the colony bees (i.e. population) are assigned around the space of search scouting in a random manner for new possible solutions. This activity is called global search (i.e. exploration). The recruited bees project randomly inside the neighborhoods. In order to establish the exploitation areas, these solutions with those of the new scouts are calculated with a number of better solutions being reserved for the succeeding learning cycle of the BA. This procedure is repeated in cycles until it is necessary to converge to the optimal global solution [10].

The Bees Algorithm detects the most promising solutions and explores in a selective manner their neighborhoods to find the global minimum of the objective function [7]. When the best solutions are selected, the BA in its basic version makes a good balance between a local (or exploitative) search and a global (or exploratory) search. Both develop random search [9]. Figure 1 summarizes the main steps of the Basic BA.

A certain number of parameters are required for the BA, called: number of the scout bees or sites (**n**), (**m**) sites selected for local search among n sites, (**e < m**) elite sites chosen from the m selected site for a more intense neighborhood search, (**nep**) recruited bees for the elite sites, (**nsp < nep**) bees recruited for neighborhood search around other **m-e** best sites, initial size of all selected patch (**ngh**) around each scout for neighborhood search (i.e. flower patch), stopping criterion for the algorithm to terminate and number of iterations [9, 10].

Two additional steps are called in local search when it does not improve any fitness enhancement in a neighborhood: the neighborhood shrinking and the site abandonment. The main objective is to enhance the computation performance and the search accuracy. This implementation is called the Standard BA [8].

2.1 Neighborhood Shrinking Approach

A large value is defined for the initial size of the neighborhoods. For each a_i of $a = \{a_1, \ldots, a_n\}$ is set as follows:

$$\begin{aligned} a_i(t) &= ngh(t) \times (max_i - min_i) \\ ngh(0) &= 1.0 \end{aligned} \tag{2}$$

where t represents the tth iteration of the BA main loop.

The initial size of flower patches is on a wide area to promote the exploration of the search space. Whilst the local search procedure gives better fitness, the size of patches is maintained unchanged, and is then gradually reduced to yield the search more exploited around the local optimal [9]. The updating formula is shown by following:

$$ngh(t + 1) = 0.8 \times ngh(t) \tag{3}$$

A new variant of BA can be obtained by applying only the shrinking procedure over Basic BA. This implementation is named the Shrinking-based BA [8]. It can be deducted from the first paper of [7].

2.2 Site Abandonment

To enhance the efficiency of the local search, the site abandonment procedure is introduced, in which there is no more enhancement of the fitness value of the fittest bee after a predetermined number (**stlim**) of successive stagnation cycles. If the abandoned site corresponds to the fitness best value, the position of the peak is registered. If no other flower patch will generate a better value of fitness in the rest of the search, the better registered fitness position is taken as the final solution [9].

To sum up, in the literature review of Bees Algorithm, three important implementations could be discovered, which are Basic-BA, Shrinking-based-BA, and Standard-BA.

3 Improved BA for Data Clustering

Many enhancements have been added to the Bees Algorithm (BA). Whereas these enhancements are logical, they do not inevitably, illustrate what happens in nature. Some BA enhancements are as follows: Neighborhood shrinking and Site abandoning by

Ghanbarzadeh [11], Fuzzy greedy selection-based BA by Pham and Darwish [12], Pheromones by Packianather et al. [13], Improved BA (IBA) by Ebrahimzadeh et al. [19] and BA adaptive neighborhood enlargement (BA-NE) by Ahmad [15].

In this work, we exploit the search ability of the BA to avoid the local optimum problem in more efficient and natural manner. The purpose is to improve the BA with memory integration to overcome repetition and unnecessarily spinning inside the same neighborhood. Thus, the search in the BA can prevent searching within infertile areas and can jump from potential local optimums.

The memory is part of the honey bees' nature, but was not included in basic BA. The strategy is to force the bees to stay away from the neighborhoods studied and experienced with no beneficial results. This will enforce the bees to visit different solutions for investigation. This needs the integration of a memory mechanism into the bees to allow them to respect the experiences and not waste their energy and time. This type of memory will be integrated to the scout and the follower bees where these bees could employ information and details about previously visited sites (i.e. positions and finesses). Different structures of arrays are used to save and handle the memory of each bee (scout and follower). The memory contains the data (i.e. information) obtained during direct communication with the environment that used to decide the way of patch visiting based on a waggle dance or depend on it. This information is very important and affects the way forager bees (both scout and follower bees) choose the patch to visit and whether it will be based on a waggle dance or not (i.e., choosing a familiar food source). Foragers depend on their own memories to discover particular locations during the visit of food patches repeatedly [16].

The memorized experiences are continuously updated and utilized to lead the further foraging of the bees, contributing to a more effective search procedure than basic BA.

The pseudo-code below shows the procedures added. It is included in the local neighborhood search for follower bees and in the global search for scout bees. This procedure decides if the new position (in different dimensions) of any bee is close to its previous one recorded in its memory.

Where r is the radius, x^d_{max} and x^d_{min} are the upper and the lower bound to the solution vector respectively, it max is the maximum number of iterations and ngh is the neighborhood size.

If the bee is a scout **then**
  $r = x1 * x^d_{max} - x^d_{min}$;
Else if the bee is a follower bee **then**
  r = x2 * ngh;
End If
While (number_iterations < itmax) **do**
  If the *new position is close to its previous position recorded in the private memory* (by using the Euclidian distance) **then**
    Find a new position randomly;
  End If
End While

Fig. 2. The pseudo code of the procedures added.

The radius is the Euclidean distance from the field center to the border. It defines the size of field. It is applied so that different patches do not congregate in the same area. As can be seen from Fig. 2, the value of radius r is based on the size of the solution space for scouts and the size of ngh for followers, which involves that r, will be greater than ngh for the scouts and less than ngh for followers.

The objective of the proposed algorithm is to get the more optimum results for clustering using the memory mechanism into the bees. Briefly, the task of the algorithm is to search for suitable centers of the clusters (C_1, C_2,..., C_k) which makes the Euclidian distance (Eq. (1)) as lower as possible.

The initial population is fixed to n scouts. Each scout constitutes a possible clustering solution as a set of k centers of cluster. The initial positions of the centers are spread in a random manner on the solution domain with uniform probability, without any prior information about the data. To construct the initial clusters, objects (points) are assigned to clusters based on Euclidean distance (1), of each data object and all clusters centers are calculated (i.e. the cluster with center closest to the objects). A data object is regarded as a member of the class with the center to which it has the minimum distance. After forming clusters, the actual clusters centroids replace the centers of original clusters to identify a particular clustering solution. Whenever new bees are to be produced, the process of initialization is used. Each scout bee assesses the fitness of candidate solution (i.e. visited site) by calculating the clustering metric (Eq. (1)), which can be inversely associated with fitness. Then, the BA enters the primary loop of four steps: Waggle dance, Local search, Global search and population update, and continue until its stopping criterion is satisfied.

At each iteration end, the algorithm will have scout and follower bees represent selected patches (clusters), and other scout bees do a random search. If the iteration is not the first one then near method and local memory use Euclidian distance to check that no repetition in visiting sites is done.

4 Experimental Results and Discussion

The main objective of our experiments is to apply the Improved BA with memory scheme (BAMS) for clustering problem. In order to evaluate and investigate the performance and the efficiency of the BA we have compared the proposed BAMS with the Basic BA [19] and several clustering algorithms available in the literature, including k-means, GA, ACO, ABC and PSO [17]. They are used for four real-life datasets (Iris, Control Chart, Crude Oil and Wood Defects), which are described in Table 1. These datasets are chosen from UCI machine learning repository [18]. Experimental results for k-means, GA, ACO, PSO, ABC, and BA were extracted from [14, 17].

To calculate the cost function, first, we determine the Euclidean distances metric between each input data and all centers of the clusters according to Eq. (1), then we define the minimum of these distances and finally, we compute the sum all minimums defined for all input data. A higher quality of clustering represents that the sum is relatively small (i.e. the smaller the value the better the clustering results). Furthermore, the mean, maximum and minimum values are mean cost function (function value), maximum or worst cost

Table 1. The characteristics of datasets.

Datasets	Number of objects	Number of features	Number of Classes
Iris	150	4	3
Wood Defects	232	17	13
Control Chart	1500	60	6
Crude Oil	56	5	3

function and minimum or best cost function respectively. To obtain statistically significant results from algorithms, all experiments were run 10 times. The parameters setting used in the clustering methods are shown in Table 2. The BAs (BA and BAMS) parameters have been empirically adjusted and the best combination of the parameter set is used. The obtained results are given in Table 3.

Table 2. Parameter setting.

Algorithm	Parameter	Value
K-means	Maximum number of iterations	1000
ABC	Number of bees (employer and onlooker)	20
	Number of sites selected for neighborhood search	10
	Number of bee recruited for best sites	5
	Number of iterations	500
ACO	Population (number of ants)	50
	Threshold of probability for maximum trail	0.95
	Probability of local search	0.01
	Evaporation rate	0.01
GA	Population size	100
	Probability of mutation	0.001
	Probability of crossover	0.8
PSO	Number of Swarm	100
	C1	1.9
	C2	2.1
	ω max	1
	ω min	0.5
	Number of iterations	500
Bees Algorithm & BAMS	Number of scout bees or sites (n)	21
	Sites selected for local search (m)	8
	Elite sites chosen from the m selected site (e)	2
	Recruited bees for elite sites (nep)	5
	Recruited bees for remaining (m-e) best sites (nsp)	2
	Number of iterations, itmax	300

Table 3. Clustering results obtained by applying the algorithms (Comparison of intra-cluster distances).

Dataset	Algorithm	Mean	Min.	Max.
Iris	k-means	107.721	97.205	124.022
	GA	97.101	97.101	97.101
	ACO	98.65729	97.12787	99.98212
	ABC	98.32564	97.54622	99.73824
	PSO	98.98634	96.06847	100.31864
	Basic-BA	**96.764**	**96.728**	**96.787**
	BAMS	**96.750**	**96.720**	**96.770**
Crude Oil	k-means	279.662	279.485	279.743
	GA	278.965	278.965	278.965
	ACO	280.12158	279.02278	281.41328
	ABC	279.65824	277.29441	279.97223
	PSO	280.69238	278.14945	281.97271
	Basic-BA	**277.339**	**277.227**	**277.558**
	BAMS	**270.893**	**267.857**	**273.929**
Control Charts	k-means	2490.267	2464.813	2528.663
	GA	2322.234	2289.450	2355.690
	ACO	2094.622	2071.113	2114.218
	ABC	2049.792	2041.334	2050.001
	PSO	2124.789	2074.225	2127.436
	Basic-BA	**1898.991**	**1860.440**	**1938.970**
	BAMS	**1880.000**	**1860.000**	**1900.000**
Wood Defects	k-means	228126.662	199306.380	270584.630
	GA	168035.200	157508.000	174784.000
	ACO	155026.17	154622.56	156001.65
	ABC	153718.67	153267.34	153992.73
	PSO	155987.86	154241.32	156095.21
	Basic-BA	162193.157	153866.531	173523.000
	BAMS	**153571.430**	**151933.260**	158928.570

For comparison of the results, we can see from the Table 3 that the Basic-BA algorithm had a mean value for clustering criterion (Eq. (1)), which was less than the mean for the other popular methods (k-means, GA, ACO, ABC, and PSO) for the Iris, Crude Oil and Control Charts datasets, but for the Wood Defects the Basic-BA was outperformed by ACO, ABC, and PSO algorithms. For example, in the Iris dataset, the average (mean) result found by Basic-BA is 96.764. Meanwhile, the result of the nearest algorithm, which is the GA, is 97.101. However, for the Wood Defect dataset, the ABC algorithm provides better performance than other compared algorithms (except the BAMS algorithm).

The results show also the superiority of the BAMS algorithm and how the addition of memory mechanism affects the results. In most cases, the BAMS method has reached the better performance in terms of the average (mean), best (min), and the worst (max) inter-cluster distances over all datasets and outperforms all clustering methods, with stable results. It allows a lower value of the clustering criterion (Eq. (1)), which means better solutions of clustering.

In fact, for every trial, the mean value of the BAMS is fewer than the minimum for the others algorithms for all datasets. For example, for the Control Charts dataset, the BAM algorithm gave 24%, 19%, 10%, 8% and 12% better results than K-means, GA, ACO, ABC and PSO algorithms respectively.

5 Conclusion

This paper presented the Bees Algorithm (BA), an optimization metaheuristic, which is inspired by the bees' forage behavior. It was proposed to look for the set of cluster centers that minimizes a given clustering metric. Due to its capability to carry out local and global search simultaneously, the method overcomes getting stuck into a local minimum.

Two variants of BA are studied for the clustering problem: the Basic BA and the proposed BA with memory scheme (BAMS).

The Experiments have demonstrated that the Basic BA demonstrated superior clustering performance than other well-known algorithms on the majority of datasets.

An enhancement to the basic BA by adding memory scheme was introduced to prevent visiting sites which are close to previously visited sites and to avoid visiting sites with the same fitness or worse. This type of memory is part of the honey bees' nature, but was not included in basic BA. The memory was added for both the follower and the scout bees. The results of applying the BA before and after the enhancements for different datasets show that the BA with memory scheme (BAMS) outperforms the basic BA and the other algorithms and the improvement percentage exceeds all previous enhancements results. Further work should be conducted to apply the BA with memory (BAMS) for others optimization problems.

References

1. Özbakır, L., Turna, F.: Clustering performance comparison of new generation meta-heuristic algorithms. Knowl. Based Syst. **130**, 1–16 (2017)
2. Jose-Garcia, A., Gomez-Flores, W.: Automatic clustering using nature-inspired metaheuristics: a survey. Appl. Soft Comput. **41**, 192–213 (2016)
3. Rokach, L., Maimon, O.Z.: Data Mining with Decision Trees: Theory and Applications. World Scientific Pub Co Inc, Hardcover (2008). 69
4. Boobord, F., Othman, Z., Abu Bakar, A.: A WK-Means Approach for Clustering. In: IAJIT (2013). ISSN:1683-3198
5. Chen, C.Y., Ye, F.: Particle swarm optimization algorithm and its application to clustering analysis. In: Proceedings of the IEEE International Conference on Networking, Sensing and Control, Taipei, Taiwan, pp. 789–794 (2004)
6. Zhang, C., Ouyang, D., Ning, J.: An artificial bee colony approach for clustering. Expert Syst. Appl. **37**(7), 4761–4767 (2010)

7. Pham, D., Ghanbarzadeh, A., Koc, E., Otri, S., Rahim, S., Zaidi, M.: The bees algorithm-a novel tool for complex optimisation problems. In: Proceedings of the 2nd Virtual International Conference on Intelligent Production Machines and Systems. Elsevier Science Ltd., Cardiff, pp. 454–459 (2006)
8. Wasim, A.H., Shahnorbanun, S., Siti, N.H.S.A.: The variants of the bees algorithm a survey. Artif. Intell. Rev. **47**(1), 67–121 (2017)
9. Pham, D.T., Castellani, M.: The bees algorithm: modelling foraging behaviour to solve continuous optimization problems. Proc. Inst. Mech. Eng. Part C.: J. Mech. **223**, 2919–2938 (2009)
10. Nemmich, M.A., Debbat, F.: Bees algorithm and its variants for complex optimisation problems. In: The 2nd International Conference on Applied Automation and Industrial Diagnostics (ICAAID 2017), Djelfa, Algeria (2017)
11. Ghanbarzadeh, A.: The bees algorithm, A Novel optimization tool. Ph.D Thesis, Manufacturing Engineering Centre, School of Engineering, Cardiff University, UK (2007)
12. Pham, D.T., Darwish, A.H. Fuzzy selection of local search sites in the bees algorithm. In: Proceedings of the 4th Virtual International Conference on Intelligent Production Machines and Systems, 1–14 July 2008, Cardiff, UI, p. 391 (2008)
13. Packianather, M.S., Landy, M., Pham, D.T.: Enhancing the speed of the bees algorithm using pheromone-based recruitment. In: Proceedings of the 7th IEEE International Conference on Industrial Informatics, 23–26 June, Cardiff, Wales, p. 789–794 (2009)
14. Ebrahimzadeh, A., Addeh, J., Ranaee, V.: Recognition of control chart patterns using an intelligent technique. J. Appl. Soft. Comput. **13**(5), 2970–2980 (2012)
15. Ahmad, S.A.: A Study of Search Neighborhood in the Bees Algorithm. Cardiff University, Cardiff (2012)
16. Grüter, C., Farina, W.M.: The honeybee waggle dance: can we follow the steps? Trends Ecol. Evol. **24**, 242–247 (2009). https://doi.org/10.1016/j.tree.2008.12.007
17. Taherdangkoo, M., Yazdi, M., Bagheri, M.H.: A powerful and efficient evolutionary optimization algorithm based on stem cells algorithm for data clustering. Cent. Eur. J. Comput. Sci. **2**(1), 47–59 (2012). https://doi.org/10.2478/s13537-012-0002-z
18. Carpenter, G., Grossberg, S.: Pattern Recognition by Self-Organizing Neural Networks. Cambridge, Massachusetts, USA (1991)
19. Pham, D.T., Otri, S., Afify, A., Mahmuddin, M., Al-Jabbouli, H.: Data clustering using the bees algorithm. In: Proceedings of the 40th CIRP International Manufacturing Systems Seminar, Liverpool, UK (2007)

RTSBL: Reduce Task Scheduling Based on the Load Balancing and the Data Locality in Hadoop

Khadidja Midoun(✉), Walid-Khaled Hidouci(✉), Malik Loudini(✉), and Djahida Belayadi(✉)

Communication in Computer Systems Laboratory, National High School of Computer Science, PO BOX 68M, 16309 Oued-Smar, Algiers, Algeria
{k_midoun,w_hidouci,m_loudini,d_belayadi}@esi.dz

Abstract. We address load balancing and data locality problems in Hadoop. These two problems limit its performance, especially, during a reduce phase where the partitioning function assigns the keys to the reducers based on a hash function. We propose in this paper a new approach to assign the keys based on the reducers' processing capability in order to ensure a good load balancing. In addition, our proposed approach called *RTSBL* takes into consideration the data locality during the partition. Our experiments prove that RTSBL achieves to up 87% improvements in the load balancing and 3× improvements of the data locality during the reduce phase in the standard Hadoop.

Keywords: MapReduce · Hadoop · Load balancing · Data locality Reduce task scheduling

1 Introduction

The advances of big data and the failure of traditional data analysis tools to process this type of data are the main reasons why companies looking for a new tool that can process a huge volume of unstructured data in a reasonable time.

Motivated by this, Google has introduced its parallel programming model MapReduce [11]. The simplicity of MapReduce makes it a popular and useful framework that is implemented by several systems as Amazon Elastic MapReduce [1], Aster MapReduce Appliance [2], Greenplum MapReduce [4], and the famous large-scale data analysis system Hadoop.

Hadoop benefits from a combination of its distributed file system called HDFS (**H**adoop **D**istributed **F**ile **S**ystem) storing data in a commodity hardware, and MapReduce as a parallel programming model for processing these data. The ability of Hadoop to hide the complexity of distribution, parallelism and fault tolerance from the programmer makes it the useful system for world's largest companies such as Facebook, Amazon, Yahoo, and the NY Times.

Hadoop uses a master-slave architecture to perform a given job. First, the Name Node, which is the master of HDFS, splits the input data into several

O. Demigha et al. (Eds.): CSA 2018, LNNS 50, pp. 271–280, 2019.
https://doi.org/10.1007/978-3-319-98352-3_29

blocks. Then, each block is stored in three copies in three different data nodes in order to ensure the reliability of the system. After that, in processing level, the master of MapReduce called JobTracker that distributes tasks into TaskTrackers and controls the job execution. To be a flexible parallel processing system for various types of applications, Hadoop runs a job in three dependent phases: the map phase, the shuffle phase and the reduce phase.

In the map phase, JobTracker assigns only one block to map the task. This later generates a set of an intermediate key/value pairs to be processed by the reduce tasks. It is required that the intermediate key/value pairs sharing the same key have to be processed by the same reduce task. To this end, the shuffle phase is responsible to sort the intermediate key/value pairs by their keys, to group them into partitions, and to shuffle each partition to only one reducer.

The load balancing and the data locality factors are playing a vital role in evaluating the effectiveness of the parallel systems. On one hand, given the fact that the running time of a given job in the parallel system is related to the running time of its last task, a good load balancing technique may significantly diminish the whole running time. On the other hand, a good data locality may improve the performance of the system by minimizing the amount of data transferred through the network. As a consequence, ignoring the effect of the data locality can cause high communication cost and at sometimes the congestion of the network.

To take profit from these two factors, (1) Hadoop evenly splits the input data into blocks to ensure load balancing of maps, and (2) it assigns a map task to the TaskTracker that has its input block to ensure data locality. Unfortunately, this is not the case for the reduce tasks where input data depends on the used partitioning function. By default, Hadoop uses a hash function as a partitioning function. According to the tests in [10], this function may cause unbalancing the reducer's load, and producing high amount of shuffled data.

Motivated by the above-mentioned problems, we propose a new approach to partition the intermediate keys across the reducers based on the nodes' capabilities and keys' distribution. Using these two factors, we can find a good tradeoff between data locality and load balancing during the reduce phase of Hadoop system.

The remainder of this paper is organized as follows. In Sect. 2 we discusses the problem statement. In Sect. 3, we present our proposed approach. In Sect. 4, we discuss the performance evaluation of our approach. In Sect. 5, we summarize some related work. In Sect. 6, we conclude our paper.

2 Problem Statement

2.1 MapReduce Overview

MapReduce framework is a parallel processing model. Its key idea is based on a typical divide and conquer parallel computing strategy. It divides the processing of a single job into several independent tasks to be executed in parallel. The processing job is acheived in two main phases: map and reduce. The execution of tasks is sequential. Therefore, the reduce phase cannot start until the map phase has finished.

Although the map function is used by a map task to process the input data, the reduce function is used by a reduce task to process the intermediate data generated by the mappers. The map and reduce tasks are executed in separated slots on TaskTrackers which is a processing node. Therefore, we define two different types of slots: map and reduce.

The number of these slots can be modified by the user. However, the definition of the map and reduce functions is the own responsibility of the user.

More precisely, an input data consists of n input items $D = \{d_1, d_2, ..., d_n\}$ that will be processed by s TaskTracker $N = \{n_1, n_2, ..., n_s\}$ through the following steps as illustrated in Fig. 1.

1. ***Input reader:*** It takes the input data from files in the basic form and converts them to (key, value) pairs. Then, the input data is evenly divided into blocks.
2. ***Map phase:*** The JobTracker schedules map tasks on nodes that have their processing data. A map task produces a list of intermediate (key, value) pairs from its input key-value pairs. $map : (K_1, V_1) \longrightarrow list(K_2, V_2)$.
3. ***Shuffle phase:*** In this phase, MapReduce is responsible for the partition process, sorting and transferring the pairs from map to reduce tasks. These operations are only based on Map's output-key. The partition process is the responsibility of *the partitioning function:* $Part : K \longrightarrow R$.
4. ***Reduce phase:*** When all maps have finished their job, the reduce phase can begin. Since it was the reducer, it has been receiving all the values of the same key of the intermediate pairs. Its output is a list (key, value) pairs. $(K_2, list(V_2)) \longrightarrow list(K_3, V_3)$.
5. ***Output writer:*** It is responsible for writing the output to stable storage.

2.2 The Load Balancing Problem

The performance of MapReduce is limited by its ability to process the data by many computing nodes in parallel within a reasonable time. The running time of a given MapReduce job is that of the slowest node. In order to benefit from parallel computation advantage, MapReduce must distribute the load evenly between the nodes. It tries to balance the workload by assigning a similar amount of input data to each mapper, and the same number of intermediate keys each to reducer. Sadly, this approach is not always efficient, and the load imbalance can occur either during the map or the reduce phases for two main reasons [7]:

(1) ***The data skew:*** The data, in this case, is not evenly distributed among the computing nodes. Generally, this type of skew affects the reduce tasks because the input blocks processed during the map phase have the same size. It occurs either when a reducer treats a larger number of clusters or whether it processes clusters bigger than other clusters that are assigned to other reduce tasks.

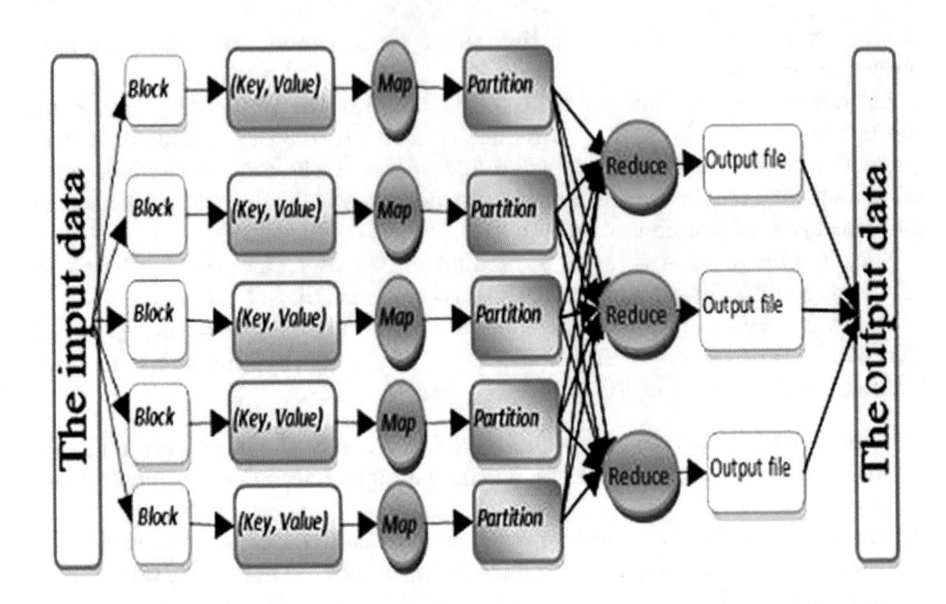

Fig. 1. MapReduce dataflow

(2) ***The workload imbalance:*** It occurs when the processing nodes have different physical characteristics e.g., the processor's speed, memory's size. This type of skew affects both tasks, map and reduce. As we noted above, the reduce phase cannot start until the map phase has finished, therefore, a map's load imbalance causes also a reducer's load imbalance. In addition, the reducer's load imbalance affects the performance of the job because its running time is that of the slowest reducer.

As a consequence, improving the load balancing can significantly reduce the job's running time.

2.3 The Data Locality Problem

The effectiveness of a parallel distributed system hinges on the manner in which tasks and data are assigned to its underlying resources. For this reason, the scheduling of the map and reduce tasks in Hadoop must respect one of the important factors which is the data locality.

The data locality can significantly reduce the utilization of the available resources such as the network bandwidth. For a given map task, JobTracker determines the TaskTracker that has its data and then sends it a map program to avoid intensive data migration. In contrast, Hadoop schedules reduce tasks on TaskTrackers upon to their request without any data locality consideration. This may cause a high usage of network bandwidth.

As a conclusion, improving the data locality for the reducers diminishes the amount of the data transferred in MapReduce's shuffle phase and therefore decreases the overhead of jobs' execution.

3 The Proposed Approach

The primary goal of RTSBL: **R**educe **T**ask **S**cheduling based on the Load **B**alancing and the Data **L**ocality in Hadoop is to ensure a better tradeoff between the load balancing and the data locality during the partition of the intermediate keys across reducers. The secondary goal is to minimize the overhead induced by that operation.

The key idea of RTSBL is to make the shuffle phase a dynamic phase. This means that the partitioning function used by RTSBL may assign a cluster to a requested reducer where the other reducers are executing their job. Therefore, the partitioning function may be executed several times in parallel with the execution of the reduce phase. This is contrary to the partition in Hadoop where the partitioning function assigns clusters to their reducers before the start of the reduce phase.

To ensure a better locality, RTSBL needs to collect information about the distribution of the highest clusters. The highest cluster is defined as the one with the size greater than or equal to a cluster's threshold τ. This threshold is an additional value that is required by RTSBL.

Algorithm 1. The algorithm of RTSBL

//JobTracker

Input : RK is the set of the intermediate keys of the highest frequencies
TT is the TaskTracker requesting a reduce task.

Description: Perform partitioning function on a set of keys. The keys are sorted in decreasing order according to their frequencies and the nodes are sorted in decreasing order according to their capabilities

- TT_i^j is a set of TaskTrackers requesting the same key k_i
- $|TTS_i|$ is the number of TT requesting k_i
- $Ratio(TT_i^j)$ is the capability ratio of TT_i to handle the key k_i.

Output : partition(k_i, TT_j)

1 **for** $k_i \in RK$ **do**

2 $j \leftarrow 0$

while $Ratio(TT_i^j, k_i) < Ratio(TT_i^{j+1}, k_i)$ **do** $j \leftarrow j+1$

end while

3 **end**

4 partition(k_i, TT_i^j);

5 **for** $TT_i \in TT$ **do**

6 Compute the new highest cluster

7 **end**

When a map phase finishes its work, all reduce slots in the TaskTrackers are idle. Therefore, each TaskTracker requests a reduce task from the JobTracker. For that, the TaskTracker sends statistics about its highest clusters to the JobTracker.

Given a cluster, C_i represented by key i and frequency fre_i. This cluster is a high cluster if $fre_i \geq \tau$. The JobTracker sorts the clusters in decreasing order of their frequencies, and the requested TaskTrackers in decreasing order of their capabilities to be scheduled based on their distribution and the capabilities of the requested TaskTrackers.

After that, the partitioning function assigns the keys with respect to their order. For each key, the partitioning function tries to achieve its better locality by assigning it to the TaskTracker with the maximum frequency of this key. If this TaskTracker is not idle, it looks to the second TaskTracker (the second maximum frequency), and so on. It is important to note that each reduce task contains only one cluster.

When all received clusters are scheduled, the JobTracker starts the communication part of the shuffle phase to transfer the partitioned data to their TaskTrackers. The remaining data are still on the local disk of their TaskTrackers, and waits their turn (when their clusters become the highest clusters).

Each requested TaskTracker starts its processing just when it receives all its necessary data. At the end of each reduce task, the TaskTracker asks for a new one using the same above-described process.

The complete algorithm is presented in Algorithm 1.

4 Performance Evaluation

We experimentally evaluate the performances of RTSBL using a simulator. This simulator allows us to generate a synthetic data with control of a skew parameter to follow a Zipf data distribution [14].

The skew parameter is an important factor to evaluate the performance of Hadoop in the context of big data. The generated data is distributed into mappers in the same way that Hadoop does. In this section, we report the results of our experiments.

We choose to evaluate RTSBL in the context of an entity resolution application because this type of application has $O(n^2)$ complexity. Hence, it is an ideal problem to be treated by Hadoop. For our experiments, we run three tests:

1. **Test 1**: We run a dataset (DS1) containing 10797117 entities using 30 homogeneous TaskTrackers.
2. **Test 2**: We run DS1 using 30 heterogeneous TaskTrackers.
3. **Test 3**: We run a skewed data set using 30 heterogeneous TaskTrackers. The skew data follows the Zipf distribution.

For each test, we compare the load balancing and the data locality qualities during the reduce phase of RTSBL with those of the standard Hadoop system. We use max-min ratio metric [12] to illustrate the balance of intermediate data distribution.

$$Max\_MinRatio = \frac{min_{1 \leq i \leq n}(Reduce\_load_i)}{max_{1 \leq i \leq n}(Reduce\_load_i)} * 100\%$$

Where *Reduce_load* is the amount of the intermediate data assigned to this reducer multiplied by its load ratio (1 in the homogeneous environment or its ratio in the heterogeneous environment). Max-min ratio metric represents the load of the slowest node compared to the fastest node.

The results reported in Fig. 2(a) and (b) show that RTSBL is able to improve up to 87% of the load balancing and up to 3.5% of the data locality during the reduce phase in the homogeneous environment and 84% and 2.69%, respectively in heterogeneous environment as shown in Fig. 3(a) and (b).

The results of the last test prove that RTSBL is still better than Hadoop despite the presence of the skewed data with an improvement of 16% of the load balancing and 3× of the data locality as shown in Fig. 4(a) and (b), respectively.

5 Related Work

Improving the load balancing and reducing the transferred data during the shuffle phase have been investigated by many researches.

Improving the Load Balancing. Gufler et al. [8] propose a new algorithm called TopCluster. The key idea of this algorithm is the monitoring of the distribution of the keys to estimate their cardinality. Based on this estimation, the keys will be distributed across the reducers. Compared to the original MapReduce algorithm, TopCluster has the advantage of balancing the load when the existing of the data skew. If it is not the case, TopCluster will produce unnecessary overhead without any improvement. In addition, its estimation produces an error. The value of this error depends on the threshold chosen by the user.

Hanif et al. [10] propose a new algorithm to find a tradeoff between the load locality and the load balancing for the reduce phase in Hadoop system for the homogeneous environment. It introduces NA-FLK metric similar to LEEN algorithm but it is adaptable to the homogeneous environment by taking into consideration the variation of nodes' capabilities. However, overlooking the variance in the intermediate key's frequencies may decrease the performance of the load balancing.

[13] proposed a new parallel programming model called MBR (**M**ap-**B**alance-**R**educe programming model). Compared to MapReduce programming model, MBR adds a new phase called Balance phase between Map and Reduce phases. The role of balance phase is receiving the Map outputs and providing Reduce with the balanced input. It has the ability to detect in advance the Reduce task that may cause an unbalanced load problem. When an unbalanced reduce task appears, a balance function will be called. The role of this function is to divide a load of this task into several loads to be processed by several reduce tasks. This Balance phase can lead a high amount of shuffled data caused by the duplication of the intermediate data.

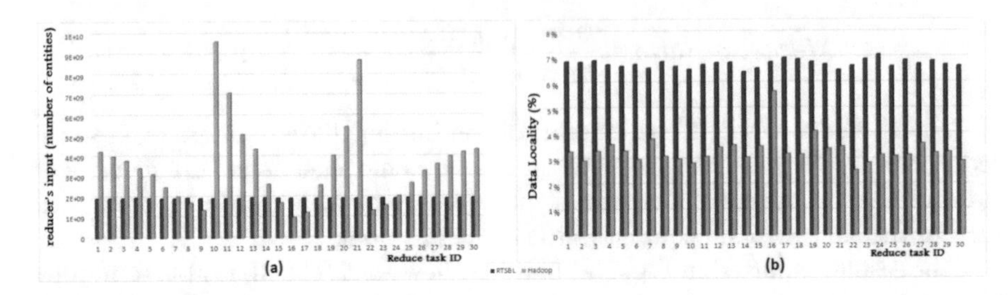

Fig. 2. Test 1: Reducers in the homogeneous environment (a) The load balancing (b) The data locality

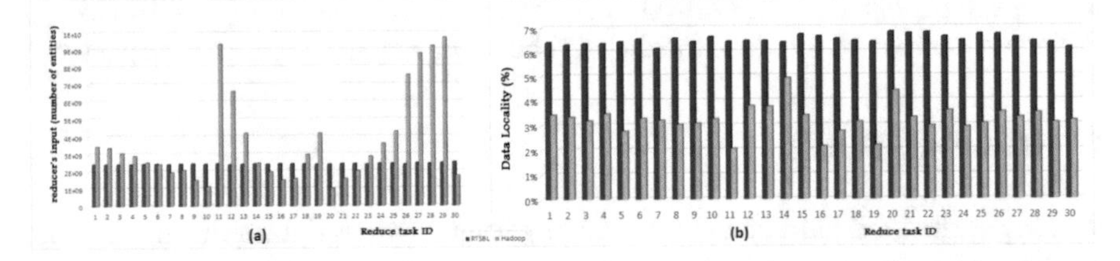

Fig. 3. Test 2: Reducers in heterogeneous environment (a) The load balancing (b) The data locality

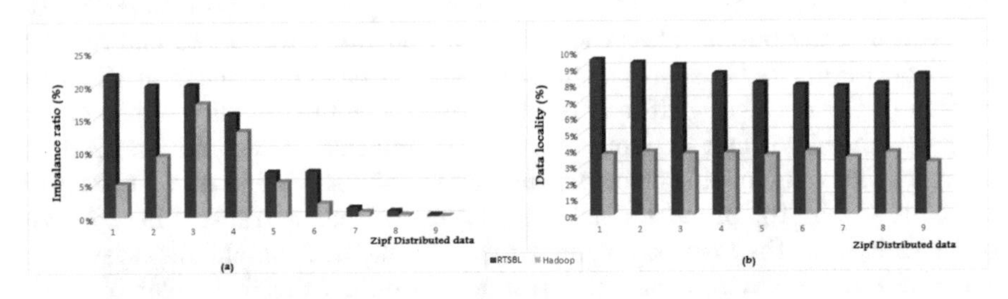

Fig. 4. Test 3: Reducers run a skewed data in heterogeneous environment (a) Imbalance ratio (b) The data locality in the system

Improving the Data Locality. [9] proposed LARST (**L**ocality-**A**ware **R**educe **T**ask **S**cheduler): a new strategy to schedule Reduce task across the node. In order to apply locality to the reduce tasks scheduling, LARST makes MapReduce aware of the location of all partitions in the network. To this end, it delays the transfer of the intermediate data until all map tasks have finished their work as a way of RTSBL. LARST suggests that a good data locality can be achieved via scheduling the reduce tasks at the maximum-node in the maximum-rack of the mappers. This condition is strictly respected and all TaskTrackers do not satisfy this condition will be rejected. When a given TaskTracker gets rejected more than α times, it will not receive any reduce task for the future. This tactic is used to avoid a scheduling delay and under-utilization of the available resources.

[15] proposed MR-Part approach to diminish the amount of the shuffled data. The key idea of MR-Part is to create a workload model for MapReduce job by using a workload sample. Then, based on this workload model, MR-Part partitions the input files containing data to be processed by the same reducer and to be assigned to the same input partition. This idea ensures a maximum of data locality in the reduce phase because the transferred data will be minimized. However, this method is good when all users use the same input data. If it is not the case, MapReduce must create a workload model for each user, which can provide a high job running time.

Improving the Load Balancing Based on the Data Locality. [6] proposed CLP partition algorithm where its goal is to improve the load balancing based on the data locality. The partition in the CLP algorithm needs two MapReduce jobs. The first job processes a random sampling of the input data to gather data distribution information. Based on this information, the second job forms a data-cluster where the input data that is responsible for generating the same intermediate key will be in the same data-cluster. According to the data locality and with respect to the load balancing, these data-clusters will be assigned to suitable computing nodes. However, the sampling error depends on the type of application, the type of the input data and its original distribution that affects the accurate data distribution information.

An extension of CLP called SRL and an extension of LEEN called LRS are used by a novel partitioner based on naive Bayes classifier proposed in [5]. This partitioner, named BAPM, classifies a given MapReduce job into three categories reduce-input-heavy, reduce-input-light, and reduce-input-zero according to the amount of intermediate data transferred through the network. Based on the category of the job, BAPM determines the algorithm used to process it (LRS or SRL). Both algorithms increase data locality and decrease data skew on reduce phase, but with reversed preference.

6 Conclusion

In this paper, we proposed a reduce task scheduler based on a tradeoff between the load balancing and the data locality. In contrast to the partitioning function used by the standard Hadoop, our new approach takes into consideration the heterogeneity of the cluster. This key idea improves the performance of Hadoop up to 80% for its load balancing and $3\times$ for its data locality during its reduce phase as proved by our experimental results.

For future work, we plan to use our proposed approach to improve the performance of Dedoop (Deduplication with Hadoop) [3]: a new tool for parallel entity resolution (ER) on cloud infrastructures using a Hadoop 0.20.2. This tool proposes an efficient approach to achieve the load balancing during the reduce phase called BlockSplit approach [16]. The major defect of this approach is its duplicate of the intermediate data in order to improve the load balancing. This duplication of intermediate data increases significantly the consumption of the network bandwidth, especially for a big data. Adapting RTSBL to Dedoop tool may handle its data locality problem resulting from its BlockSplit approach.

References

1. Amazon elastic MapReduce. http://aws.amazon.com/elasticmapreduce/. Accessed 10 Jan 2018
2. Aster MapReduce appliance. http://www.asterdata.com/product/deployment/appliance.php. Accessed 10 Jan 2018
3. Dedoop tool. https://dbs.uni-leipzig.de/howto_dedoop. Accessed 10 Jan 2018
4. Pivotal greenplum database. http://gopivotal.com/pivotal-products/pivotal-data-fabric/pivotal-analytic-database. Accessed 10 Jan 2018
5. Chen, L., Lu, W., Wang, L., Bao, E., Xing, W., Yang, Y., Yuan, V.: Optimizing MapReduce partitioner using naive bayes classifier. In: 2017 IEEE 15th International conference on Dependable, Autonomic and Secure Computing, 15th International Conference on Pervasive Intelligence & Computing, 3rd International Conference on Big Data Intelligence and Computing and Cyber Science and Technology Congress (DASC/PiCom/DataCom/CyberSciTech), pp. 812–819. IEEE (2017)
6. Chen, Y., Liu, Z., Wang, T., Wang, L.: Load balancing in MapReduce based on data locality. In: International Conference on Algorithms and Architectures for Parallel Processing, pp. 229–241. Springer (2014)
7. Grolinger, K., Hayes, M., Higashino, W.A., L'Heureux, A., Allison, D.S., Capretz, M.A.M.: Challenges for MapReduce in big data. In: 2014 IEEE World Congress on Services (SERVICES), pp. 182–189. IEEE (2014)
8. Gufler, B., Augsten, N., Reiser, A., Kemper, A.: Load balancing in MapReduce based on scalable cardinality estimates. In: 2012 IEEE 28th International Conference on Data Engineering (ICDE), pp. 522–533. IEEE (2012)
9. Hammoud, M., Sakr, M.F.: Locality-aware reduce task scheduling for MapReduce. In: 2011 IEEE Third International Conference on Cloud Computing Technology and Science (CloudCom), pp. 570–576. IEEE (2011)
10. Hanif, M., Lee, C.: An efficient key partitioning scheme for heterogeneous MapReduce clusters. In: 2016 18th International Conference on Advanced Communication Technology (ICACT), pp. 364–367. IEEE (2016)
11. Dean, J., Ghemawat, S.: Mapreduce: simplified data processing on large clusters. ACM **51**, 107–113 (2008)
12. Jain, R., Chiu, D.-M., Hawe, W.R.: A quantitative measure of fairness and discrimination for resource allocation in shared computer system. In: Eastern Research Laboratory, vol. 38, Digital Equipment Corporation Hudson, MA (1984)
13. Li, J., Liu, Y., Pan, J., Zhang, P., Chen, W., Wang, L.: Map-balance-reduce: an improved parallel programming model for load balancing of MapReduce. Future Gener. Comput. Syst. (2017)
14. Lin, J., et al.: The curse of ZIPF and limits to parallelization: a look at the stragglers problem in MapReduce. In: 7th Workshop on Large-Scale Distributed Systems for Information Retrieval, vol. 1, pp. 57–62. ACM, Boston (2009)
15. Liroz-Gistau, M., Akbarinia, R., Agrawal, D., Pacitti, E., Valduriez, P.: Data partitioning for minimizing transferred data in MapReduce. In: International Conference on Data Management in Cloud, Grid and P2P Systems, pp. 1–12. Springer (2013)
16. Mestre, D.G., Pires, C.E.S.: Improving load balancing for MapReduce-based entity matching. In: 2013 IEEE Symposium on Computers and Communications (ISCC), pp. 000618–000624. IEEE (2013)

CARP: Cost Effective Load-Balancing Approach for Range-Partitioned Data

Djahida Belayadi(✉), Khaled-Walid Hidouci, and Khadidja Midoun

Ecole Nationale Supérieure d'Informatique, Algiers, Algeria
{d_belayadi,w_hidouci,k_midoun}@esi.dz

Abstract. One of the important issues in range partitioning schemes is data skew. Tuples distribution across nodes may be skewed (some nodes have many tuples, while others may have fewer tuples). Processing skewed data not only slows down the response time, but also generates hot nodes. In such a situation, data may need to be moved from the most-loaded partitions to the least-loaded ones in order to achieve storage balancing requirements. Early works from the State-of-The-Art focused on achieving load balancing. However, today's works focus on reducing the load balancing cost. This latter involves reducing the cost of maintaining partition statistics. In this context, we propose to improve one of the best load balancing work, that is the one of Ganesan et al., to reduce the cost of maintaining the statistics of load balancing. We introduce the concept of fuzzy system image. Both nodes and clients have approximate information about the load distribution. They can nevertheless locate any data with almost the same efficiency as using exact partition statistics. Furthermore, maintaining load distribution statistics do not require exchanging additional messages as opposed to the cost of efficient solutions from the State-of-The-Art (which requires at least $\mathbb{O}(\log n)$ messages).

Keywords: Data skew · Load balancing · Parallel database
Range partitioning

1 Introduction

Range partitioning maps tuples to partitions according to a partitioning key. This scheme is the most common type of partitioning in parallel databases and it is often used with dates [1]. A key requirement in such systems is that the data has to be uniformly partitioned on all the nodes. This requirement is challenging to enforce when the input data is skewed. Data skew is a well-known concern in range partitioning where only few partitions (nodes) may be more loaded than others. In that case, data migration approaches are an appealing solution. The out-of-range data has to be moved from the most loaded area to least loaded one in order to satisfy the storage balancing requirement.

Data movement must be followed by a change in the partition statistics (partition boundaries, data size, etc.). All the nodes/clients of the system must be

O. Demigha et al. (Eds.): CSA 2018, LNNS 50, pp. 281–290, 2019.
https://doi.org/10.1007/978-3-319-98352-3_30

aware of these changes in order to decide whether to call the balancing algorithm and addressing the right node. One of the dominant measures that we want to optimize in such a situation is communication cost. The focus is on the solutions that reduce the cost of maintaining data distribution information.

Ganesan et al., [2] proposed an online load balancing algorithm on a linearly ordered nodes. Their algorithm called ADJUSTLOAD guarantees a low imbalance ratio between the maximum and the minimum loads among nodes. Although, the ADJUSTLOAD algorithm is easy to state, each balancing operation may require global load information, that may be expensive in terms of operations costs. Their algorithm uses a data structure called skip graphs [3] to maintain load information and ensure efficient range queries. Each balancing operation may require global information with a cost of $\mathbb{O}(\log n)$ messages. Moreover, a change of partition boundaries of neighboring nodes in load-balancing will require a change in the two skip graphs.

In this paper, we improve the Ganesan et al., work by reducing the cost of maintaining partition statistics. We propose CARP, a **C**ost effective load-balancing **A**pproach for **R**ange-**P**artitioned data. Our algorithm uses the same primitive operations, NBRADJUST and REORDER as in Ganesan et al. The primitive NBRADJUST transfers the surplus of data from the current node to one of its neighbors. The primitive REORDER changes the nodes order to achieve the storage balancing requirements. As a result, the partition boundaries change as well as the data size of each partition. However, our algorithm is not based on skip graphs to maintain load partition statistics. The key point of our contribution is the fuzzy system image, where both nodes and clients have approximate information on data distribution statistics. Node/client image is an estimate of partition boundaries and data size related to the other nodes. After a balancing operation, the participating nodes may change their own boundaries. Those nodes do not need to inform the other ones by these changes. The clients use their image to route the queries. As a result, clients may address the wrong node when their images are outdated. Nevertheless, whenever an interaction happens between two peers (node or client), they exchange their images in order to correct each other. Our solution outperforms the State-of-The-Art methods in terms of communication cost. There is no additional cost for maintaining load statistic as in Ganesan et al.

In our solution, we are targeting range-partitioned databases that have to deal with the frequent end users requests that are continuously supplied in real time from multiple sources (e.g., Wireless Sensor Networks where data is inserted continuously in a parallel database). Note that in this case, the major risk that can be encountered is data skew. This would influence negatively the performance of parallel processing (such as range queries). We propose to improve Ganesan et al.'s work and other works that use skip graphs to avoid the additional cost related to maintaining the partition statistics. This allows heavy loaded nodes to dynamically adjust the boundaries of their ranges by migrating data to less loaded ones. We avoid using a central site or indexes to maintain load information.

This paper is organized as follows: we present the problem formulation and the load balancing solution of Ganesan et al. in Sects. 2 and 3, respectively. We describe our CARP solution in Sect. 4. We experimentally evaluate it in Sect. 5. Finally, we discuss the related work in Sect. 6.

2 Problem Formulation

In this section, we define a simple abstraction of a parallel database and make some considerations:

- Let $V = \{N_1, N_2, \ldots, N_n\}$ be a set of n heterogeneous nodes, each node stores a collection of keys. The key space is partitioned into n ranges, with boundaries $R_1 \leq R_1 \leq ... \leq R_n$. The node N_i manages a range $[R_{i-1}, R_i]$. We consider that the nodes are logically ordered by their ranges. This ordering defines left and right relations between them.
- Let $U = \{C_1, C_2, \ldots, C_m\}$ be a set of m clients. The clients insert, delete, or search for objects. They can also perform range queries.
- Both nodes and clients do not have the whole information about the data distribution across the partitions. Instead, they have their own image about the global load distribution, which may differ from the real one. The clients uses this image to send the insert, delete or range queries. The nodes use their images to know if they should call the balancing algorithm and to which neighbor they should send the surplus data.
- Each insert or delete operation is followed by the execution of the load-balancing algorithm. Data may have to be moved from node N_i to node N_j. node N_i sends its image with the set of migrated keys. This image describes the view of node N_i about the other nodes loads and boundaries.

3 Load Balancing Solution of Ganesan et al.

Ganesan et al., suggested an inspiring algorithm for reducing data skew. It guarantees a small imbalance ratio σ between the largest and the smallest load. This ratio is always bounded by a small constant which is 4,24. The algorithm uses two operations:

- NBRADJUT: the node N_i transfers its surplus of data to one of its neighbors (N_{i+1} or N_{i-1}, it depends on the least loaded neighbor). Data migration may change the boundaries of N_i and the neighbor receiving data.
- REORDER: the least loaded node (N_r) among all the nodes transfers its entire content to one of its neighbors and change its logical position to share data with the node performing the load balancing algorithm.

In both operations, a node requires non-local information (neighbors' loads, position of the most and least loaded node). A given node attempts to shed its load whenever its load increases by a factor δ. For some constant c, Ganesan et al., define a sequence of thresholds $T_i = c\delta^i$, for all $i \geq 1$. The node N_i attempts to trigger the AdjustLoad procedure whenever its load $L(N_i)$ is greater than its threshold T_i. When $\delta = 2$, they call their algorithm the Doubling Algorithm. The AdjustLoad procedure works as well when $\delta > \phi = (\sqrt{5}+1)/2 = 1,62$, the golden ratio. They call their algorithm that operates at that ratio, the Fibbing Algorithm. They prove that the AdjustLoad procedure running on that ratio would guarantee the imbalance ratio σ of $\delta^3 = 4.24$.

Ganesan et al., use two skip graphs. Skip graphs are circular linked lists [4], in which every node has $\log n$ pointers. Routing between two nodes needs $\mathbb{O}(\log n)$ messages. The first skip graph is used to get neighbors' loads (one message) and to route range queries to the appropriate node. The second skip graph is used to get the positions of the most and least loaded node in the system ($\mathbb{O}(\log n)$ messages for locality plus costs of updating the two skip graphs).

4 CARP Overview

In this section, we present the algorithm for balancing data over nodes with a minimum operation cost. We present the *FuzzyBalance* procedure which uses the neighbor item exchange and the node reorder. Before that, we describe the mechanism of the fuzzy image used to maintain the global information, unlike Ganesan et al., that use skip graphs.

4.1 Node's Image

Consider that the node's image is encoded in a table $node\_image[1,n]$, where each element "$node\_image[i]$" stores an estimate of the information about node N_i. Each node has an index i so that $i \in [1,n]$, N_{i+1} is the logical successor and N_{i-1} is the logical predecessor. We denote "$node\_image[i].Load$" the current load of N_i, "$node\_image[i].LowerBound$" its lower bound, the upper bound of N_i is "$node\_image[i].UpperBound$". Each node N_i maintains the timestamps of the last update, this value is stored in "$node\_image[i].LastUpdate$".

4.2 Client's Image

Each client has its own image about all the other nodes. Consider that the client's image is encoded in a table $client\_image[1,m]$, where each entry "$client\_image[i]$" stores the information about node N_i. For each node, the client knows its load "$client\_image[i].Load$", lower bound "$client\_image[i].LowerBound$", upper bound "$client\_image[i].UpperBound$" and address "$client\_image[i].Add$". This image is not necessarily correct, it is updated every time the client sends a request to an incorrect node.

4.3 FuzzyBalance Operation

We present the *FuzzyBalance* procedure, which uses the two universal load balancing primitives: Nodes Reorder (Reorder) and Neighbor Item Exchange

(NBRADJUT) as in Ganesan et al.'s work. However, we use the concept of fuzzy image to get global load information instead of using the skip graphs. A node N_i executes *FuzzyBalance* algorithm whenever its load increases beyond a threshold T_i. The algorithm access to the node's image to check if data can be shared with the lightly loaded neighbors (less than half of N_i's load), if so, it performs the primitive NBRADJUT to average out the load with it, otherwise, it gets the position of the least-loaded node N_r from the local image, if the load of N_r is less than a quarter of N_i's load, N_r sends all of its data to the lightly loaded neighbors and changes its position to take over half of N_i's load. If N_i is unable to perform neither item exchange nor node reorder, we conclude that the system load is balanced.

The Algorithm 1 describes the *FuzzyBalance* operation. Note that at each message exchange between peers (nodes/clients) whether it is a client request or a data migration between nodes. The image is piggybacked in the message. This is why we do not need to implement a data structure for maintaining load statistics or to set up a central node. This concept significantly reduces the costs of load balancing.

Algorithm 1. *FuzzyBalance* (N_i, *node_image*)

```
1  Let N_j be the lightly loaded node between N_{i+1} and N_{i-1};
2  if (node_image[i].Load/2 ≥ node_image[j].Load) then
3  |   NB = (node_image[i].Load − node_image[j].Load)/2;
4  |   Send the NB tuples to N_j;
5  |   node_image[i].Load = node_image[i].Load − NB;
6  |   Inode_image[j].Load = node_image[j].Load + NB;
7  |   Update N_i boundaries;
8  |   FuzzyBalance (N_i, node_image[i]);
9  |   FuzzyBalance (N_j, node_image[j]);
10 else
11 |   Find r so that: ∀k ∈ [1, n], node_image[k].Load ≥ node_image[r].Load;
12 |   if (node_image[i].Load/4 ≥ node_image[r].Load) then
13 |   |   Let N_j be the lightly loaded node between N_{r+1} and N_{r−1};
14 |   |   Send node_image[r].Load tuples to N_j;
15 |   |   N_j changes its position to be N_i's neighbor;
16 |   |   Send node_image[i].Load/2 to N_r
   |   |   node_image[i].Load = node_image[i].Load − node_image[i].Load/2);
17 |   |   node_image[r].Load = node_image[i].Load/2);
18 |   |   Update the images of the concerned nodes;
19 |   |   FuzzyBalance (N_i, node_image[i]);
20 |   |   Rename nodes appropriately after the REORDER;
21 |   else
22 |   |   The system is balanced;
23 |   end
24 end
```

5 Experimental Evaluation

In this Section, we present the results from our simulation of $CARP$ approach on a networks of 8 nodes and 2 clients. Processing node software and client software were executed on machines with Intel(R) Core(TM) i7-5500U CPU@2.40 GHz and 8 GiB of RAM. Both nodes and clients were connected through a Gigabit Ethernet network. Algorithms are implemented in C language using the Message Passing Library (Open MPI).

In the experiments, we present the algorithm for the insert-only case. This case is simpler to analyze and provides general ideas on how to deal with the general case. The system is studied under a simulation model that we call $HOTSPOT$. All insert operations are directed to a single hot node. We use a sequence of $5 * 10^4$ frequent insert operations.

5.1 Imbalance Ratio

First of all, we evaluate the imbalance ratio σ. We measure the imbalance ratio as the ratio between the largest and smallest loads after each insert operation. As the thresholds in the system are an infinite (increasing geometric sequence, as in Ganesan et al.'s work), we measure the imbalance ratio with three values of the factor δ, $(\delta = \phi, \delta = 2, \delta = 4)$. ϕ is the golden ratio, $\phi = (\sqrt{5}+1)/2 = 1,62$. Figure 1 shows the imbalance ratio (Y-axis) against the number of insert operations (X-axis).

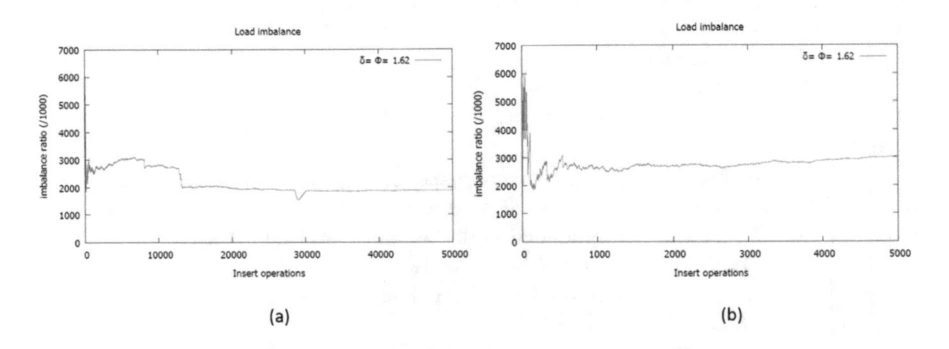

Fig. 1. The imbalance ratio when $\delta = 1.62$.

The graph in Fig. 1(a) shows that the imbalance ratio is always bounded by a constant 6 (in Fig. 1(b), the curve is enlarged to visualize this constant) and it converges to 1 after $2 * 10^4$ operations. The spikes in the curve correspond to an $FuzzyBalance$ invocation. The curve presents several variations at the beginning because the nodes are in the growing phase where data are loaded, which explains the variation of the ratio. At the beginning, all the insert operations are sent towards the node N_1. The successive invocations of the $FuzzyBalance$ algorithm

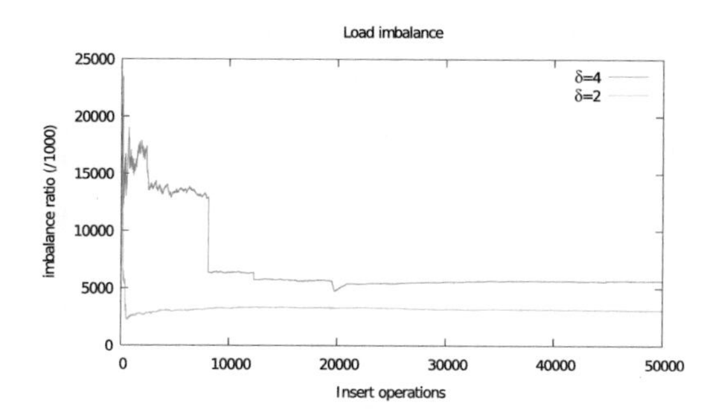

Fig. 2. The imbalance ratio ($\delta = 2$ and $\delta = 4$).

leads to a variation of the partition boundaries (the interval of N_1 will be reduced so that the data will be directed towards the neighbor).

The graphs in Fig. 2 are given for $\delta = 2$ and $\delta = 4$. Our algorithm, unlike the algorithm of Ganesan et al., does not present a big ratio difference between $\delta = 1.62$ and $\delta = 2$. As in the first experiment, the growing phase presents big variations in the ratio, thereafter, the ratio values converge to 2. However, the results of the third experiment $\delta = 4$ are different from the previous ones, the ratio values are larger and converge towards 5.

5.2 Image Adjustment

After a balancing operation, two nodes at least change their boundaries due to the data migration, which leads to changing the image of these two nodes. The client with an outdated image can address a wrong node that has changed its bounds. In our set of experiments, we measure the number of times the client sends a query to a wrong node and hence made an addressing error and received an IAM. The results shown in Fig. 3 present a rapid increase of the number of addressing errors in the growing phase (from 1 to 1000 insert operations). This is explained by the fact that there is frequent invocation of the balancing algorithm, and therefore a frequent change of the bounds.

5.3 Performance Analysis

For balancing loads among nodes, we need to minimize the movement cost as much as possible. After measuring the imbalance ratio and the image adjustment, we next measure the data movement cost.

Figure 4(a) plots the cumulative number of tuples migrated by the *FuzzyBalance* algorithm (Y-axis) against the number of insert operations (X-axis) during a run when $\delta = 1,62$ and $\delta = 2$. We observe that in the growing phase, the number of migrated tuples for different δ values is rising, this is because keeping the system tightly balanced causes a larger number of rebalancing operations. Figure 4(b) plots the data movement cost when $\delta = 4$.

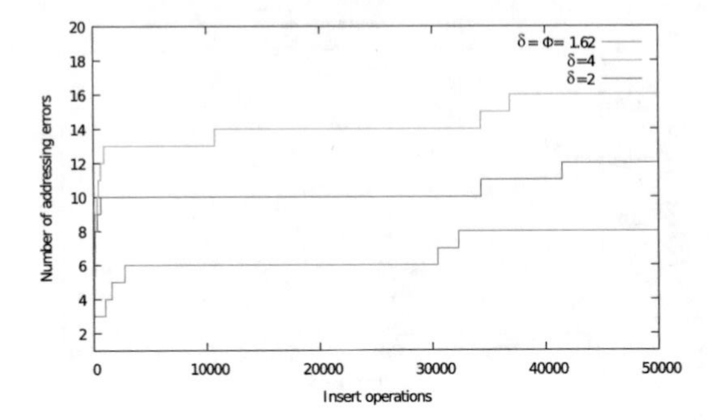

Fig. 3. The number of addressing errors ($\delta = \phi = 1.62$, $\delta = 2$, and $\delta = 4$)

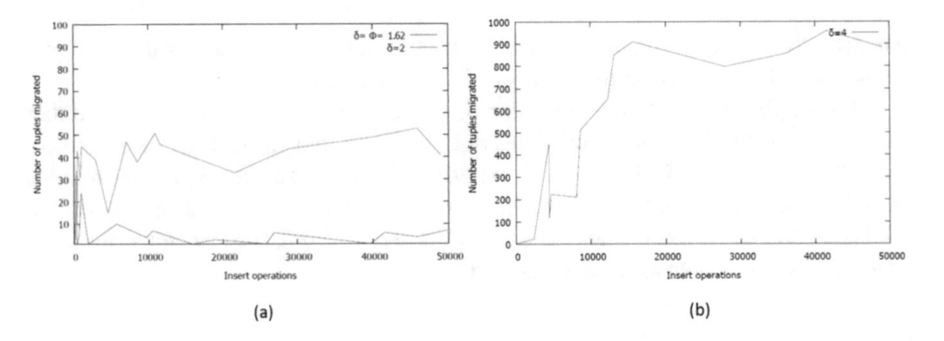

Fig. 4. Data movement cost when (a): $\delta = 1.62$, $\delta = 2$ and (b): $\delta = 4$.

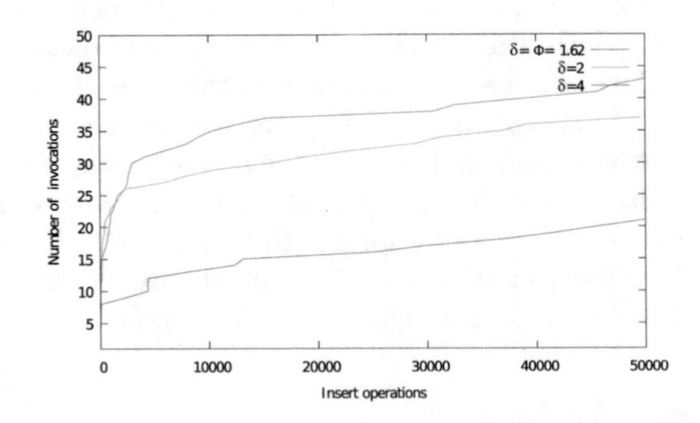

Fig. 5. Number of invocations when $\delta = 1.62$, $\delta = 2$ and $\delta = 4$.

Figure 5 illustrates the number of invocations of the *FuzzyBalance* (Y-axis) against the number of insert operations (X-axis) during a run. The observation we make is that the number of invocations of the algorithm increases for the three values of δ during the growing phase. The number of procedure invocations presented in the Fig. 5 was measured when $\delta = 1.62$, $\delta = 2$ and $\delta = 4$. We observe that the number of invocations is relatively small. This comes back to the fact that we are supporting some imbalance situations.

6 Related Work

In this section, we describe the current research related to our work that achieves load balancing method while supporting range queries.

In P2P networks, A number of recent load balancing approaches have been proposed [5–10]. Structured P2P networks are an efficient tool for storage and location of data since there is no central server which could become a bottleneck. The basic load balancing approach in structured P2P networks is consistent hashing [5]. Unfortunately, consistent hashing approach destroys data order by randomizing placement of data items which is not applicable to some applications. For example, to support range searching in a database application, the items need to be in a specific order. SkipNet [11] and skip graph [3] are data structures for range searching. Both of them are adapted from Skip Lists [4]. The computational cost of object search in a skip graph and SkipNet networks is $\mathbb{O}(\log n)$, where n is the number of nodes in the network.

Chawachat et al., [7] also proposed an improvement of the Ganesan et al.'s approach, there algorithm guarantees a ratio of 7.464 between the maximum and minimum loads among nodes. In their work, the same load balancing primitives are used as in Ganesan et al.'s work. Despite the effectiveness of their work, the cost of maintaining load balancing information is also high, because their solution is based on skip graph.

The above State-of-The-Art approaches deal with the problem of data skew in a range-partitioned data system. The conclusion we could make is that the cost of maintaining load statistics is at least $\mathbb{O}(\log n)$. Our solution provides load balancing mechanism to handle the data skew for applications where append operations and range queries are dominant. We provide a strategy for maintaining the global load statistics without any extra communication cost.

7 Conclusion

In this paper, we presented an effective on-line load balancing algorithm that deals with the problem of data skew so that the partition boundaries change dynamically and the ratio between the most and least loaded nodes is close to 1. Our experimental results that we set in our laboratory show that our approach does not need extra cost of maintaining partition statistics as opposed to the cost of efficient solutions from the State-of-the-Art. Although our proposal was presented in the context of balancing storage load, it can be generalized to balance execution load. All that is required is an ability to partition load evenly across two machines.

References

1. Bensberg, C., Becker, J., Mueller, C., Thumfart, A.: Dynamic range partitioning, US Patent App. 14/463,060, 19 August 2014
2. Ganesan, P., Bawa, M., Garcia-Molina, H.: Online balancing of range-partitioned data with applications to peer-to-peer systems. In: Proceedings of the Thirtieth International Conference on Very Large Data Bases, vol. 30, pp. 444–455. VLDB Endowment (2004)
3. Aspnes, J., Shah, G.: Skip graphs. ACM Trans. Algorithms (TALG) **3**(4), 37 (2007)
4. Pugh, W.: Skip lists: a probabilistic alternative to balanced trees. Commun. ACM **33**(6), 668–676 (1990)
5. Felber, P., Kropf, P., Schiller, E., Serbu, S.: Survey on load balancing in peer-to-peer distributed hash tables. IEEE Commun. Surv. Tutorials **16**(1), 473–492 (2014)
6. Mirrezaei, S.I., Shahparian, J.: Data load balancing in heterogeneous dynamic networks. arXiv preprint arXiv:1602.04536 (2016)
7. Chawachat, J., Fakcharoenphol, J.: A simpler load-balancing algorithm for range-partitioned data in peer-to-peer systems. Networks **66**(3), 235–249 (2015)
8. Antoine, M., Pellegrino, L., Huet, F., Baude, F.: A generic API for load balancing in structured P2P systems. In: 2014 International Symposium on Computer Architecture and High Performance Computing Workshop (SBAC-PADW), pp. 138–143. IEEE (2014)
9. Takeda, A., Oide, T., Takahashi, A., Suganuma, T.: Efficient dynamic load balancing for structured P2P network. In: 2015 18th International Conference on Network-Based Information Systems (NBiS), pp. 432–437. IEEE (2015)
10. Mizutani, K., Inoue, T., Mano, T., Akashi, O., Matsuura, S., Fujikawa, K.: Stable load balancing with overlapping ID-space management in range-based structured overlay networks. Inf. Media Technol. **11**, 1–10 (2016)
11. Harvey, N.J.A., Jones, M.B., Saroiu, S., Theimer, M., Wolman, A.: Skipnet: a scalable overlay network with practical locality properties. Networks **34**(38) (2003)

Parallel Clustering Validation Based on MapReduce

Soumeya Zerabi[1(✉)], Souham Meshoul[1], and Bilel Khantoul[2]

[1] Department of Computer Science and Its Applications, Abdelhamid Mehri Constantine 2 University, Constantine, Algeria
{soumeya.zerabi,souham.meshoul}@univ-constantine2.dz

[2] Department of Computer Science, Batna 2 University, Batna, Algeria
khantoul_b@yahoo.fr

Abstract. In this work, we developed and experimentally validated a novel model for external clustering validation to deal with huge data sets using Conditional Entropy index. The model allows clustering validation in a parallel and a distributed manner using Map-Reduce framework, it is termed MR-Centropy. The aim is to be able to scale with increasing dataset sizes when ground truth clustering is available. The proposed MR-Centropy is a three-jobs process where each job consists of Map and Reduce functions. Three jobs were necessary to gather all the statistics involved in the computation of the Conditional Entropy index. Each step in the proposed framework is done in parallel. Numerical tests on real and synthetic datasets demonstrate the effectiveness of our proposed model.

Keywords: Big data · Clustering · MapReduce · External clustering validation Conditional entropy index

1 Introduction

One of the important data mining tasks is clustering which is an effective method used in many different fields such as biology, business, climate, spatial data analysis, web mining, etc. Clustering [1] is a technique of partitioning the data set into subsets called clusters, without a known prior information, where the similarity measure between the data objects in the same cluster must be maximized and the similarity measure between the data objects from different clusters must be minimized. A wealth of clustering algorithms has been proposed in the literature among which the most popular are K-means [2], DBSCAN [3], STING [4], BRICH [5] and many others.

One of the most important challenges in cluster analysis is to evaluate clustering algorithms results in order to test the goodness of partitions after clustering [6]. A number of clustering validation indexes have been proposed in the literature [7–10], they are usually classified into two types which are: internal and external clustering validation indexes. The first type uses the external information, which is not present in the data for clustering validation to compare between clustering solutions with other clustering [11], the most popular indexes in this type are: purity, F-measure, Conditional Entropy, Jaccard coefficient, etc. However, the second one assumes that the access to a ground truth grouping is not available in most real applications. Therefore, these indexes have

O. Demigha et al. (Eds.): CSA 2018, LNNS 50, pp. 291–299, 2019.
https://doi.org/10.1007/978-3-319-98352-3_31

to use the notions of intra-cluster similarity or compactness. The indexes widely used in this type are: Silhouette index [12], Davies Bouldin index [13], Dunn index [14], etc.

When the size of the data set is very large, traditional clustering validation indexes become computationally expensive in terms of time and space. For these reasons, the parallelization of clustering validation indexes is necessary.

The idea of a parallel computing framework is to provide simultaneous calculations of big tasks that can be divided into smaller ones.

In order to develop parallel applications easier, Google introduced in 2004 a programming paradigm called MapReduce [15] and it has become popular as an alternative model to data parallel programming over the past few years. MapReduce uses the Map and Reduce functions and provides scalability, load balancing and fault tolerance.

Therefore, the aim for this paper is to propose a new parallel and distributed model of external clustering validation indexes especially Conditional Entropy named MR_Centropy in order to tune up traditional index for a big data context using the MapReduce framework.

The rest of this paper is organized as follows. Section 2 describes Conditional Entropy index and some related works in clustering validation indexes based on MapReduce. Section 3 gives an overview about MapReduce framework. Section 4 exposes the design of the proposed model. Section 5 shows the experiments performed with real and synthetic datasets and discusses the obtained results. Finally, a conclusion and future work are given in Sect. 6.

2 Background

In this section, we describe a formal definition of the Conditional Entropy then, we summarize some works related to the external clustering validation indexes based on MapReduce.

In the following, let's denote $C = \{C_1,\ldots,C_r\}$ a set of obtained cluster composed of r clusters, $R = \{R_1,\ldots,R_k\}$ a set of ground truth (real) partition composed of k clusters.

Conditional Entropy

Conditional Entropy (*CE*) is a popular and a commonly used index in clustering validation. It measures the "purity" of clusters based on a given class labels [16].

The Conditional Entropy of R with respect to cluster C_i is defined as follows [17]:

$$H(R/C_i) = -\sum_{j=1}^{k} \frac{n_{ij}}{n_i} log\left(\frac{n_{ij}}{n_i}\right) \tag{1}$$

The Conditional Entropy of a given clustering C is defined as the weighted sum of conditional entropies of all clusters [17].

$$H(R/C) = \sum_{i=1}^{r} \frac{n_i}{n} H(R/C_i) \tag{2}$$

With:

n_{ij}: the number of common points between the obtained clusters C_i and ground-truth (real) partition R_j.
n_i: the number of points in each obtained cluster C_i.
n: the number of points in a dataset.
k: the number of clusters in ground truth partition R_j.
r: the number of clusters in the obtained clusters C_i.

The Conditional Entropy should be minimized for a perfect clustering solution and its values ranges in [0, log k].

To the best of our knowledge, very few attempts if any have been suggested to solve the problem of clustering validation in a big data context. In [18, 19], we proposed new parallel and distributed models for external validation index for Purity and F-measure based on MapReduce. Following this effort, we tackle in this paper the issue of revisiting *CE* index computation using parallel and distributed paradigms.

3 MapReduce Framework

Apache Hadoop [20, 21] is an open source framework used to process large data sets [22]. It includes two main sub-components: the Hadoop Distributed File System (HDFS) used for storing data and the MapReduce framework used for processing data as depicted in Fig. 1.

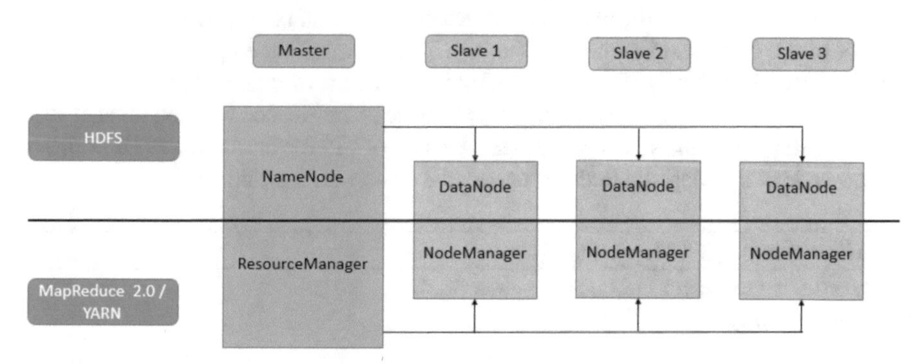

Fig. 1. Hadoop components [23]

HDFS [24] is a file system designed for providing a cost effective and reliable storage for huge amounts of structured and unstructured data. Additionally, it ensures data replication in order to reduce network congestion and to increase system performance by ensuring fault tolerance [22].

HDFS is based on a master/slave architecture, where the master node is called NameNode and the slave nodes is called DataNode.

MapReduce is a programming model used for processing and generating large data-sets stored in HDFS [15]. It offers fault tolerance, scalability, load balancing and simplicity. Those features explain its wide adoption.

MapReduce framework works as follows [26]: the Map function takes a set of key/values pairs and returns a set of intermediate key/value pairs then all values with the same key are grouped and submitted by the framework to the same reducer. The Reduce function produces one or more output key/value pairs. These outputs are stored in HDFS. For an in-depth insight into HDFS-MapReduce one can refer to [25].

4 Proposed MapReduce Model for Conditional Entropy

Taking into account the previous description of MapReduce principle and in order to develop a parallel architecture for Conditional Entropy index, it is necessary to define MapReduce necessary jobs for the computation of *CE* index. In our model we propose a three jobs process.

- The first MapReduce job calculates n_{ij} which is the number of points that are common to the obtained cluster C_i and the real cluster R_j.
- The second MapReduce job finds for each obtained cluster C_i the list of the number of common points for this cluster and each real cluster R_j.
- The third MapReduce job is used to calculate the overall value of Conditional Entropy.

First MapReduce Job

The input of the Map function is a file of <key, value> pairs stored on HDFS. First of all, the dataset is split into blocks, each block is stored as lines and sent to a Mapper as series of ***<key***, ***value>*** pairs, where ***key*** indicates the line number, and ***value*** denotes the content of the corresponding line i.e. (R_j, C_i). The intermediate output is a list of ***<key, value>*** pairs where ***key*** is (R_j, C_i) and ***value*** is "1". A list of intermediate ***<key, value>*** pairs with the same ***key*** are sent to the same Reducer. The Reduce function outputs $<(R_j, C_i), n_{ij}>$ pairs where the value n_{ij} indicates the sum of "1" for the same key (R_j, C_i). The pseudo-code of the Map and Reduce functions is shown in Algorithm 1.

Algorithm 1. First MapReduce job

---

function map1 (key: LineNumber, value: (R_j,C_i))
Emit((R_j,C_i), 1);
end function

function reduce1 (key: (R_j, C_i) , Iterator values: 1)
int $n_{ij} = 0$;

for each v in values **do**

n_{ij} += v;

end for

Emit((R_j,C_i),n_{ij});

end function

Second MapReduce Job

The input of the Map function in this job is the output of the previous job, this function extracts C_i from the key of the output of the first job and emits the pair $<C_i,n_{ij}>$. The

Reduce function generates a list of n_{ij} for all pairs with the same value of C_i. The pseudo-code for the Map and Reduce functions is shown in Algorithm 2.

Algorithm 2. Second MapReduce job

function map2 (key: (R_j,C_i), value: n_{ij})
for each v in key **do**
extract (C_i)
end for
Emit ((C_i), n_{ij});
end function

function reduce2 (key:(C_i), Iterator Values:n_{ij})
Emit ((C_i), <n_{11},n_{12},…n_{ij}>);
end function

Third MapReduce Job

At each Mapper, we calculate the local Conditional Entropy $H(R/C_i)$ of each cluster C_i. The intermediate output of this Map function is a list of *<key, value>* pairs where *key* indicates "CE" for all the records and value indicates both the Conditional Entropy $H(R/C_i)$ of each cluster C_i and the number of points in this cluster n_i. These intermediate results are sorted and partitioned by key on the Mappers and then sent to the reducers. The Reduce function calculates the value $H(R/C)$ of all the clustering. The pseudo-code of the Map and Reduce functions is shown in Algorithm 3.

Algorithm 3. Third MapReduce job

function map3 (key: (C_i) , value: < n_{11},..,n_{ij}>)
for each v in key **do**
n_i=n_{11}+…+n_{ij}
calculate $H(R/C_i)$ using Equation 1.
end for
Emit ((CE) , ($H(R/C_i)$,n_i);
end function

function reduce3 (key: ("CE") , Iterator values: ($H(R/C_i)$,n_i)
for each v in values **do**
calculate ($H(R/C)$) using Equation2.
end for
Emit(CE,$H(R/C)$);
end function

5 Results

In this section, we describe and discuss the experimental results performed for our proposed model.

5.1 Experimental Environment

We have implemented and executed our proposed model of MR_Centropy in JDK 1.7.0_25 and have used the platform Hadoop 2.2.0 for the MapReduce framework. The local machine has the following characteristics: Intel® CoreTM i5 4200U CPU @1.60 GHz x4, 4 Go RAM.

5.2 Experimental Datasets

To evaluate our MR_Centropy model, we used four (04) real datasets taken from the UCI repository [27] and three (03) synthetic datasets used by [28] (see Table 1).

Table 1. Description of the datasets.

Data sets	Term	No. of points	No. of clusters	Type
Wisconsin	RD1	683	2	real
Iris	RD2	150	3	real
RUspini	RD3	75	4	real
Dermatology	RD4	366	6	real
D_10d10c	SD1	436	10	synthetic
D_2d10c	SD2	520	10	synthetic
D_10d4c	SD3	733	4	synthetic

5.3 Results

We evaluated the correctness of our proposed model for three cases depending on the similarity between the obtained clustering C_i and the real partitioning R_j. Let us denote Nbr_R_j the number of points for each R_j, Nbr_C_i the number of points for each C_i.

Case 1

In the first case, there is a low similarity between the obtained clustering and the real partitions. Table 2 shows the partition of points in C_i and R_j.

Case 2

In the second case, we apply k-means algorithm to obtain the partitions of points in C_i. We assume three trials.

Case 3

In this case, there is a strong relationship between the obtained clustering and true partitions as illustrated on Table 3.

Table 2. Case 1, partitioning of points for obtained clustering and true partitions.

Datasets	Nbr_R_j	Nbr_C_i
RD1	{444,239}	{280,403}
RD2	{50,50,50}	{40,70,40}
RD3	{20,23,17,15}	{30,15,10,20}
RD4	{112,61,72,49,52,20}	{60,80,80,50,55,41}
SD1	{18,83,57,26,67,50,12,72,39,12}	{26,39,46,44,46,50,38,46,50,51}
SD2	{67,15,19,53,83,64,65,68,68,18}	{51,49,39,45,66,56,52,58,56,48}
SD3	{252,244,166,71}	{180,250,270,33}

Table 3. Case 3, partitioning of points for obtained clustering and true partitions.

Datasets	Nbr_R_j	Nbr_C_i
RD1	{444,239}	{444,239}
RD2	{50,50,50}	{50,50,50}
RD3	{20,23,17,15}	{20,23,17,15}
RD4	{112,61,72,49,52,20}	{112,61,72,49,52,20}
SD1	{18,83,57,26,67,50,12,72,39,12}	{18,83,57,26,67,50,12,72,39,12}
SD2	{67,15,19,53,83,64,65,68,68,18}	{67,15,19,53,83,64,65,68,68,18}
SD3	{252,244,166,71}	{252,244,166,71}

Table 4 shows the results of obtained MR_Centropy values where applying our proposed model to the above cases.

Table 4. Overview of the obtained MR_Centropy values.

Datasets	Case 1	Case 2			Case 3
		Trial 1	Trial 2	Trial 3	
RD1	0.933999	0.921674	0.921674	0.921674	0.0
RD2	0.975698	0.366683	0.0	0.535376	0.0
RD3	1.483111	0.0	0.240642	0.0	0.0
RD4	2.132256	0.526786	0.663147	0.597107	0.0
SD1	2.849623	0.362630	0.121528	0.480738	0.0
SD2	3.005500	0.414502	0.502389	0.419303	0.0
SD3	1.453981	0.728568	0.106085	0.092581	0.0

5.4 Discussion

From Table 4, we can see that the obtained values of Conditional Entropy are correctly aligned with the related objectives of each case. For instance, in case 3, the minimum obtained value of Conditional Entropy is zero for all the datasets which indicates a strong relationship between C_i and R_j. This behavior is almost the same in the rest of the cases.

6 Conclusion

In this paper, we proposed a novel parallel and distributed model for Conditional Entropy namely MR_Centropy using the MapReduce framework. This model was verified experimentally both on synthetic and real datasets. The results show that the new model works in a correct manner.

Future work includes the application to very large datasets in order to prove the scalability of our model, the proposition of models for internal validation indexes besides external validation indexes and the use of different frameworks like Spark and Storm.

Acknowledgment. This work has been supported by the National Research Project CNEPRU under grant N: B*07120140037.

References

1. Davidson, I., Ravi, S.S., Shamis, L.: A SAT-based framework for efficient constrained clustering. In: The Proceedings of the 10th SIAM International Conference on Data Mining, pp. 94–105 (2010)
2. MacQueen, J.: Some methods for classification and analysis of multivariate observations. In: 5th Berkeley Symposium on Mathematics, Statistics and Probability, pp 281–296 (1967)
3. Ester, M., Kriegel, H.-P., Sander, J., Xu, X.: A density based algorithm for discovering clusters in large spatial databases with noise. In: Proceedings of the Second International Conference on Knowledge Discovery and Data Mining (KDD), Portland, pp. 226–231 (1996)
4. Wang, W., Yang, J., Muntz, R.: STING: a statistical information grid approach to spatial data mining. In: Proceedings of the 23rd International Conference on Very Large Data Bases (VLDB), pp. 186–195. Morgan Kaufmann Publishers, Athens (1997)
5. Tian, Z., Raghu, R., Miron, L.: BIRCH: an efficient data clustering method for very large databases. In: Proceedings of the Conference of Data Management, pp. 103–114. ACM SIGMOD, Montreal (1996)
6. Xiong, H., Li, Z.: Clustering validation measures. In: Aggarwal, C.C., Reddy, C.K. (eds.). Data Clust. Algorithms Appl., vol. 43(3), pp. 571–605. CRC, Boca Raton (2014)
7. Santibanez, M., Valdovinos, R.-M., Truebam, A., Rendon, E., Alejo, R., Lopez, E.: Applicability of cluster validation indexes for large data sets. In: The 12th Mexican International Conference on Artificial Intelligence, pp. 187–193. IEEE, Mexico (2013)
8. Campo, D.N., Stegmayer, G., Milone, D.H.: A new index for clustering validation with overlapped clusters, pp. 549–556. Elsevier (2016)
9. Halkidi, M., Batistakis, Y., Vazirgiannis, M.: On clustering validation techniques. J. Intell. Inf. Syst. **17**(2/3), 107–145 (2001)
10. Wu, J., Xiong, H., Chen, J.: Adapting the right measures for k-means clustering. In: Proceedings of the 15th ACM SIGKDD International Conference on Knowledge Discovery and Data Mining, Paris France, pp. 877–886 (2009)
11. Liu, Y., Li, Z., Xiong, H., Gao, X., Wu, J., Wu, S.: Understanding and enhancement of internal clustering validation measures. IEEE Trans. Cybernet. **43**(3), 982–993 (2013)
12. Rousseeuw, P.J.: Silhouettes: a graphical aid to the interpretation and validation of cluster analysis. J. Comput. Appl. Math. **20**(1), 53–65 (1987)
13. Davies, D.L., Bouldin, D.W.: A cluster separation measure. IEEE Trans. Patt Anal. Mach. Intell. **2**, 224–227 (1979)

14. Dunn, J.: Well separated clusters and optimal fuzzy partitions. J. Cybernet. Syst. **4**(1), 95–104 (1974)
15. Dean, J., Ghemawat, S.: MapReduce: simplified data processing on large clusters. Commun. ACM **51**(1), 107–113 (2008)
16. Rendón, E., Abundez, I., Arizmendi, A., Quiroz, E.M.: Internal versus external cluster validation indexes. Int. J. Comput. Commun. **5**(1), 27–34 (2011)
17. Zaki Mohamed, J., Wagner, M.J.R.: Data Mining and Analysis, 1st edn. Cambridge University Press, Cambridge (2014)
18. Zerabi, S., Meshoul, S., Merniz, A., Melal, R.: Towards clustering validation in big data context. In: Proceedings of the 2nd International Conference on Big Data, Cloud and Applications, pp. 28–33. ACM, Tetouan (2017)
19. Zerabi, S., Meshoul, S.: External clustering validation in Big Data context. In: Proceedings of the 3nd International Conference on Cloud Computing Technologies and Applications. IEEE, Rabat Morocco (2017)
20. Apache Hadoop. http://hadoop.apache.org/. Accessed 12 Aug 2017
21. White, T.: Hadoop: The Definitive Guide Storage and Analysis at Internet Scale, 4th edn. O'Reilly Media, Sebastopol (2015)
22. Oussous, A., Benjelloun, F.Z., AitLahcen, A., Belfkih, S.: Big data technologies: a survey. J. King Saud Univ. Comput. Inf. Sci. (2017)
23. Chullipparambil, C.P.: Big data analytics using hadoop tools. Ph.D. thesis San Diego State University (2016)
24. White, T.: Hadoop: The Definitive Guide, 3rd edn. O'Reilly Media, Sebastopol (2012)
25. Ibrahim, A., Hashem, T., Anuar, N.B., Gani, A., Yaqoob, I., Xia, F., Khan, S.U.: MapReduce: review and open challenges. Scientometrics **109**, 389–422 (2016)
26. Ha, L.K., Hyansik, C., Bongki, M., Lee, Y.J., Chung, Y.D.: Parallel data processing with MapReduce: a survey. SIGMOD Rec. **40**(4), 11–20 (2011)
27. Machine learning repository. http://archive.ics.uci.edu/ml/datasets.html
28. Handl, J., Knowles, J.: Improvements to the scalability of multi objective clustering. IEEE Congr. Evol. Comput. **3**, 2372–2379 (2005)

AMAM: Adaptive Multi-Agents Based Model for Negative Key Players Identification in Social Networks

Nassira Chekkai[1(✉)], Souham Meshoul[1], Imene Boukhalfa[1], Badreddine Chekkai[1], Amel Ziani[2], and Salim Chikhi[1]

[1] Abdelhamid Mehri-Constantine 2 University, Nouvelle ville-Ali Mendjli, 25000 Constantine, Algeria
{nassira.chekkai, souhame.meshoule}@univ-constantine2.dz, boukhalfa_imene@hotmail.com, slchikhi@yahoo.com

[2] University of Badji Mokhtar, 23000 Annaba, Algeria
z_amel1911@live.fr

Abstract. Social Network Analysis (SNA) is an active research topic. It arises in a broad range of fields. One important issue in SNA is the discovery of key players who are the most influential actors in a social network. Negative Key Player Problem (KPP-NEG) aims at finding the set of actors whose removal will break the social network into fragments. By another way, Multi-Agents Systems (MAS) paradigm suggests suitable ways to design adaptive systems that exhibit desirable properties such as reaction, learning, reasoning and evolution. A fortiori, the intrinsic nature of social networks and the requirements of their analysis could be efficiently handled using a MAS framework. Within this context, this paper proposes a multi-agents based-model AMAM for KPP-NEG. We first represent the social network in terms of a weighted graph. Then, a set of agents cooperate in order to identify the most important nodes. Simulation and computational results are demonstrated to confirm the effectiveness of our approach.

Keywords: Key players · KPP-NEG · Social networks · Multi-agent system Adaptation · Weighted graphs

1 Introduction

In recent years, social networks have become a very active area of research. They are made up of a set of members called actors or players, and a set of relationships that bring these social actors together. Social Networks Analysis (SNA) is a set of theories, tools, and processes for understanding the relationships and structures of a network [1]. Indeed, SNA is an interdisciplinary academic field which is becoming extremely important in various fields.

O. Demigha et al. (Eds.): CSA 2018, LNNS 50, pp. 300–310, 2019.
https://doi.org/10.1007/978-3-319-98352-3_32

In addition, SNA makes use of basic concepts of graph theory with the objective of studying the key issues in social networks. A social graph includes a set of nodes or points that represent the social actors and a set of edges or ties that connect them; with a variety in the nature of these connections. In a weighted social graph, weights are modeled by scores attributed to the edges; to illustrate the trust or similarity between the corresponding actors.

Key player identification is a basic issue in SNA. The latter consists of finding those actors who are considered important in their social network, in regard to some criteria [2]; such as their connectivity and the paths passing through them. The key players are identified according to their influential value which reflects how important an actor is in the social network [3].

In this paper, we are interested in KPP-NEG which is related to the maximally disrupt communication and finds applications in several fields; such as marketing [4] where the biggest challenge is how to find a small segment of players who are able to influence other segments, by their positive or negative opinions about products or services [5]. KPP-NEG has also been applied in psychology to study the orientation of actions in ergonomics area [6]. According to the same view, this problem has been studied in sociology; where a community is considered as a set of followers who follow the key actors. In the military context, KPP-NEG is used for locating an efficient set of enemies to follow up [7]. In the same line of thought, the key players have been detected in criminal networks to extract the set of the most dangerous criminals. In the public health domain, the key players were identified whenever we need to select a subset of population members to be immunized or vaccinated; in order to optimally contain an epidemic by avoiding its propagation [7].

We attempt to identify the set of negative players by taking into consideration the mutual trust between pairs of actors and its adaption. Thus, we represent the social network by a weighted graph and we propose an adaptive multi-agents based-approach aiming at identifying the set of the most influential cut points in regard to some conditions.

The rest of the paper is organized as follows: Sect. 2 reviews the related work and gives our motivation. Section 3 presents the basic concepts upon which our approach AMAM is built. Section 4 details the process of AMAM by describing the role of each agent. Section 5 provides the adaptation of our model. In Sect. 6, we present our experimental results. Finally, we conclude our work and give future directions in Sect. 7.

2 Related Work and Motivation

Research in KPP has been highly driven by the goal of enhancing the quality of interactions within the social network and improving information sharing. Thus, a large number of approaches, from different axes, have studied the effect of identifying the sets of the key players. Indeed, the existing solutions for solving this problem use different methods such us centrality, fragmentation and entropy as summarized in Table 1.

Table 1. Comparison between the existing methods for key players identification

Measures/References		Feature	Pros	Cons
Centrality	Degree of Centrality (DC) [9]	Uses the number of weights of direct neighbors that a given node is connected to	The nodes having high degree can influence a lot of other nodes	The indirect connections and the global structure of the network are not taken into account
	Closeness Centrality (CC) [10]	Assesses the length of the paths from a node to all other nodes in the social network	Identifies the nodes with beneficial paths with other nodes	The quickest path is not always the efficient one when the weights represent social information
	Betweenness Centrality (BC)	Measures the number of the shortest paths passing through a node [11]	The nodes occurring the most often on the shortest paths linking the remaining nodes, can propagate information productively within the network	There are other paths linking the remaining nodes, even if they are longer
	Eigenvector Centrality (EC)	Detects the nodes having more effective ties to higher scoring nodes [12]	Estimates the influence of a node on the transmission of information within the network, by taking into account the global structure of this network	Calculates the values for symmetric structures only (i.e. undirected graphs)
Fragmentation [13, 14]		Measures the drop in network efficiency after deleting the important nodes	Simple to implement. It takes into consideration the influence of a node on the network connection	Identifying individual key players without simultaneously detecting the set of key players

(*continued*)

Table 1. *(continued)*

Measures/References	Feature	Pros	Cons
Entropy [15]	Quantifies the influence of a node on the degree of uncertainty in the transmitted information within a network	Can quantify errors in the diffused information and messages within the network	The actors are acting in the social network to produce fruitful relations and reactions rather than making errors

From Table 1, it can be clearly seen that the centrality measure is not optimal for solving the KPP-NEG because the deletion of the central nodes may not guarantee the fragmentation of the network or even its disconnectivity [8]. In addition, the fragmentation measure aims at detecting each node separately; while the removal of one node at a time may not generate the maximum fragmentation. However, the deletion of the same node simultaneously with the other nodes can cause a significant disruption in the network.

Thus, we propose, in this paper, a new approach that allows the identification of the set of the negative key players by detecting the set of the most efficient articulation points.

3 Basic Concepts

3.1 Articulation Points

An articulation point of a graph is a point whose removal increases the number of components [16]. In fact, a vertex in a connected undirected graph is an articulation point if removing it and all edges incident to it results in a non-connected graph.

3.2 Adaptive Agents

The aim of a multi-agent systems is to find a solution to a given problem by attributing some tasks to a set of processes, called agents, working together. Hence, the process of the problem solving is divided into subunits for which an intelligent agent is operating. An agent is considered adaptive if it is capable of responding to other agents and/or its environment to some degree [17]. In other words, an adaptive agent has the ability to adapt its structure, knowledge and reaction when its environment changes.

Four basic forms of adaptation can be depicted:

- *Reaction:* a direct, predetermined response to a particular event or an environmental signal.
- *Reasoning:* ability to make inferences.
- *Learning:* change that occurs during the lifetime of an agent.
- *Evolution:* change that occurs over successive generations of agents [17].

4 AMAM Process

The multiagent systems paradigm offers a suitable framework to design adaptive systems that exhibit the aforementioned abilities. Hence, we propose an agent system to deal with the tackled problem. Each agent is responsible of a task in the process of identifying the set of the key cut points.

4.1 Graph Agent

This agent represents the social network by a graph. In fact, it models the actors by nodes and the relationships between them by edges and each trust score between two actors by a weight on the edge linking the corresponding nodes. It should be mentioned that the calculation of these nodes depends on the type of the social network then the existing interactions. Later, these nodes can adapt in time by increasing or decreasing according to the environment changes.

We consider, in (Fig. 1), a weighted graph representing a social network of 21 players. Each node corresponds to a player, and each edge weight estimates the trust between two players in the range [0, 1] where:

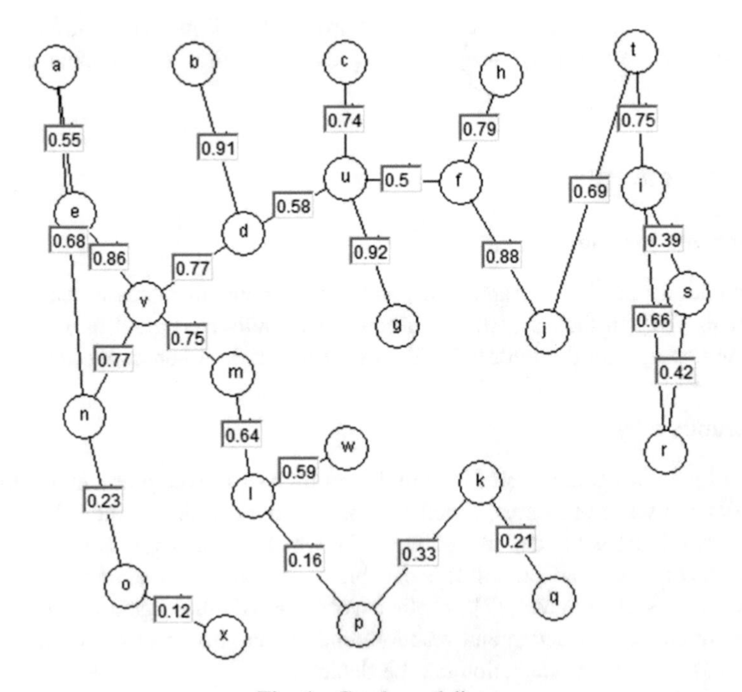

Fig. 1. Graph modeling

- Weights between [0, 0.4] represent a weak interaction.
- Weights between [0.5, 0.7] represent an average interaction.
- Weights between [0.8, 1] represent a good interaction.

4.2 Articulation-Points Agent

Once representing the social network by a graph, the Articulation-Points agent operates in a second step to find the set of articulation points using the following algorithm.

```
procedure ArticulationPoint(node v)
 visited(v) = true
 num (v) = low(v) = counter++
 for each edge(v,w) do
  if (!visited(w)) then
   DFS(w)
   Low(v) = min(low(v), low(w))
    if low(w) >= num (v) then
     print "v is an articulation point"
    end if
    else
     low(v) = min(low(v), num (v)
    end if
  end for
end
```

Algorithm 1. Articulation points detection

In Algorithm 1, we first draw a Spanning Tree (ST) for the given graph starting from any of its nodes. For that, we traverse the graph using the Depth First Search (DFS) and every time we follow an edge, we mark it. Once traversing, we mark the edges form the tree. Then, we associate to each edge a number according to the traversal number *num(v)*. Afterwards, we search for each node v, a child w such that $low(w) \geq num(v)$ where: $low(w)$ is the least number $num(v)$ in a DFS traversal of a graph such that node v can be reached by a possibly empty path of tree edges followed by at most one back edge. Then, $low(v)$ is the minimum of:

$low(v) = \min\{num(v)$, for all children w, $low(w)$ among all the tree edges (v, w), $num(w)$ among all back edges $(v, w)\}$.

In fact, a node v is an articulation point, for every child w of v, in these cases:

- If num(v) $\leq$ low(w), then v is an articulation point, otherwise v is the root in the ST.
- If v is the root of the ST, then it has two or more children in this ST.

It should be noted that this algorithm takes linear time (*O(N)* complexity) since it depends only on the number of nodes and the number of edges. Additionally, the removal of each articulation point results in the fragmentation of two bi-components at least.

4.3 Cut-Points Agent

Once detecting the set of articulation points within the network, Cut-Points agent operates in a last step in order to identify the minimum set "*S*" of cut points having the maximum influence on the network in terms of connection of the social communities. These latter are represented by the graph fragments. Thus, Cut-Points agent controls the articulation points under two conditions in order to select the most efficient cut points, as follows:

- **The set S connects the maximum number of fragments with similar sizes:** The more a node brings components together; the better the network fragmentation after its deletion.
- **The set S includes the nodes whose edges have maximum weights:** Our approach takes into account the weights values; by focusing on the greatest ones. We also use the standard deviation to choose the key players whose trust scores are well distributed.

The mathematical formulation of the Cut-Points agent task is given as follows:

Denote an undirected graph $G = (V, E)$ with a set of vertices $V = (v1, v2, \ldots, vn)$ and a set of edges $E = \{uv : u, v \in V\}$; with $W_v \in [0, 1]$ the weights on the edges. Noting that P is the set of articulation points, we aim to detect a subset S of cut points where $S \subseteq P \subseteq V$. The nodes included in S have the greatest average weights *A(S)* and the smallest standard deviation between the weights *D(S)*. Additionally, the set S must be linked to a set of fragments $F = (f1, f2, \ldots, fp)$ having a maximum cost *CF*. *CF* is calculated by using two values $\sigma 1$ and $\sigma 2$, where $\sigma 1$ is the number of fragments linked to a vertex v to be included in S, and $\sigma 2$ is the standard deviation between the sizes of these fragments.

$$CF = \sigma 1 + (0.5 - \sigma 2) \tag{1}$$

Taking into consideration that our network is dynamic and can evolve over time, we squash the values of $\sigma 1$ and $\sigma 2$ in the range [0, 0.5]. This is to keep identifying the set of negative key players; while verifying the two conditions regarding the number and the size of the fragments at the same time. We have normalized them using the following formula:

$$Normalized = \frac{(Original - MIN) * (max - min)}{(MAX - MIN) + min} \tag{2}$$

Where:

[*MAX*, *MIN*]: is the original interval
[*min*, *max*]: is the new interval
Original: is the value in the original interval
Normalized: is the value in the new interval.

It should be mentioned that:

- The maximum number of fragments is equal to $|V - S|$: in the worst case each node is a fragment.
- The minimum number of fragments is equal to 0.
- The maximum difference between the sizes of two fragments is equal to $|V - S|$ as the generated components include the rest of nodes after deleting the cut points. Thus, the size of the largest component must include all of the remaining nodes of the graph G.
- The minimum size between two fragments is equal to 0.

The global definition of the proposed method is described in the following formula.

$$\begin{cases} \mathrm{A(S)} \geq \propto \\ \mathrm{D(S)} \leq \beta \\ \arg\max\left\{\sum_{i \in P} CF\right\} \end{cases} \tag{3}$$

Where: $\propto$ and β are constant values.

In Formula 3, we identify the set S of cut points in a weighted graph of a social network by extracting those articulation points whose weights average is greater than a threshold $\propto$ and the standard deviation between the weights of these nodes is less than a threshold β. Furthermore, these cut points are the most trusted ones as they join the set of components having the maximum cost CF. The method terminates when the size of the set S is equal to a value K fixed a priory.

It should be noted that we have identified the connected components using the depth first search algorithms [18]. In addition, the weights of the edges that have been normalized using *sigmoid function* that varies between [0, 1]. It is defined by:

$$y = \frac{1}{1 + e^{-x}} \tag{4}$$

Where x is the original weight and y is the normalized one.

5 Adaption Model

In order to define the efficiency of each key player and estimate their influence on the social communities and on the global system, Cut-points agent uses a table that indicates when to change a key player once his total efficiency will fell much. This adaptation allows a periodic evolution of our AMAM model. Table 2 shows an example of the influence of the key players in a social network. Since each key player is modeled by a node, the influence of this latter was calculated using the average of its edges' weights representing the trust degrees with the neighbors.

Table 2. Table of key players.

NEG-KPP	Influenced communities (graph components)					Total efficiency
Community	1	2	3	4	5	
KPP 1	89%	92%				90.5%
KPP 2		91%	90%			90.5%
KPP 3			92%			92%
KPP 4				88%	89%	88.5%
KPP 5	98%				92%	95%

In addition, AMAM evolves dynamically as the social network can receive new actors. Furthermore, the actors can leave the network and the weights of the edges (trust scores) evolve over time. Thus, Graph agent has the capacity of adapting to these changes by adding new nodes, deleting nodes and updating the trust scores.

6 Experimental Results

We have simulated our approach using the multi-agents Platform JADE [19]. We have measured the efficiency of our approach using fragmentation utility [20] given in (Eq. 5). This measure varies between 0 and 1. The more the value is close to 1, the better the fragmentation (several small components), and the more the value is close to 0, the worse the fragmentation (a lot of players are still connected).

$$F = 1 - \frac{\sum_k S_k(S_k - 1)}{n(n-1)} \tag{5}$$

We have tested our approach on mobile tourism recommender systems that are mobile social networks. We have used OpinRank dataset [21] that contains full reviews for cars and hotels collected from Tripadvisor (259,000 reviews) and Edmunds (42,230 reviews). The results are summarized in the following table.

Table 3. Comparaison between AMAM, Efficiency method and Degree centrality technique using OpinRank dataset

Random KPP	Efficiency method [14]	Degree centrality [9]	AMAM
39%	73%	56%	84%

From Table 3 it can be seen that our approach performs more better than Efficiency method, this is because Efficiency technique has focused on the geodisique distances by detecting the nodes that appear on the shortest path linking two nodes. Also the degree centrality can not guarantee the network fragmentation as another path connecting them may exist, even with a larger size.

7 Conclusion

In this paper, we presented a new approach AMAM for KPP-NEG problem in social networks. Our method is based on graph theory and multi agent technology. To this end, three types of agents have been used where each one handles a specific task. A graph agent is dedicated to modeling the social network in terms of a weighted graph in which each social concept finds its equivalent in graph theory. Articulation-Points agent is used as the detector of all articulation points within the social graph. Then, Cut-Points agent is responsible for selecting the most efficient articulation points that represent the set of cut points, according to the connected components and the trust scores. Experimental results have confirmed the effectiveness of the proposed approach on social networks connections and interactions.

Our future work follows three directions. We, first, look for using centrality measures together with the fragmentation in order to get more beneficial results. Secondly, we plan to apply our methods on directed social graphs that allow the modeling of trust between each pair of users in both directions.

References

1. Hoppe, B., Reinelt, C.: Social network analysis and the evaluation of leadership networks. Leadersh. Q. **2**(4), 600–619 (2010)
2. Arroyo, D.O., Akbar Hussain, D.M.: An information theory approach to identify sets of key players. In: Proceedings of the 1st European Conference on Intelligence and Security Informatics (EuroISI 2008), pp. 15–26 (2008)
3. Lin, P., Chen, L., Yuan, M., Nie, P.: Discover the misinformation broadcasting in on-line social networks. J. Inf. Sci. Eng. **31**(3), 763–785 (2015)
4. Domingos, P., Richardson, M.: Mining the network value of customers. In: Proceedings of the Seventh ACM SIGKDD International Conference on Knowledge Discovery and Data Mining (KDD 2001), pp. 57–66, New York, NY, USA (2001)
5. Watts, D., Dodds, P.: Influentials, networks, and public opinion formation. J. Consum. Res. **34**(4), 441–458 (2007)
6. Ouimet, G.: Pour une psychologie du changement L'incontournable décodage de la culture. Direction de la recherche, Editor (HEC Montréal), Canada (2005)
7. Borgatti, P.: Identifying sets of key players in a network. Comput. Math. Organ. Theor. **12** (1), 21–34 (2006)
8. Lindquist, M.J., Zenou, Y.: Key players in co-offending networks. IZA, Margard Ody, P.O. Box 7240, D-53072 Bonn, Germany (2014)
9. Arulselvan, A., Commander, C.W., Pardalos, P.M., Shylo, O.: Managing network risk via critical node identification. In: Gulpinar, N., Rustem, B. (eds.) Risk Management in Telecommunication Networks 2009. Springer, Heidelberg (2009)
10. Qi, X., Fuller, E., Wu, Q., Wu, Y., Zhang, C.Q.: Laplacian centrality: a new centrality measure for weighted networks. Inf. Sci. **194**, 240–253 (2012)
11. Opsahl, T., Agneessensb, F., Skvoretz, J.: Node centrality in weighted networks: generalizing degree and shortest paths. Soc. Netw. **32**(3), 245–251 (2010)
12. Avrachenkov, K.E., Mazalov, V.V., Tsynguev, B.T.: Beta current flow centrality for weighted networks. In: Proceedings of the 4th International Conference (CSoNet 2015), Beijing, China (2015)

13. Justification and Application of Eigenvector Centrality. https://www.math.washington.edu/~morrow/336_11/papers/leo.pdf. Accessed 24 Jul 2018
14. Latora, V., Marchiori, M.: How the science of complex networks can help developing strategies against terrorism. Chaos Solitons Fractals **20**(1), 69–75 (2004)
15. Eugene, K.Y., Alex, N.C.L., Alvin, W.S.: Characterizing terrorist networks using efficiency method. USC3001 Term Paper, 2005–2006
16. Arroyo, D.O.: Discovering sets of key players in social networks. In: Computational Social Network Analysis, pp. 27–47. Springer, Heidelberg (2010)
17. Tarjan, R.: Depth first search and linear graph algorithms. SIAM J. Comput. **1**(2), 146–160 (1972)
18. Everton, S.F.: Network topography, key players and terrorist networks. Connect. J. **32**(1), 12–19 (2012)
19. Java Agent Development Framework «JADE». http://jade.tilab.com/. Accessed 22 Feb 2016
20. Victor, P., Cornelis, C., De Cock, M., Teredesai, A.M.: Key figure impact in trust-enhanced recommender systems. AI Commun. **21**, 127–143 (2008)
21. OpinRank Dataset. http://kavita-ganesan.com/entity-ranking-data/#.WrKVgOzwbIV. Accessed 21 Mar 2018

Temperature Sensor Faults Monitoring in a Heat Exchanger Using Evolving Fuzzy Classification

Meryem Mouhoun(✉) and Hacene Habbi

Applied Automation Laboratory, M'hamed Bougara University of Boumerdès, Av. de l'indépendance, Boumerdès, Algeria
meriem-mouhoun@hotmail.fr, habbi_hacene@hotmail.com

Abstract. In this paper, an advanced evolving clustering strategy is used to design a fuzzy model-based sensor faults detection mechanism for a pilot parallel-type heat exchanger. The change in the process operating mode is detected by an incremental unsupervised clustering procedure based on participatory learning. Real experimental data is used to construct signals for fuzzy residual generation. The resulting residuals are then processed by the evolving classifier to supervise the heat exchanger operation. The obtained results clearly show the ability of the evolving fuzzy classifier to early detect the considered temperature sensor faults.

Keywords: Fault detection · Evolving clustering · Fuzzy modeling
Heat exchanger

1 Introduction

Fault detection and diagnosis (FDD) is an essential task in process industry. The early detection of abnormal operating situations is of major importance to keep systems safe and working as smoothly as possible. The basics of FDD schemes are the methods of detecting faults and those leading to identify their causes. For instance, some applications of FDD strategies in process industry use temperature and/or pressure measurements at various locations in a system to detect and diagnose common faults [1–3]. Sensors might put in evidence every change in process dynamics provided that they keep reliable, which cannot be entirely guaranteed in faulty situations [4, 5]. There are several techniques that have been developed for fault detection and diagnosis in industrial processes. Their use depends on the type of system to which they are applied and the available degree of knowledge about it. Methods used in that field of interest are mostly model-based or data-based. Model-based FDD performance is closely related to the quality of the available process model [6, 7]. Data-based FDD approach is built upon available data obtained directly from the monitored process. In general, there is no need to a priori knowledge about the monitored process when data-based FDD mechanisms are used [8–10].

O. Demigha et al. (Eds.): CSA 2018, LNNS 50, pp. 311–321, 2019.
https://doi.org/10.1007/978-3-319-98352-3_33

In recent literature, evolving clustering approaches have been introduced and applied to different research fields such as fault diagnosis [10, 16, 20–22], and process identification [23, 24]. Learning structures and models from a stream of data using recursive clustering algorithms represent the main feature of the evolving clustering based methods. The proposed approaches showed different results with promising performances, but also introduced compromises that might be processed depending on the application study.

Heat exchangers are important devices in industry. In thermal plants, they are used for cooling and heating which might involve many complex heat transfer phenomena. The nonlinearity of the process and its slow transient response are the most desirable characteristics for testing an advanced fault detection mechanism [11–13, 25]. The present work addresses the problem of designing a fault detection system for a parallel flow heat exchanger based on evolving fuzzy classifier with particular emphasis on temperature sensor faults. The fault detection design procedure is based on a fuzzy model-based residual generator (FRG) and a Gaussian evolving clustering-based (GEC) classifier. Experimental data collected from the real heat exchanger is used in the design.

This paper is organized as follows. A brief review of Gaussian evolving clustering (GEC) approach is first given in Sect. 2. Section 3 describes the heat exchanger process under study and the fuzzy modeling procedure. Section 4 presents the designed fuzzy residual generator used for temperature sensor faults detection. Detailed sensor faults detection results obtained through evolving GEC-based data classification is provided in Sect. 5. Then conclusions are formulated in Sect. 6.

2 Gaussian Evolving Clustering Approach

The Gaussian evolving (GEC) method is based on the concept of participatory learning [14], which has been developed to determine the expected number of clusters and the corresponding cluster center in data stream. This principle was used at first in clustering [15].

GEC-based method is capable to process and learn from stream of data which gives the algorithm greater robustness in the presence of outliers and noise. Data processing is performed at each new observation, new cluster must be created, an old cluster should be modified, or redundant cluster must be eliminated. In this Section, the formulation of Gaussian evolving clustering method is briefly discussed according to the concept studied in [9, 16, 17].

In GEC algorithm, each cluster is represented by a multivariable Gaussian distribution. Mahalanobis distance measure is employed to define elliptical cluster shape as follows:

$$D(x^k, v_i^k) = (x^k - v_i^k)(\Sigma_i^k)^{-1}(x^k - v_i^k)^T \qquad (1)$$

where x is an $1 \times m$ input vector, v is an $1 \times m$ center vector, and Σ is an $m \times m$ symmetric positive-definite matrix which represents the dispersion matrix.

The GEC algorithm makes use of an updating cluster structure mechanism which is based on compatibility measure. This index includes some measure of similarity between input data samples and existing clusters.

The compatibility measure, denoted as $\rho_i^k \in [0, 1]$, is calculated at each step k by:

$$\rho_i^k = F(x^k, v_i^k) = \exp\left[-\frac{1}{2}D(x^k, v_i^k)\right] \tag{2}$$

The compatibility measure indicates how an observation is compatible with the current cluster structure. The observation with $\rho_i^k = 1$ becomes complete, equal to a cluster center, and $\rho_i^k = 0$ provides no new information. The threshold value for the compatibility measure can be determined as follows:

$$T_\rho = \exp\left[-\frac{1}{2}X_{m,\lambda}^2\right] \tag{3}$$

where $X_{m,\lambda}^2$ is the value of Chi-Square distribution, λ is an upper unilateral confidence interval, and m is the degree of freedom which depends on the input space dimension.

The arousal index is the output of an arousal mechanism used on a revision of the dynamics compatibility measure. Here, it is denoted as $a_i^k \in [0, 1]$, and is evaluated for each observation based on the occurrence value O_i^k which can be found using the following expression:

$$O_i^k = \begin{cases} 0, & \textit{for}\, D(x^k, v_i^k) < X_{m,\lambda}^2 \\ 1, & \textit{otherwise} \end{cases} \tag{4}$$

The occurrence value O_i^k indicates the violation of the arousal threshold. Let nv_i^k denotes the number of violations of arousal threshold in a sequence w which slides along k, then:

$$nv_i^k = \begin{cases} \sum_{j=0}^{w-1} O_i^{k-j} & k > w \\ 0 & \text{otherwise} \end{cases} \tag{5}$$

The probability of getting from 0 to w violations corresponds to the cumulative sum of these evaluated probabilities which is computed by using a binomial calculator. The discrete probability distribution of observing nv_i^k threshold violations on a window of size w is denoted by $P(NV_i^k = nv)$ with $nv = 1, \ldots., w$. The binomial probability is given by:

$$P(NV_i^k = nv) = \begin{cases} \binom{w}{vv} \lambda^{nv}(1-\lambda)^{w-nv}, & nv = 1, \ldots., w \\ 0 & \text{otherwise} \end{cases} \tag{6}$$

The binomial distribution gives the probability of observing nv threshold violations in a sequence of w observations. The cumulative probability distribution function of sequence defined by Binomial distribution with probability of success λ is used to compute the compatibility threshold, which is expressed by:

$$a_i^k = p(NV_i^k < nv) \tag{7}$$

The shape of clusters will change at each step k, the dispersion matrix of each cluster Σ_i^k is estimated at each step as:

$$\Sigma_i^{k+1} = \left(1 - G_i^k\right)\left(\Sigma_i^k - G_i^k\left(x^k - v_i^k\right)\left(x^k - v_i^k\right)^T\right) \tag{8}$$

where G_i^k is defined by:

$$G_i^k = \beta\left(p_i^k\right)^{1-a_i^k} \tag{9}$$

and β is the basic learning rate which is usually set small to avoid undesirable far cluster centers. The threshold value of the arousal index is set as:

$$T_a = 1 - \lambda \tag{10}$$

the λ upper unilateral is same as in threshold value definition used for the compatibility measure.

Based on the above description, the cluster structure is updated using a compatibility measure and an arousal index. If the compatibility measure of current observation is less than threshold for all clusters i.e. $p_i^k < T_p \forall i = 1, \ldots, c_k$, and the arousal index of the cluster with higher compatibility is greater than the threshold i.e. $a_i^k > T_a$ for $i = \arg\max_i p_i^k$, then a new cluster is created, otherwise, the cluster center with the highest compatibility is adjusted as follows:

$$v_i^{k+1} = v_i^k + G_i^k\left(x^k - v_i^k\right) \tag{11}$$

Then, GEC algorithm proceeds by revising the current cluster structure by merging every two similar clusters. The distance between the updated or created clusters and all remaining clusters centers is computed at each step by:

$$D\left(v_j^k, v_i^k\right) = \left(v_j^k - v_i^k\right)\left(\Sigma_i^k\right)^{-1}\left(v_j^k - v_i^k\right)^T \tag{12}$$

$$D\left(v_i^k, v_j^k\right) = \left(v_i^k - v_j^k\right)\left(\Sigma_j^k\right)^{-1}\left(v_i^k - v_j^k\right)^T \tag{13}$$

The compatibility of two clusters i and j are given by $p_i^k\left(v_j^k, v_i^k\right)$ and $p_j^k\left(v_i^k, v_j^k\right)$ since we usually have $\left(\Sigma_i^k\right) \neq \left(\Sigma_j^k\right)$. Those quantities are evaluated as follows:

$$p_i^k\left(v_j^k, v_i^k\right) = \exp\left[-\frac{1}{2}D\left(v_j^k, v_i^k\right)\right] \tag{14}$$

$$p_j^k\left(v_j^k, v_i^k\right) = \exp\left[-\frac{1}{2}D\left(v_i^k, v_j^k\right)\right] \tag{15}$$

The clusters i and j are merged if one of the following conditions is met:

$$p_i^k\left(v_j^k, v_i^k\right) > T_p \tag{16}$$

$$p_j^k\left(v_i^k, v_j^k\right) > T_p \tag{17}$$

It must be noted that this algorithm has essentially three control parameters: the learning rate β which is used to update the center of cluster v_i^k and its dispersion matrix Σ_i^k which is taken in the interval $[0.08, 0.2] * I_2$; the confidence level λ which is used to define thresholds of compatibility T_p and of arousal T_a; and the windows size w which is required to define the number of samples to be used to compute the arousal index.

3 Heat Exchanger Process Description and Modeling

The thermal process under study is a parallel-type gas-liquid heat exchanger with the set of components depicted in Fig. 1. There are three subsystems in the process configuration: an electric heater of air (E), and two circuits for air and water circulation. The hot air coming from the electric heater with temperature T14 enters the heat exchanger with the temperature T16 after flowing through the air circulation pipe. Motor-driven valves are installed to control the air evacuation and recycling while a pump of variable speed controls the water flow in the system. The water entering the heat exchanger with the temperature T33 is heated up to the temperature T34 by thermal energy transfer from hot air. The main process measurable variables are: the heating power P [kW]; the air temperature after the heater T14 [°C] and before the heat exchanger T16 [°C]; the water temperature at the inlet and the outlet of the exchanger T33 and T34 [°C], respectively; the position of the air recycling valve Vr [%] and the air evacuation valve Ve [%]; the air flow rate Qa [m^3/s], and the water flow rate Q [m^3/s].

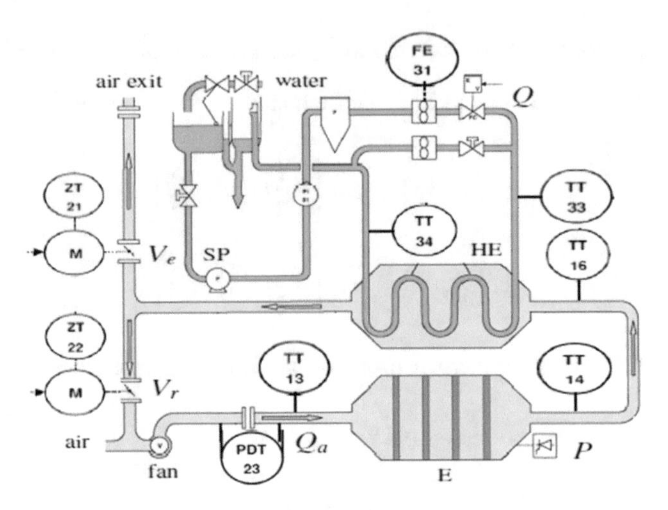

Fig. 1. Schematic of the heat exchange process [18].

The fault detection mechanism presented in this work is based on a data-driven multivariable fuzzy model of the pilot heat exchanger which was developed in [26]. The model is built by means of fuzzy identification from experimental data. With regard to practical considerations, it was found that the normal input-output behaviors of the water and air circuits depend mainly on T34, T16, T14, P and Vr. Therefore, a supervision scheme involving these five measurements is specified based on the following nonlinear model structure:

$$T_{34}(k+1) = \psi_1(T_{34}(k), T_{16}(k), P(k), Vr(k)) \tag{18}$$

$$T_{16}(k+1) = \psi_2(T_{16}(k), T_{14}(k), P(k), Vr(k)) \tag{19}$$

where ψ_1 and ψ_2 are unknown nonlinear functions which are approximated by means of Takagi-Sugeno-type (TS) fuzzy modeling. For process modeling, real data from the heat exchanger is generated in fault-free mode [19, 26]. The power P is manipulated over its whole operating domain from 0 to 10 kW, while the air valves Vr and Ve are varied in the range [0–100%]. For wide-range operation, dynamic excitations in P, Vr and Ve are applied by using an amplitude-modulated pseudo-random binary signals. The obtained T34 and T16 TS-type fuzzy prediction models of the heat exchanger are detailed in [26], where a complete study on modelling reliability and performance is provided. Those models will be used as a basement for the design of the temperature sensor fault detection system proposed in the present contribution.

4 Fuzzy Residual Generation

Residual signals generation is an important step in the fault detection design procedure. For temperature sensor fault detection, residuals are generated from the comparison of measured process output temperatures and estimated temperatures provided by the water and air temperature fuzzy prediction models (18) and (19). The considered faults are bias of sensor for T_{34} measurement, denoted by $\mathcal{F}_{T_{34}}$, and bias of sensor for T_{16} measurement, denoted by $\mathcal{F}_{T_{16}}$. The recorded signals are denoted as $r_{T_{34}}$ and $r_{T_{16}}$ which correspond to differences between the real process outputs $y(k) = [T_{34}(k)T_{16}(k)]^T$ and the fuzzy temperature prediction model outputs $\widehat{y}(k) = \left[\widehat{T}_{34}(k)\widehat{T}_{16}(k)\right]^T$. Hence, the residual vector is computed within a time window as follows [19]:

$$r(k) = y(k) - \widehat{y}(k) \tag{20}$$

The temperature sensor faults are considered by adding offsets with magnitude of 4 °C to the measured values of water temperature T_{34} and to the measured values of the air temperature T_{16} at different time instants. The resulting data stream to be processed by the GEC classifier for fault diagnosis purpose corresponds to temperature residual data which are illustrated in the following Section.

5 GEC-Based Fault Diagnosis Results

To assess the performance of the GEC-based classifier, simulations are performed using the real data collected from the heat exchanger process of Fig. 1. As mentioned above, the manipulated data correspond to residual signals provided by the air and water temperature sensors. The data sets are composed of 2000 data samples of residual signal $r_{T_{34}}$ and $r_{T_{16}}$ as shown in Figs. 2 and 5. At the starting phase, the heat exchanger is working under normal operation. Faults on either water temperature sensor or air temperature sensor are applied at different instants. The impact of inducing faults on each

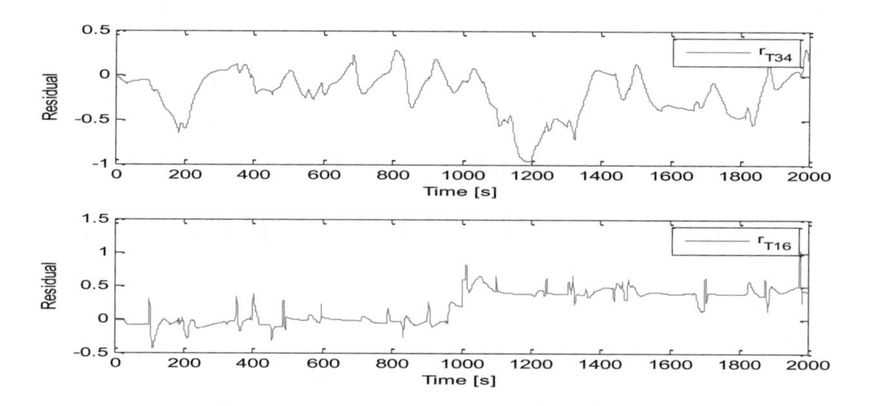

Fig. 2. Residual signals associated to air temperature sensor fault.

recorded residual signal can be clearly seen from the plots. The fault scenario was simulated, the algorithm GEC starts by associating the center of the first cluster to the initial values of the residuals signals. The corresponding dispersion matrix is chosen as $0.041 * I_2$, the parameters of the classifier are set to $\lambda = 0.01$, $X^2_{m,\lambda} = 9.210$, $\omega = 10$, and $\beta = 0.01$.

Figures 3 and 4 show the compatibility index and the arousal index obtained in this case study. From this plots, it is clear that, in the presence of one cluster, the compatibility index is staying greater than the threshold T_r and the arousal index less than the threshold T_a, At time instant k = 1012, the two predefined thresholds are violated and then a new cluster is created. The detected cluster centers are v = [−0.0751 0.7480]. The failure mode is detected after 24 s of its occurrence which is an acceptable result.

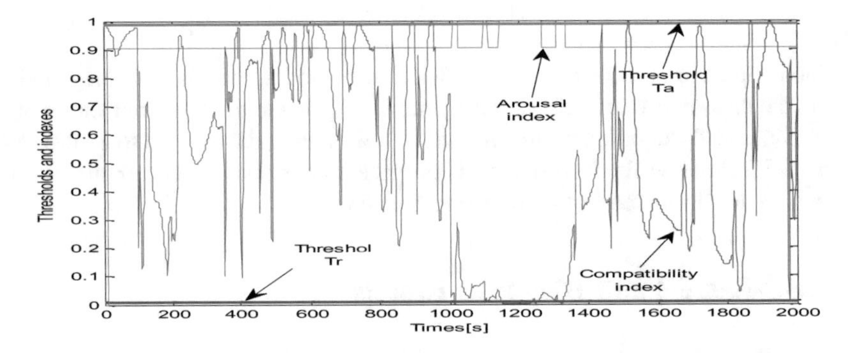

Fig. 3. Compatibility and arousal indexes before air temperature sensor fault occurrence.

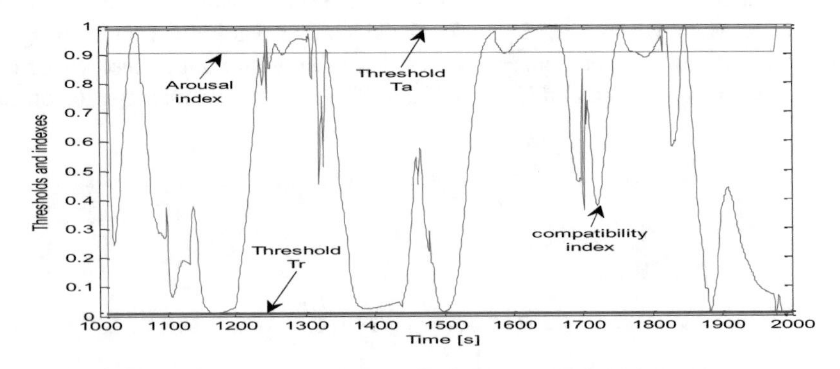

Fig. 4. Compatibility and arousal indexes after air temperature sensor fault occurrence.

Furthermore, to evaluate the feasibility of the GEC-based classifier, a single fault on water temperature sensor denoted as $\mathcal{F}_{T_{34}}$ is also considered. Figure 5 depicts the recorded residual signals for that case study. Based on the same concept of evaluating the compatibility and arousal indexes, Fig. 6 illustrates the resulting plots after fault occurrence. As can be seen, at time instant k = 1500, a faulty situation is introduced. The bias of magnitude 4 °C clearly induced thresholds violation, which resulted in formation of a new cluster with center v = [0.7363 0.0682]. Evaluating the obtained results leads to a detection delay of 2 s which is very acceptable.

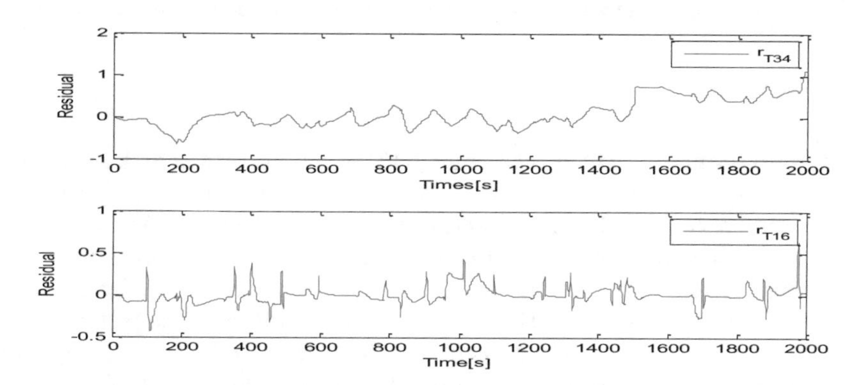

Fig. 5. Residual signals associated to water temperature sensor fault.

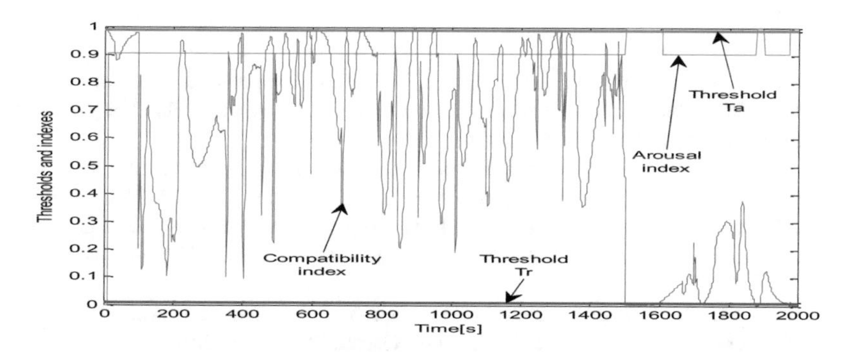

Fig. 6. Compatibility and arousal indexes after water temperature sensor fault occurrence.

As can be noticed, the obtained results show satisfactory performance of the GEC-based approach to sensor fault detection in the studied heat exchange process. From feasibility viewpoint, the evolving clustering based fault detection approach presents interesting merits while compared to the cumulative sum (CUSUM) based decision logic studied in [18, 19]. Acceptability of the GEC-based results is made with respect to the same case studies described in [19], where it can be clearly seen that the fault in T16 sensor is detected with a delay of 24 s which is same as the CUSUM-based mechanism. In addition, the fault induced in T34 sensor is detected after 2 s from its effective occurrence time, which is better than the recorded detection delay of 4 s obtained by using the CUSUM logic. Here, it should be noticed that the detection scheme did not induce any false alarms. This, of course, represents a promising result which might lead to further improvements when extended to other fault case studies tackled in [19]. Obviously, the procedure of applying the evolving fuzzy classifier to the real heat exchanger data requires specific assessment of the monitored thermal plant to make the fault detection design efficient and robust. In addition, the setting of the control parameters has an important impact on the results. This should be handled through different tests with taking into account the specifications of the monitored process.

6 Conclusion

In this paper, Gaussian evolving clustering (GEC) algorithm is used to develop a fault detection scheme for sensor faults in a parallel heat exchanger process. Experimental data collected from the real process is employed to design a GEC-based fuzzy classifier which serves as a basement to detect and evaluate faults present in the system. This data-based method is able to ensure continuous monitoring of process variables through the supervision of data clusters. The obtained results clearly show the performance of The GEC-based classifier and its feasibility for fault detection in heat exchanger processes. Future works might focus on detailed analysis of the impact of the control parameters on the fault detection system with considering other types of faults.

References

1. Costa, B.S.J., Angelov, P.P., Guedes, L.A.: Fully unsupervised fault detection and identification based on recursive density estimation and self-evolving cloud-based classifier. Neurocomputing **150**, 289–303 (2015)
2. Ma, H., Hu, Y., Shi, H.: Fault detection and identification based on the neighborhood standardized local outlier factor method. Ind. Eng. Chem. Res. **52**(6), 2389–2402 (2013)
3. Bezerra, C.G., Costa, B.S.J., Guedes, L.A., Angelov, P.P.: An evolving approach to unsupervised and real-time fault detection in industrial processes. Expert Syst. Appl. **63**, 134–144 (2016)
4. Yan, R., Ma, Z., Kokogiannakis, G., Zhao, Y.: A sensor fault detection strategy for air handling units using cluster analysis. Autom. Constr. **70**, 77–88 (2016)
5. Du, Z., Fan, B., Chi, J., Jin, X.: Sensor fault detection and its efficiency analysis in air handling unit using the combined neural networks. Energy Build. **72**, 157–166 (2014)
6. Isermann, R.: Model-based fault-detection and diagnosis–status and applications. Annu. Rev. Control **29**(1), 71–85 (2005)
7. Beghi, L.A., Cecchinato, F., Peterle, M., Rampazzo, F., Simmini, F.: Model-based fault detection and diagnosis for centrifugal chillers. In: 3rd Conference on Control and Fault-Tolerant Systems (SysTol), pp. 158–163. IEEE (2016)
8. Filev, D.P., Chinnam, R.B., Tseng, F., Baruah, P.: An industrial strength novelty detection framework for autonomous equipment monitoring and diagnostics. IEEE Trans. Ind. Inform. **6**(4), 767–779 (2010)
9. Lemos, A., Caminhas, W., Gomide, F.: Adaptive fault detection and diagnosis using an evolving fuzzy classifier. Inf. Sci. **220**, 64–85 (2013)
10. Inacio, M., Lemos, A., Caminhas, W.: Fault diagnosis with evolving fuzzy classifier based on clustering algorithm and drift detection. In: Mathematical Problems in Engineering (2015)

11. Dragan, D.: Fault detection of an industrial heat-exchanger: a model-based approach. Strojniški vestnik-J. Mech. Eng. **57**(6), 477–484 (2011)
12. Tudón-Martínez, J.C., Morales-Menendez, R., Garza-Castañón, L.E.: Fault diagnosis in a heat exchanger using process history based-methods. Comput. Aided Chem. Eng. **28**, 169–174 (2010)
13. Habbi, H., Kidouche, M., Kouadri, A., Zelmat, M.: Design and real-time implementation of a fuzzy residual generator for process fault detection in a co-current heat exchanger. In: AIP Conference Proceedings, vol. 1107, no. 1, pp. 91–95. AIP (2009)
14. Yager, R.R.: A model of participatory learning. IEEE Trans. Syst. Man Cybern. **20**(5), 1229–1234 (1990)
15. Silva, L., Gomide, F., Yager, R.: Participatory learning in fuzzy clustering. In: 14th IEEE International Conference on Fuzzy Systems, FUZZ 2005, pp. 857–861. IEEE (2005)
16. Lemos, A.P., Caminhas, W.M., Gomide, F.A.: Fuzzy multivariable gaussian evolving approach for fault detection and diagnosis. In: IPMU, pp. 360–369 (2010)
17. Lemos, A., Caminhas, W., Gomide, F.: Multivariable Gaussian evolving fuzzy modeling system. IEEE Trans. Fuzzy Syst. **19**(1), 91–104 (2011)
18. Habbi, H., Kidouche, M., Zelmat, M.: Nonlinear identification and fault diagnosis using multiple-model approach. In: First International IMPACT 2010 Conference on Dynamics of Systems, Material and Structures, Djerba. Tumisia (2010)
19. Habbi, H., Kidouche, M., Kinnaert, M., Zelmat, M.: Fuzzy model-based fault detection and diagnosis for a pilot heat exchanger. Int. J. Syst. Sci. **42**(4), 587–599 (2011)
20. Inacio, M., Lemos, A., Caminhas, W.: Evolving fuzzy classifier based on clustering algorithm and drift detection for fault diagnosis applications. In: Annual Conference of the Prognostics of Prognostics and Health Management Society (2014)
21. Jianu, O., Wang, W.: A self-evolving fuzzy classifier for gear fault diagnosis. Int. J. Mech. **14**(05), 90–96 (2014)
22. Alippi, C., Roveri, M., Trovò, F.: A self-building and cluster-based cognitive fault diagnosis system for sensor networks. IEEE Trans. Neural Netw. Learn. Syst. **25**(6), 1021–1032 (2014)
23. Rocha Filho, O.D., Oliveira Serra, G.L.: Evolving fuzzy clustering algorithm based on maximum likelihood with participatory learning. In: IEEE Conference Evolving and Adaptive Intelligent Systems (EAIS), pp. 65–72. IEEE (2016)
24. Maciel, L., Gomide, F., Ballini, R.: MIMO evolving participatory learning fuzzy modeling. In: IEEE International Conference on Fuzzy Systems, (FUZZ-IEEE), pp. 1–8. IEEE (2012)
25. Zhu, X., Shu, L., Zhang, H., Zheng, A., Han, G.: Preliminary exploration: fault diagnosis of the circulating-water heat exchangers based on sound sensor and non-destructive testing technique. In: 8th International Conference on Communications and Networking, (CHINACOM), pp. 488–492. IEEE, China (2013)
26. Habbi, H., Kidouche, M., Zelmat, M.: Data-driven fuzzy models for nonlinear identification of a complex heat exchanger. Appl. Math. Model. **35**(3), 1470–1482 (2011)

A Novel Artificial Bee Colony Learning System for Data Classification

Fatima Harfouchi(✉) and Hacene Habbi

Applied Automation Laboratory, M'hamed Bougara University of Boumerdès, Av. de l'indépendance, Boumerdès, Algeria
harfouchi.fatima03@hotmail.com, habbi_hacene@hotmail.com

Abstract. Training artificial neural networks (ANNs) is a common hard optimization problem. The process of neural nets training is generally defined on synaptic weights and thresholds of artificial neurons with the aim to find optimal or near-optimal values. Artificial bee colony (ABC) optimization has been successfully applied to several optimization problems, including the optimization of weights and biases of ANNs. This paper addresses the problem of feed-forward ANNs training by using a novel ABC variant named cooperative learning artificial bee colony algorithm (CLABC), which we have developed in our previous work. The performance of the CLABC-trained feed-forward ANN is validated on different classification problems, namely the XOR problem, the 3-bit parity, 4-bit encoder-decoder and Iris benchmark problems. The results are compared to other advanced optimization methods.

Keywords: Neural networks · Training · Artificial bee colony algorithm (ABC) Swarm intelligence · Cooperative learning

1 Introduction

Artificial neural networks (ANNs) have been used in several scientific and engineering problems as universal approximators [1], including linear and nonlinear modeling, prediction, image processing, classification and combinatorial optimization problems. However, finding systematically an optimal neural network structure with suitable weights and thresholds remains a challenging issue. The ANN based strategies provide different ways to solve complex problems that are hard to solve by conventional techniques. Good solutions can be achieved, but again this depends on the adopted training strategy. Many algorithms are proposed for the optimization and improvement of ANNs learning. They are generally classified into local optimization and global optimization algorithms [2]. Local minimization algorithms, such as the gradient descent, are fast, but converge for local minima. On the other hand, global minimization algorithms use heuristic strategies to escape local minima [3].

To overcome some of these limitations, population based optimization techniques have been investigated. Let us mention for instance particle swarm optimization (PSO) [4], genetic algorithms (GA) [5], firefly algorithm [6], differential evolution (DE) [7] and artificial bee colony (ABC) algorithm [8, 9].

O. Demigha et al. (Eds.): CSA 2018, LNNS 50, pp. 322–331, 2019.
https://doi.org/10.1007/978-3-319-98352-3_34

The ABC algorithm is one of the most popular swarm algorithms which was introduced by Karaboga in 2005. Since ABC is simple in concept and easy to implement, it has rapidly gained the attention of researchers and applied to a variety of scientific and engineering problems in several fields [8–16]. The basic ABC model was subjected to many improvements to enhance its computational performance and robustness [17–19]. On the particular issue of ANNs design, let us mention for instance the study proposed in [8], where ABC was used to train feed-forward ANNs as an alternative global search approach to back-propagation algorithms. Compared results with gradient-descent based back-propagation (GD-BP) algorithm, Levenberg-Marquardt based back-propagation (LM-BP) algorithm and the global search GA algorithm have shown clear superiority of ABC model. Improving ABC-based ANNs training is still devoting substantial interest in the literature. Enhancements are particularly introduced in order to deal with the challenging issues of exploration and exploitation which have considerable impact on ABC convergence.

The cooperative learning artificial bee colony (CLABC) algorithm which has been recently developed in our research study is a novel ABC variant which relies on the concept of cooperative foraging [20]. An extensive comparative study on low, medium and high-dimensional numerical benchmark problems including several ABC variants and advanced metaheuristics have been conducted to show the feasibility and the performance of the proposed CLABC model. To further assess the performance of the novel method, applications to engineering problems are currently investigated. The present work addresses in fact the problem of data classification using ANNs. As will be shown, the problem is formulated as a search optimization problem defined on the weights and thresholds of artificial neurons. Structural and parameter settings are set to allow the adaptation of the algorithmic structure of CLABC to the training problem.

The paper is organized as follows. In Sect. 2 a brief description of ANNs basics is recalled. Section 3 reviews the artificial bee colony algorithm. The cooperative learning ABC algorithm is described in Sect. 4. Section 5 shows the numerical experiments and discusses the compared results. Conclusions are formulated in Sect. 6.

2 Basics of Neural Systems

Artificial neural network is an information processing model of a human brain. The basic structure is depicted in Fig. 1. A three-layer feed-forward neural network is considered which consists of two fully connected layers of neurons: one hidden layer and one output layer. Thresholds of nodes, the weight values between input nodes, hidden nodes and output nodes are randomly initialized. For example, the network's structure labeled as n-p-m, stands for a neural net with n input nodes, p hidden nodes and m output nodes. The output of the ith is given by:

$$y_i = f_i\left(\sum\nolimits_{j=1}^{n} w_{ij}x_j + b_i\right) \tag{1}$$

where y_i is the output of the node, x_j is the jth input to the node, w_{ij} is the connection weight between the node and input x_j, b_i is the threshold (or bias) of the node, and f_i is the node transfer function. The transfer function is usually chosen as a nonlinear function (Heaviside, sigmoid, Gaussian, etc.).

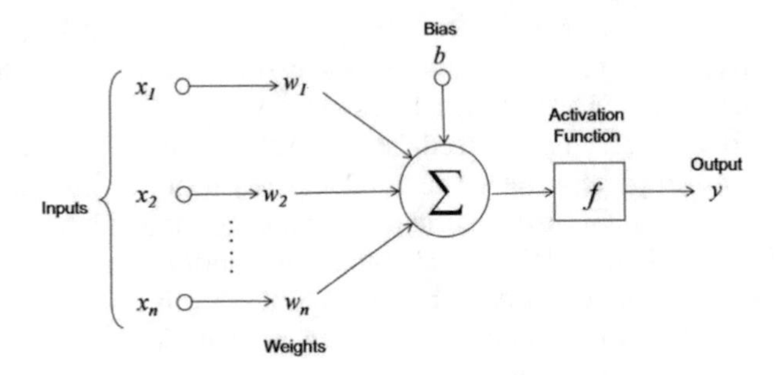

Fig. 1. Processing unit of an artificial neuron

The sigmoid function is defined as:

$$y = f(net) = \frac{1}{1 + e^{-net}} \tag{2}$$

The process of ANN training is formulated as a nonlinear optimization problem on neurons synaptic weights and thresholds. In supervised learning, the objective function stands for the error between the actual output and the desired one. In this paper, the mean square error (MSE) is taken as objective function to minimize. The training MSE is expressed by:

$$E(w(t), b(t)) = \frac{1}{N} \sum\nolimits_{i=1}^{m} \sum\nolimits_{j=1}^{N} \left(y_i - o_i\right)^2 \tag{3}$$

where $E(w(t), b(t))$ is the error at the tth iteration, $w(t)$ and $b(t)$ are the weights and thresholds associated to the connections at the tth iteration, y_i and o_i are the actual and desired outputs of the ith output node, and N is the number of patterns.

3 Artificial Bee Colony Optimization

The artificial bee colony (ABC) algorithm is a swarm based meta-heuristic algorithm which imitates the foraging behavior of honey bees [21]. In ABC concept [22], the honey bee colony model employs three types of bees: employed bees, onlooker bees and scouts. An employed bee searches around the current food source to find a new source position with better nectar amount. If the nectar amount of the discovered position is higher than that of the previous one, the bee saves the new position in her memory and forgets the old one. Each employed bee is associated with a food source, in other words, the number of employed bees is equal to the number of food sources. After all employed bees complete their search processes, they share their information with onlooker bees which are waiting in the hive. An onlooker bee chooses a food source according to a probabilistic greedy selection mechanism. Therefore, food sources with better profitability will get higher probability to be selected by the onlookers. Similar to employed bees, each onlooker bee produces a modification on the position in her memory and investigates the nectar amount of the generated candidate source. If a position cannot be improved further through a predetermined limit tolerance called "limit", then that food source is assumed to be abandoned. The corresponding employed bee becomes then a scout. The abandoned position will be replaced with a new food source found by the scout.

In the initialization phase, the algorithm generates randomly a population of SN food sources (solutions) in the admissible search domain as follows:

$$x_i^j = x_{\min}^j + rand(0,1)\left(x_{\max}^j - x_{\min}^j\right) \tag{4}$$

where $i = 1, 2, \ldots, SN$, $j = 1, 2, \ldots, D$, and D is the number of optimization parameters; $x_{\min}^j$ and $x_{\max}^j$ are lower and upper bounds for the dimension j, respectively. After initialization, the population of food sources is subjected to repeated cycles of the search processes of employed bees, onlooker bees and scout bees.

In the employed bees' phase, each employed bee generates a new candidate solution around its current position by the following equation:

$$v_i^j = x_i^j + \phi_i^j\left(x_i^j - x_k^j\right) \tag{5}$$

where $i = 1, 2, \ldots, SN$, $j \in \{1, 2, \ldots, D\}$ and $k \in \{1, 2, \ldots, SN\}$ are randomly chosen indexes; k has to be different from i, and ϕ_i^j is a random number in the range $[-1, 1]$. Greedy selection between the old and the updated food source position is performed by the employed bee based on fitness value evaluation. This valuable information about the position and the quality of the food sources are shared with the onlooker bees.

In the onlooker bees' phase, the onlookers receive the information about the food sources shared by employed bees and choose a food source to exploit depending on a probability related to its nectar value which is computed by:

$$p_i = \frac{fit_i}{\sum_{n=1}^{SN} fit_n} \tag{6}$$

where fit_i is the fitness value of the solution i. It should be noted that onlooker bees also use Eq. (5) to generate new candidate solutions.

In the scout bees' phase, the ABC algorithm deals with food sources abandonment. Any solution that cannot be improved through a predefined number of generations will be abandoned and replaced by a new position that is randomly determined by a scout bee by using Eq. (4).

4 Cooperative Learning Artificial Bee Colony Optimization with Multiple Search Strategy

Relying on the idea that employed foragers or onlooker foragers of a bee swarm might have to evolve in separate groups during the search process, a novel ABC variant, referred to as CLABC, is introduced in our recent research work [20]. The search strategy in the proposed CLABC algorithm is based on multiple solution search equations which are integrated in different ways at the employed and onlooker stages. In the present study, the employed bees are divided into 3 groups; each group forms a sub-population of potential solutions and uses search equations different from the other groups. The onlookers are arranged in one group. The solution updating equations are given as follows [20]:

$$v_i^j = x_i^j + \phi_i^j (x_i^j - x_k^j) \tag{7}$$

$$v_i^j = x_k^j + \phi_i^j (x_i^j - x_{k1}^j) \tag{8}$$

$$v_i^j = x_i^j + r_i^j (x_i^j - x_{mean}^j) + r1_i^j (x_k^j - x_{k1}^j) \tag{9}$$

$$v_i^j = x_i^j + r_i^j (x_i^j - x_{mean}^j) + r1_i^j (x_k^j - x_{k1}^j) \tag{10}$$

$$v_i^j = x_k^j + \phi_i^j (x_{k1}^j + x_{k2}^j) \tag{11}$$

$$v_i^j = x_{best}^j + \phi_i^j (x_k^j - x_{k1}^j) \tag{12}$$

where $\phi_i^j \in [-1, 1], r_i^j, r1_i^j \in [0, 1]$ are random numbers generated for each dimension at each processing time. $x_k^j, x_{k1}^j, x_{k2}^j$, are the jth dimensions of solutions selected randomly from the sub-populations. k, k_1, k_2, are not equal to each other and to i. x_{best}^j denotes the jth dimension of the best solution obtained by the sub-population so far. x_{mean}^j is the average solution corresponding to the jth dimension for all solutions in the sub-populations. At employed phase, search Eqs. (7) and (8) are called by the first group of foragers to generate new candidate solutions. Only one of these two search equations is used to perform solution updating based on an increasing probability defined by $(iter/MCN)^{1/3}$, where $iter$ is the current generation number and MCN is the maximum number of cycles. If the random value between 0 and 1 is smaller than or equal to the predefined probability, then Eq. (7) is used. Otherwise, search mechanism described by Eq. (8) is performed. Based on the same definition of probability given above, (9) and (10) are used for the second group, and (11) and (12) for the last group.

Besides, in order to increase the convergence speed of the algorithm, employed bees are managed to start by changing one random dimension as basic ABC does. Then, when the number of iterations reaches $iter = MCN/4$, the algorithm will have to optimize randomly one or all parameters simultaneously. While reaching the end of the optimization process, i.e. between $iter = MCN/2$ and $iter = MCN$, the employed bees will update all parameters of the solution.

Subsequently, after completion of the employed bee phase and based on greedy selection, the onlookers will start to change one random parameter of the solution by using Eq. (8). When $iter = MCN/4$, the onlooker bees will have to update randomly all parameters of the solution by using Eq. (7) or only one random dimension through Eq. (8).

5 Numerical Experiments and Results

In order to illustrate the performance of the proposed strategy, a number of benchmark neural systems based classification problems are considered in the numerical experiments. More precisely, we emphasized to test the proposed training strategy on the XOR problem, the 3-bit parity problem, the 4-bit encoder-decoder problem and Iris classification problem. The objective function to be minimized is chosen as defined in Eq. 3. The considered test problems are described in the next subsection.

The ANN to be trained is a three layer feed-forward neural network with one hidden layer, one input layer and one output layer. In the predefined network structure, bias nodes are applied and sigmoid functions are selected as hidden neurons' activation functions.

A. **Benchmark Classification Problems**

(1) **The Exclusive-OR (XOR) Problem:** This is the first problem considered in the experiments. It is a non linear separable problem where Boolean functions that maps two binary inputs to a single binary output are defined. The input-output samples for this problem are: (0 0; 0 1; 1 0; 1 1) → (0; 1; 1; 0). In the simulations, we use a 2-2-1 feed-forward neural network with six connection weights and three biases which gives a total of 9 parameters (XOR9).

(2) **3-Bit Parity Problem:** The second problem to solve calculates the modulus 2 of summation of three inputs. It is the indicator function that returns 1, when the number of binary units is odd, and 0 otherwise. The input-output samples for 3-Bit parity problem are: (0 0 0; 0 0 1; 0 1 0; 0 1 1; 1 0 0; 1 0 1; 1 1 0; 1 1 1) → (0; 1; 1; 0; 1; 0; 0; 1). In the experiments, we use a 3-3-1 feed-forward neural network structure. It has twelve (12) connection weights and four (4) biases, i.e. sixteen parameters to optimize.

(3) **4-Bit Encoder/Decoder Problem:** This problem has four distinct input samples, each of them has only one bit tuned on. The output is a duplication of the inputs. The input-output samples for this problem are: (0 0 0 1; 0 0 1 0; 0 1 0 0; 1 0 0 0) → (0 0 0 1; 0 0 1 0; 0 1 0 0; 1 0 0 0). This is quite close to real world pattern classification problems, where small changes in the input samples cause small changes in the output samples. In the simulations, we use a 4-2-4 feed-forward neural network structure. It has eighteen connection weights and six biases, totally 22 parameters.

(4) **Iris Classification Problem:** The Iris dataset is indeed one of the most popular data used in classification problems [23]. It consists of 4 attributes, 150 input-output samples and 3 classes. In the experiments, we used 4-5-3 and 4-15-3 feed-forward neural network models with biases.

B. Parameter Settings

To evaluate the performance of the proposed CLABC-trained ANN, the obtained results are compared to gradient-descent based back-propagation (GD-BP) algorithm, Lenvenberg-Marquant based back-propagation (LM-BP) algorithm and other advanced methods, namely the Bat algorithm (BA), the Cuckoo search (CS) algorithm, the ABC algorithm and the modified Bat algorithm (MBA). For each benchmark problem, experiments were repeated 30 times and training processes were stopped when the mean squared errors become equal to or less than 0.01 ($MSE \leq 0.01$), or when the maximum cycle number (MCN) has been reached. The control parameters of GD-BP, LM-BP, BA, CS, ABC and MBA are taken same as in [24]. The control parameters of CLABC are chosen same as those used with ABC [24]. The colony size is set to 50 and the abandoned limit is chosen as $SN * D$.

The parameter ranges, the dimension of the problems (D), the maximum cycle number (MCN), epoch number and the network structures are presented in Table 1.

Table 1. Setting parameters of the benchmark problems

Problem	Range	NN structure	D	MCN	Epoch
XOR	[− 10, 10]	2-2-1 + Bias(3)	9	100	500
3-bit parity	[− 10, 10]	3-3-1 + Bias(4)	16	1000	1600
Enc. Dec.	[− 10, 10]	4-2-4 + Bias(6)	22	1000	2100
Iris	[− 10, 10]	4-5-3 + Bias(8)	43	500	–
		4-15-3 + Bias(18)	123	500	–

C. Results and Discussion

Comparative results of GD-BP, LM-BP, BA, CS, ABC, MBA and CLABC trained ANN for solving XOR, 3-bit parity, 4-bit encoder-decoder and Iris classification problems are given in Tables 2 and 3, respectively. The results are shown in terms of mean of mean squared errors (MMSE), standard deviation of mean squared errors (SDMSE), mean of epoch numbers (ME), mean of cycle numbers (MC), success rate (SR) and average classification percentage. The best results are marked in bold.

From Table 2, it can be concluded that CLABC achieves the best results among the compared algorithms. It gives the best mean of mean squared error (MMSE), and the fastest rate of convergence in all test cases. In addition, CLABC, CS, ABC and MBA algorithms achieve 100% success rate for all test problems. Superiority of CLABC over BA, CS and MBA on Iris data classification problems is also noticeable as can be seen from Table 3. The obtained results demonstrate clearly the goodness of the CLABC in those case studies. The performance of CLABC-trained feed-forward ANN is induced by the modifications introduced at the employed and onlooker stages leading to convenient balance between exploration and exploitation through different solution updating equations involved in the novel cooperative foraging strategy.

Table 2. Performance of CLABC-trained feed-forward ANNs on XOR9, 3-bit parity and 4-bit encoder/decoder benchmark classification problems

Algorithm		XOR9	3-bit parity	Enc./Dec.
GD-BP	MMSE(SDMSE)	0.212(0.0369)	0.2493(0.0025)	0.0809(0.0756)
	ME	500	1600	2020
	SR	0	0	2
LM-BP	MMSE(SDMSE)	0.0491(0.0646)	0.0209(0.043)	0.0243(0.0424)
	ME	13	21.333	83.3667
	SR	66.66	86.66	73.33
BA	MMSE	0.006275(0.00345)	0.006924(0.00219)	0.007601(0.00183)
	MC	60	325.1	300.1667
	SR	86.6667	100	100
CS	MMSE(SDMSE)	0.005913(0.00235)	0.006565(0.00212)	0.007285(0.00181)
	MC	30.96667	212.1	219.86667
	SR	100	100	100
ABC	MMSE(SDMSE)	0.006956(0.00240)	0.006679(0.00282)	0.008191(0.00186)
	MC	32	179.07	185
	SR	100	100	100
MBA	MMSE(SDMSE)	0.004656(0.00287)	0.005432(0.00299)	0.006864(0.00234)
	MC	**26.233**	137.4	184.37
	SR	100	100	100
CLABC	MMSE(SDMSE)	**5.60e−05(7.38e −06)**	**5.41e−05(6.78e −20)**	**1.53e−04(8.36e −10)**
	MC	28	**52**	**111**
	SR	**100**	**100**	**100**

Table 3. Performance of CLABC-trained feed-forward ANNs on Iris classification problems

Algorithm	Iris (4-5-3)			Iris (4-15-3)		
	MMSE	SDMSE	Average classification %	MMSE	SDMSE	Average classification %
BA	0.029381	0.007464	96.33333	0.030534	0.009083	91
CS	0.024712	0.010707	97.66667	0.026972	0.008258	97.33333
MBA	0.023088	0.006785	98.66667	0.026620	0.006345	98.66667
CLABC	**0.012432**	**0.003030**	**99.88889**	**0.011701**	**0.002927**	**99.88889**

6 Conclusion

Cooperative learning ABC (CLABC) algorithm is a newly proposed ABC variant which relies on the concept of cooperative foraging among bee swarms. The performance of the CLABC has been demonstrated in our recently published work on solving several benchmark numerical functions. In this paper, the novel CLABC is used to train feed-forward artificial neural networks. The algorithmic structure of the CLABC is managed to handle the problem of ANN training which is basically regarded as a hard optimization problem. To assess the performance of CLABC-based ANN design, benchmark classification problems are investigated, namely the XOR problem, the 3-bit parity problem, the 4-bit encoder-decoder problem and Iris classification problems. The numerical results show clearly the superiority of the CLABC-trained ANN system on solving the considered test problems in comparison with back-propagation algorithms and other advanced methods, namely BA, CS, ABC and MBA algorithms. Fast convergence and higher quality of solutions are the main features of the proposed ANN training strategy. Future works will focus on the application of the novel swarm optimization strategy to design intelligent systems including different types of ANNs.

References

1. Haykin, S.: Neural Networks and Learning Machines. Pearson, Upper Saddle River (2009)
2. Camargo, L.C., Tissot, H.C., Pozo, A.T.R.: Use of backpropagation and differential evolution algorithms to training MLPs. In: 31st International Conference of the Chilean Science Society (SCCC), pp. 78–86. IEEE (2012)
3. Shang, Y., Benjamin, W.: Global optimization for neural network training. Computer **29**, 45–54 (1996)
4. Mendes, R., Cortez, P., Rocha, M., Neves, J.: Particle swarms for feedforward neural network training. In: Proceedings of the International Joint Conference on Neural Networks, vol. 2, pp. 1895–1899. IEEE (2002)
5. Che, Z.G., Chiang, T.A., Che, Z.H.: Feed-forward neural networks training: a comparison between genetic algorithm and back-propagation learning algorithm. Int. J. Innov. Comput. Inf. Control **7**, 5839–5850 (2011)
6. Brajevic, I., Tuba, M.: Training feed-forward neural networks using firefly algorithm. In: Proceedings of the 12th International Conference on Artificial Intelligence, Knowledge Engineering and Data Bases (AIKED 2013), pp. 156–161 (2013)
7. Ilonen, J., Kamarainen, J.K., Lampinen, J.: Differential evolution training algorithm for feed-forward neural networks. Neural Process. Lett. **17**, 93–105 (2003)
8. Karaboga, D., Akay, B., Ozturk, C.: Artificial bee colony (ABC) optimization algorithm for training feed-forward neural networks. MDAI **7**, 318–319 (2007)
9. Karaboga, D., Ozturk, C.: Neural networks training by artificial bee colony algorithm on pattern classification. Neural Netw. World **19**, 279–292 (2009)
10. Habbi, H., Boudouaoui, Y., Karaboga, D., Ozturk, C.: Self-generated fuzzy systems design using artificial bee colony optimization. Inf. Sci. **295**, 145–159 (2015)
11. Habbi, H.: Artificial bee colony optimization algorithm for TS-type fuzzy systems learning. In: 25th International Conference of European Chapter on Combinatorial Optimization, pp. 26–28 (2012)
12. Saffari, H., Sadeghi, S., Khoshzat, M., Mehregan, P.: Thermodynamic analysis and optimization of a geothermal Kalina cycle system using artificial bee colony algorithm. Renew. Energy **89**, 154–167 (2016)
13. Habbi, H., Boudouaoui, Y.: Hybrid artificial bee colony and least squares method for Rule-Based systems learning. Waset Int. J. Comput. Control Quantum Inf. Eng. **08**, 1968–1971 (2014)
14. Secui, D.C.: A new modified artificial bee colony algorithm for the economic dispatch problem. Energy Convers. Manag. **89**, 43–62 (2015)

15. Habbi, H., Boudouaoui, Y., Ozturk, C., Karaboga, D.: Fuzzy rule-based modeling of thermal heat exchanger dynamics through swarm bee colony optimization. In: International Conference on Advanced Technology and Sciences, ICAT 2015, pp. 4–7 (2015)
16. Boudouaoui, Y., Habbi, H., Harfouchi, F.: Swarm bee colony optimization for heat exchanger distributed dynamics approximation with application to leak detection. In: Handbook of Research on Emergent Applications of Optimization Algorithms, pp. 557–578. IGI Global (2018)
17. Zhu, G., Kwong, S.: Gbest-guided artificial bee colony algorithm for numerical function optimization. Appl. Math. Comput. **217**, 3166–3173 (2010)
18. Kiran, M.S., Hakli, H., Gunduz, M., Uguz, H.: Artificial bee colony algorithm with variable search strategy for continuous optimization. Inf. Sci. **300**, 140–157 (2015)
19. Akay, B., Karaboga, D.: A modified artificial bee colony algorithm for real-parameter optimization. Inf. Sci. **192**, 120–142 (2012)
20. Harfouchi, F., Habbi, H.: A cooperative learning artificial bee colony algorithm with multiple search mechanisms. Int. J. Hybrid Intell. Syst. **13**, 113–124 (2016)
21. Karaboga, D.: An idea based on honey bee swarm for numerical optimization. Technical report-tr06, Computer Engineering Department, Engineering Faculty, Erciyes University (2005)
22. Karaboga, D., Basturk, B.: A powerful and efficient algorithm for numerical function optimization: artificial bee colony (ABC) algorithm. J. Glob. Optim. **39**, 459–471 (2007)
23. Dua, D., Taniskidou, E.K.: UCI machine learning repository (2017). School of Information and Computer Science, University of California, Irvine, CA. http://archive.ics.uci.edu/ml
24. Tuba, M., Alihodzic, A., Bacanin, N.: Cuckoo search and bat algorithm applied to training feed-forward neural networks. In: Recent Advances in Swarm Intelligence and Evolutionary Computation, vol. 585, pp. 139–162. Springer, Cham (2015)

Author Index

O. Demigha et al. (Eds.): CSA 2018, LNNS 50, pp. 333–334, 2019.
https://doi.org/10.1007/978-3-319-98352-3

Printed in the United States
By Bookmasters